Epithelial Transport

Epithelial Transport

A guide to methods and experimental analysis

Edited by

Nancy K. Wills, Luis Reuss and Simon A. Lewis

Department of Physiology and Biophysics,
University of Texas Medical Branch, Galveston, Texas, USA

CHAPMAN & HALL
London · Weinheim · New York · Tokyo · Melbourne · Madras

Published by Chapman & Hall, 2–6 Boundary Row, London SE1 8HN, UK

Chapman & Hall, 2–6 Boundary Row, London SE1 8HN, UK

Chapman & Hall GmbH, Pappelallee 3, 69469 Weinheim, Germany

Chapman & Hall USA, 115 Fifth Avenue, New York, NY 10003, USA

Chapman & Hall Japan, ITP-Japan, Kyowa Building, 3F, 2-2-1 Hirakawacho, Chiyoda-ku, Tokyo 102, Japan

Chapman & Hall Australia, 102 Dodds Street, South Melbourne, Victoria 3205, Australia

Chapman & Hall India, R. Seshadri, 32 Second Main Road, CIT East, Madras 600 035, India

First edition 1996

Softcover reprint of the hardcover 1st edition 1996

Typeset in 10/12pt Palatino by AFS Image Setters Ltd, Glasgow

ISBN-13: 978-94-010-7168-0 e-ISBN-13: 978-94-009-1495-7

DOI:10.1007/ 978-94-009-1495-7

A catalogue record for this book is available from the British Library

Printed on acid-free paper, manufactured in accordance with ANSI/NISO Z39.48-1992 and ANSI/NISO Z39.48-1984 (Permanence of Paper)

Contents

Contributors

Guillermo A. Altenberg MD
Department of Physiology and Biophysics
University of Texas Medical Branch
Galveston
Texas 77555-0641
USA

M. Cereijido MD, PhD
Centro de Investigation y de Studios Avanzados
Fuselage, Biofisica y Neurociencias
Apartheid Postal 14-740
07000 Mexico DF
Mexico

Chris Clausen PhD
Department of Physiology and Biophysics
State University of New York at Stony Brook
Stony Brook
New York 11704-8861
USA

R. G. Contreras PhD
Centro de Investigacion y de Estudios Avanzados
Fisologia, Biofisica y Neurociencias
Apartado Postal 14-740
07000 Mexico DF
Mexico

Calvin U. Cotton PhD
Department of Physiology
Case Western Reserve
Cleveland
Ohio 44106-4948
USA

E. Radford Decker PhD
Department of Pharmacology
Texas Biotechnology Corporation
Houston 77030
Texas
USA

M. R. García-Villegas MD
Centro de Investigacion y de Estudios Avanzados
Fisologia, Biofisica y Neurociencias
Apartado Postal 14-740
07000 Mexico DF
Mexico

L. González-Mariscal MD
Centro de Investigacion y de Estudios Avanzados
Fisologia, Biofisica y Neurociencias
Apartado Postal 14-740
07000 Mexico DF
Mexico

Norman J. Karin PhD
Department of Integrative Biology
University of Texas, Houston Health Science Center
Houston
Texas 77030
USA

Simon A. Lewis PhD
Department of Physiology and Biophysics
University of Texas Medical Branch
Galveston
Texas 77555-0641
USA

Alicia McDonough PhD
Department of Physiology and Biophysics
University of Southern California, School of Medicine
1333 San Pablo
Los Angeles
CA 90033-1026
USA

John W. Mills PhD
Department of Biology
Clarkson University
Potsdam
NY 13676
USA

Austin K. Mircheff PhD
Department of Physiology and Biophysics
University of Southern California, School of Medicine
1333 San Pablo
Los Angeles
CA 90033-1026
USA

Roger O'Neil PhD
Department of Integrative Biology
University of Texas, Houston Health Science Center
Houston
Texas 77030
USA

Luis Reuss MD
Department of Physiology and Biophysics
University of Texas Medical Branch
Galveston
Texas 77555-0641
USA

J. Valdés PhD
Centro de Investigacion y de Estudios Avanzados
Fisologia, Biofisica y Neurociencias
Apartado Postal 14-740
07000 Mexico DF
Mexico

Nancy K. Wills PhD
Department of Physiology and Biophysics
University of Texas Medical Branch
Galveston
Texas 77555-0641
USA

Min I.N. Zhang MD
Department of Integrative Biology
University of Texas, Houston Health Science Center
Houston
Texas 77030
USA

Preface

Epithelial cells function to exchange substances between the body and the external world or body fluid compartments. The complexity of these processes demands the orchestration of many biological mechanisms. Perhaps for this reason, the field of epithelial transport is populated by a diverse array of experimental approaches and methodologies.

This book presents an introduction to some of the major questions concerning epithelial function and the experimental methods that can be used to answer them. It is intended for advanced undergraduate and graduate students as well as post-doctoral fellows and established investigators who desire a working knowledge of this field. The first section of the book (Chapters 1–3) focuses on structure – function issues, progressing from the general organizational features of epithelia (Chapter 1) to an overview of the molecules which mediate epithelial ion transport (Chapter 2), and next to the processes involved in the establishment and maintenance of epithelial cell polarity (Chapter 3). Chapters 4 and 5 describe electrical and other approaches for determining basic epithelial properties. The advantages and pitfalls of specific methods for measuring epithelial ion and water flow, as well as membrane fractionization and approaches involving cell culture, are provided in Chapters 6–11. Finally, the last three chapters provide a synopsis of recent progress in the use of pharmacological and molecular methods to understand the role of intracellular messenger systems, cytoskeleton and genetic factors in epithelial function.

The idea for this book was conceived during a conversation with Dr Susan Hemming. We are grateful for her support and to Rachel Young, Alison Conneller and Nigel Balmforth who brought it to fruition. We also wish to thank our esteemed colleagues (and leading specialists) who kindly agreed to contribute chapters to this volume. Our thanks to Dr Karl Karnaky for reading a preliminary version of the manuscript and to Lynette Durant and LiJun Mo for their help with the wordprocessing and artwork.

Nancy K. Wills
Luis Reuss
Simon A. Lewis

1

Epithelial structure and function

Simon A. Lewis

The necessity to maintain and regulate extracellular fluid volume and composition is one of the greatest challenges faced by both vertebrates and invertebrates. Such a requirement for homeostasis is subserved by a diverse collection of organ systems which can selectively absorb from, or excrete into, the external environment the necessary amounts of non-electrolytes (sugars, amino acids, water) and electrolytes (sodium, potassium, chloride, bicarbonate etc). A feature shared by all of these organ systems is that they are lined by a layer of closely packed cells which are tightly joined to each other by a structure called tight junctions (for models of tight junction architecture, see Chapter 3 and Gumbiner, 1987). The cell ensemble is given additional structural rigidity (on one side) by a glycoprotein matrix called the basement membrane. This configuration of a layer of cells bonded together by a hoop of protein and attached to a basement membrane is the basic structural characteristic of an epithelium. What distinguishes an epithelial cell from any other cell type in the body is the ability to segregate different but complementary transport proteins between two distinct membrane domains (epithelial cell polarity, Chapters 3 and 4). It has been proposed that the properties of these two domains are, in part, maintained by the tight junctions. This two unit-structure (polarized cell and tight junction) allows an epithelium to

Epithelial Transport: A guide to methods and experimental analysis.
Edited by Nancy K. Wills, Luis Reuss and Simon A. Lewis.
Published in 1996 by Chapman & Hall, London. ISBN 978-94-010-7168-0

perform vectorial transport of electrolytes and non-electrolytes.

There are two pathways that electrolytes and non-electrolytes can follow when moving across this planar array of cells called an epithelium. The first is across the tight junctions and then along the lateral intercellular spaces surrounding each epithelial cell (the intercellular or **paracellular** pathway), and the second is across one of the membranes, into the cell, followed by exit across the other membrane (this is the **transcellular** pathway). Given its structure, the epithelium can also act as a barrier to the movement of substances between two fluid compartments. Without exception, one of these compartments is in close contact with the extracellular fluid and in many instances the opposing compartment is directly connected to the external environment.

Epithelia can perform selective absorption (movement into the extracullular fluid – ECF) and/or secretion (movement away from the ECF) of non-electrolytes and electrolytes against concentration and/or electrical gradients. This movement demands the use of energy and requires different but complementary transport proteins in the two cell membrane domains (Chapter 2). In other words, if one membrane possesses a transporter that moves a substance into the cell, then the series membrane should contain a transporter which moves the substance out of the cell into the opposing compartment. Because differing epithelial organs perform different transport functions, it is obvious that the transport properties of individual membranes must differ between various epithelia and organ systems. For example, the transporters found in the proximal sections of the renal tubules or small intestine are quite different from the transporters found in the distal segments of the renal tubule or the intestinal tract. Also, the transport properties of the intestinal crypt cells differ from those of intestinal villus cells.

The ability to restrict or enhance the movement of electrolytes and non-electrolytes varies amongst epithelia. Some epithelia are able to absorb or secrete copious quantities of electrolytes, non-electrolytes and water but are not able to establish or support large chemical or osmotic gradients between adjacent fluid compartments. Other epithelia are efficient electrolyte transporters, but are relatively impermeable to water. Consequently, these epithelia have the ability to generate and support osmotic and either electrolyte or non-electrolyte concentration gradients.

In the following sections, we first overview epithelial structure and nomenclature. Next, we categorize epithelia into two broad groupings ('tight' and 'leaky') and list the physiological correlates that encompass these groupings. This is followed by an overview of some of the common properties shared by epithelia. Examples are given of models of electrolyte and non-electrolyte transport by epithelia. Lastly, we review possible sites of electrolyte and non-electrolyte transport regulation.

1.1 EPITHELIAL STRUCTURE

Epithelia consist of three basic building blocks: cells; **tight junctions** (which bind the cells together and sometimes restrict the movement of substances between the epithelial cells); and the basement membrane (which acts as a structural support and site of attachment for the epithelial cells). An epithelium and these structures are shown in Figure 1.1, and Table 1.1 gives the commonly used names for these components according to the epithelium being studied. Briefly, the **apical membrane** is separated from the **basolateral membrane** by the tight junctions. The tight junctions consist of a very close apposition of the lateral membranes to adjacent cells, with the membranes being anchored together by interwoven strands of protein which completely encircle the cells (Chapter 3 and Cereijido, 1992). Other junctional structures include: the **gap junctions** which allow for cell-to-cell communication (usually between like cells); **desmosomes** (spot junctions) for structural integrity; and the **zona adherens**, which is the major site of attachment of actin filaments to the lateral membrane. The space between the cells is the **lateral intercellular space**, and the series combination of the tight junction and the lateral intercellular space is the **paracellular pathway**. Last, the basal aspect of all cells rests on the basement membrane.

The distance between the lateral membranes of apposing cells depends upon the type of epithelium, and can be as narrow as 10 nanometers (nm) and as wide as 1–3 microns (μm). This lateral intercellular space has important implications for ion and water transport across epithelia. Its presence can result in overestimates of the basolateral membrane resistance (Chapters 5 and 6).

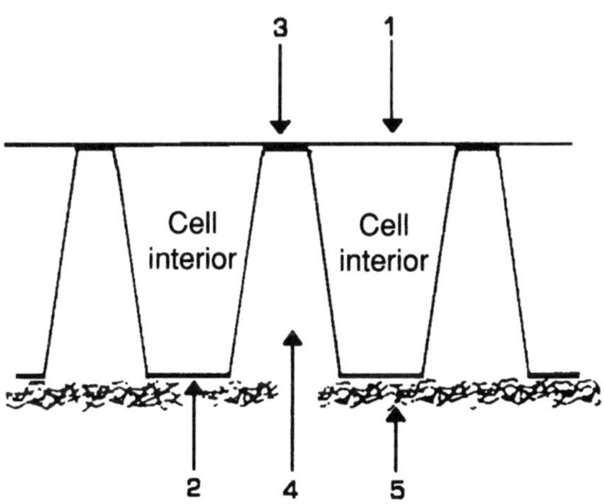

Figure 1.1 An epithelial cell layer: (1) apical membrane; (2) basolateral membrane; (3) tight junction; (4) lateral intercellular space; (5) basement membrane.

Table 1.1 Nomenclature for structural components of an epithelium: the most commonly used terms (top of table) and tissue specific nomenclature

Tissue	*Luminal compartment*	*ECF compartment*	*Apical membrane*	*Basolateral menbrane*	*Tight junction*
	mucosal	serosal	apical	basolateral	tight junction
	apical solution	basolateral solution	brush border	basolateral	zona occludens
Renal	tubular luminal	peritubular contraluminal	brush border	basolateral	paracellular
Corneal epithelium	tear	stromal	apical	basolateral	tight junction
Liver	canaliculus	sinusoidal space	canalicular	sinusoidal	
Gastric	secretory	nutrient			
Frog skin	pond; outside	blood; inside	outside	inner	

The microscopic structure of apical membranes can vary from being smooth (e.g. the epithelium of the mammalian urinary bladder) to extensively folded (e.g. a brush-border membrane of renal proximal tubule, small intestine and choroid plexus). The actual surface area of the apical membrane can be equal to the nominal area of a flat disk (i.e. 1 cm^2 of tissue area = 1cm^2 of nominal apical membrane area) or can be considerably amplified, as is the case of the small and large intestine, gall bladder or renal proximal tubule, where 8–10 cm^2 of apical membrane = 1 cm^2 of nominal area. (The methods to measure such amplification are outlined in Chapter 6.)

For the basolateral membrane, if the cells are cuboidal, the minimal surface area is about five times that of the nominal area of a disk. Therefore, neglecting microscopic amplification, the basolateral membrane area is about five times larger than the apical membrane area. Such a simple geometry has been found for the mammalian urinary bladder, where under conditions of moderate stretch the apical membrane area is about one fifth that of the basolateral membrane area. For cells from the mammalian renal proximal tubules, small and large intestine and gallbladder, the basal and lateral membranes (i.e. the basolateral membrane) are significantly infolded and thus increase the actual surface area by a factor of two to four.

Although most transporting epithelia are monolayers, a number are composed of two to seven cell layers. For example, corneal epithelium and frog skin epithelium each have ca. seven cell layers and the sweat duct has two cell layers. In the case of corneal epithelium, electrical measurements which are capable of measuring the surface area of the apical and basolateral membrane in a non-invasive manner yield area amplification factors of 2.6 for the apical membrane (reasonably folded) and 94 for the basolateral membrane. This large number for the basolateral surface area tells us that the seven cell layers are all electrically coupled to each other, i.e. the total basolateral membrane of the corneal epithelium is composed of the basolateral membrane of the first cell layer plus the total surface area of the remaining six cell layers (Chapter 6).

The demarcation line between the apical and basolateral membrane is the tight junctions. This junctional complex serves three very important functions:

1. It acts as a barrier to the migration of apical-membrane proteins to the basolateral membrane and vice versa. Thus it is commonly believed that the tight junction acts as a fence which helps to maintain the polarity of the two membrane domains of the epithelium.
2. It binds the epithelial cells together and thus provides structural integrity to the epithelium.
3. It constitutes an extracellular (paracellular) route for the movement of

electrolytes and non-electrolytes. Whereas the movement of hydrophilic substances through the cell requires the existence of specific transport proteins in both apical and basolateral membranes, the movement of similar hydrophilic substances through the tight junctions is based on size and charge.

The precise structure of the tight junctions has not been resolved. Freeze-fracture electron microscopy studies have shown that the tight junction proteins are a two-dimensional lattice, in which the apposing cell membranes 'kiss' along the length of the lattice's ridges. It has been proposed that these ridges (and not the space between these ridges) impede the movement of hydrophilic substances from the apical bathing solution to the lateral intercellular space. Claude and Goodenough (1973) proposed that the resistive properties of tight junctions could be explained by the average number of parallel strands in the lattice. Thus, the more strands in series, the greater the ability of the tight junction to restrict the movement of electrolytes and non-electrolytes. This simple hypothesis was tested and found not to be valid. A re-evaluation of the original hypothesis requires the consideration of not the mean strand number (which would equal the mean resistor value for a series arrangement of resistors) but rather the inverse of the mean value of the inverse strand number (in effect the normalized resistance of a parallel arrangement of resistors; see Chapter 6). The possibility that the tight junctions can be regulated as a function of transport across an epithelium is addressed in Chapter 13.

1.2 CATEGORIZING EPITHELIA

Given the number of different types of epithelia and the variety of transport functions that epithelia can perform, it would be helpful to categorize epithelia and as such develop a common language when discussing epithelial function. A number of different schemes can be used. For example, epithelia could be categorized according to whether they absorb or secrete substances, the molecular species of substances transported, the histology of the epithelium (e.g. transitional, cuboidal, columnar or squamous) or the ability to restrict the movement of ions between the two compartments that the epithelium separates.

Frömter and Diamond (1972) realized that epithelia from a variety of sources possessed not only different transport functions but, in addition, a variable ability to restrict the movement of ions between two compartments, i.e. there was a wide range in the reported transepithelial resistance among epithelia. As a general rule, there is an inverse relationship between transepithelial resistance and ionic permeability. The extremes of this transepithelial resistance (R_t) are illustrated by the mammalian proximal tubule and the mammalian urinary bladder. The proximal tubule has a R_t 5 Ωcm^2 and that of mammalian urinary bladder is 20 000–50 000 Ωcm^2. Since the transcellular resistance (R_c) is in parallel with the paracellular resistance (R_{pc}), which is a series arrangement of the lateral intercellular space resistance (R_{lis}) and the tight junction resistance (R_j), this large difference in transepithelial resistance can be accounted for by either a great variability of the transcellular resistance (e.g. 6 Ωcm^2 to 100 000 Ωcm^2) with a constant and large value for the paracellular resistance (e.g. 100 000 Ωcm^2) or, conversely, by a high and near constant cellular resistance (100 000 Ωcm^2) but a variable value for the paracellular resistance (6 cm^2 to 100 000 Ωcm^2). The transepithelial resistance can be formalized as follows:

$$R_t = \frac{R_c \cdot R_{pc}}{R_c + R_{pc}} \tag{1.1}$$

where R_t, R_c and R_j are transepithelial, cellular and tight junctional resistances respectively.

For this equation, it is important to note that the value of R_t is always less than the lower of the two values on the right-hand side of the equation. Knowing this relationship and using electrophysiological techniques (Chapter 5), Frömter and Diamond devised a simple yet elegant experiment to determine whether the limiting resistance for an epithelium (in this case the *Necturus* gallbladder epithelium) was the transcellular or the tight junctional pathway. The method for determining the low resistance pathway was to pass a transepithelial current pulse across the epithelium and to scan the epithelial surface for **current sinks** (tight junctions and apical membrane) using a microelectrode under microscopic observation. The site (cell surface or tight junction) over which the highest current density is recorded is a current sink, i.e. has the lowest resistance. In the gallbladder, it was determined that the tight junction had a lower resistance than the cellular pathway. This method is qualitative, i.e. it does not permit one to determine the absolute resistance of the tight junctions compared with the cellular pathway. Determination of absolute values for these parallel resistances requires the use of microelectrodes and transepithelial recording techniques (Chapter 5). From such methods, it was determined that the variability of the tight junctional resistance accounted for the large variability of the transepithelial resistance among different types of epithelia. This does not imply that for all epithelia the transcellular resistance is large and constant, but rather that there is a greater range of paracellular resistance values (from 6 Ωcm^2 to at least 100 000 Ωcm^2) than of transcellular resistance (from 1000 Ωcm^{2}to 80 000 Ωcm^2).

Frömter and Diamond divided epithelia into two categories based on the resistive properties of the tight junction. Epithelia with low resistance tight junctions (R_j < 100 Ωcm^2) were termed **leaky epithelia**; those with high resistance tight junctions (R_j > 500 Ωcm^2) were termed **tight epithelia**. In addition to the differences in tight junctional resistance, there are a number of other physiological properties of epithelia which conveniently fall into one category or the other, some of which are a consequence of the junctional resistance (Table 1.2).

As will be demonstrated in Chapter 5, it is not surprising that, with identical 'physiological' bathing solutions on both sides, the transepithelial voltage (V_t) is lower in leaky epithelia than in tight epithelia. This lower voltage has two sources. First, the low junctional resistance attenuates the voltage produced by active ion transport in the epithelial cells. Second, some leaky epithelia actively transport ions in an electroneutral manner, i.e. the cells have coupled ion transport systems which either cotransport (a cation and an anion) or exchange (a cation for another cation) and thus do not produce an electrical current.

Similarly, the inability of a leaky epithelium to maintain an ion concentration gradient is not surprising. First, the junctional electrical resistance is low (i.e. the junction is highly permeable to ions) and therefore the generation of an ion concentration gradient is attenuated by the back diffusion of ions (through the low resistance junction) from the compartment of high ion concentration to the compartment with low ion concentration. Second, a leaky epithelium cannot maintain an ion concentration gradient (if this gradient also generates an isosmotic driving force) because of its high permeability to water. The high water permeability is thought to reside exclusively or predominantly in the cellular pathway, although there is some disagreement on this point (for a review see Whittenbury and Reuss, 1991).

The arguments used to explain the low voltage and inability to maintain an ion concentration gradient or osmotic gradient across leaky epithelia

Table 1.2 Characteristics of tight and leaky epithelia

Type	R_t *(Ω.cm^2)*	V_t *(mV)*	*Concentration ratio*	L_p *(cm/sec.osmol/kg)*	*Osmotic ratio*
Leaky	6–100	0–11	1–12	4×10^{-3} to 4×10^{-5}	1
Tight	500–70 000	10–120	70 to 1×10^{6}	2×10^{-5} to 1×10^{-6}	>1

R_t = transepithelial resistance
V_t = spontaneous transepithelial potential, which is a result of active transepithelial ion transport (see Chapter 5 for details)
Concentration ratio = steady-state concentration gradient which the epithelium can establish and maintain using active transport processes
L_p = water permeability of the epithelium
Osmotic ratio = osmotic gradient which the epithelium can maintain between the two compartments, and which is a balance between net transport of osmotically active substances and the relative water permeability (L_p) of the epithelium.

can be turned around to explain the ability of tight epithelia to generate large transepithelial voltages (V_t), maintain large ion concentration gradients and maintain large osmotic gradients. Tight epithelia generate large transepithelial voltages because the junctions have a high electrical resistance and, as a consequence, the voltage generated by the cells is not attenuated to the same degree as in the leaky epithelia (Chapter 5). In addition, in tight epithelia the cells generally perform electrogenic ion transport, i.e. active transport of net charge across the two series membranes. Since in tight epithelia the junctions are relatively impermeable to solutes, there will be less passive dissipation of ion concentration gradients generated by the active transport system of the epithelium. In addition, since the epithelium has also a low water permeability, the rate of water movement due to the osmotic gradient established by the ion concentration gradient will be far less than in leaky epithelia. In conclusion, since the cell water permeability is low in tight epithelia, they can generate and maintain significant osmotic gradients.

While leaky epithelia are very efficient transporters of large quantities of electrolytes, non-electrolytes and water, tight epithelia transport lower amounts of electrolytes, non- electrolytes and water. Nevertheless, tight epithelia can establish and maintain large electrolyte and non-electrolyte gradients. In addition, unlike most leaky epithelia, transport by tight epithelia is under hormonal control.

1.3 COMMON PROPERTIES OF EPITHELIAL CELLS

Although different epithelia have diverse functional properties, they share a number of membrane and cytoplasmic features. These major common features are summarized in Figure 1.2.

In nearly all epithelia so far studied, a Na^+,K^+-ATPase is located in the basolateral membrane. (The exception to this rule is that choroid-plexus and retinal-pigmented epithelia have the Na^+,K^+-ATPase in their apical (brush border) membrane.) The Na^+,K^+-ATPase transports three Na^+ from the cytoplasm to the serosal solution in exchange for two K^+ transported from the serosal solution to the cell cytoplasm and hydrolyzes one ATP molecule per cycle. The pump is electrogenic, i.e. in each cycle a positive charge is moved from the cytoplasm to the serosal solution. This process is termed **primary active transport** and is capable of transporting Na^+ and K^+ against their respective electrochemical gradients. For further details, see Chapter 2.

The basolateral membrane voltage is negative (cell interior with respect to the serosal solution). Under normal conditions, this potential is in the range of –30 to –80 mV. More details are given in Chapter 5.

The intracellular ion concentrations (or activities) of most epithelia are as follows. K^+ concentration is higher in the cell than in the extracellular

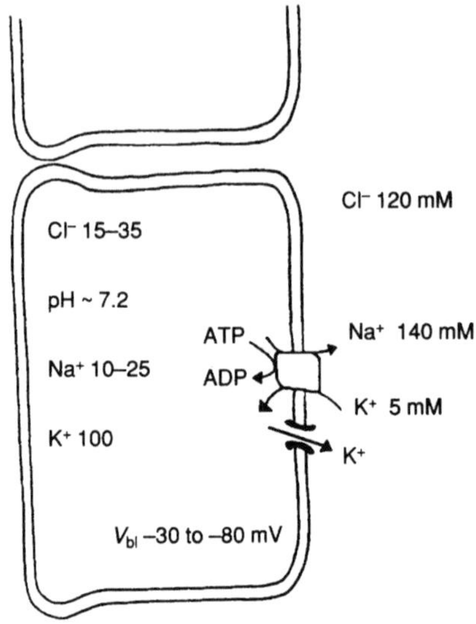

Figure 1.2 Idealized epithelial cell demonstrating some of the properties common to all epithelia.

fluid (in mammalian epithelia, > 100 mM compared with 5 mM), and is not in electrochemical equilibrium across the basolateral membrane. Since the high cell K^+ concentration is dependent upon the continuing operation of the Na^+,K^+-ATPase, this implies that K^+ is actively accumulated by this ATPase. Na^+ concentration is much lower in the cell than in the extracellular fluid (10–25 mM and 140 mM respectively). Since the cell membrane voltage is negative, then cell Na^+ is not in equilibrium. Inhibition of the Na^+,K^+-ATPase causes intracellular Na^+ to increase, implying that the low cell [Na^+] is maintained by the Na^+,K^+-ATPase. Cell pH is either the same as or slightly more acid (e.g. pH 7.2 to 7.4) than the extracellular fluid (pH 7.4). Because this cell H^+ is not in electrochemical equilibrium across the basolateral membrane, there must be a mechanism for active (energy-requiring) extrusion, from cell to extracellular space, of acid equivalents across one of the epithelial cell membranes (see Chapter 2 for possible transporters). Cell Cl^- is either in equilibrium or above equilibrium, across the basolateral membrane, compared with the extracellular fluid (e.g. 15–35 mM and 121 mM, respectively). Hence, energy is utilized to move Cl^- into the cell (see Chapter 2 for discussion of mechanisms). The sum of the cell monovalent cation concentrations greatly exceeds the sum of the monovalent anions. Hence, there must be a high level of non-permeant polyvalent anions (largely proteins) in the cells. The methods available for measuring intracellular ion concentrations are addressed in Chapter 7.

Under physiological conditions, the basolateral membrane of all studied epithelia is predominantly K^+ conductive (underlain by K^+ channels; Chapters 2 and 10), with small conductances to Na^+ and in some cases Cl^-. Since cell [K^+] is higher than predicted for electrochemical equilibrium, there is a net electrochemical gradient favoring K^+ movement from the cell interior to the extracellular space. The basolateral membrane potential is then a K^+ diffusion potential attenuated by the finite conductances of this membrane to Na^+ and Cl^- and by an intraepithelial current loop if the apical membrane is not K^+ selective.

The properties of the apical membranes vary considerably among epithelia. In addition, vectorial transcellular transport requires the expression of complementary but intrinsically different transport system in the basolateral membrane.

1.4 MODEL FOR SOLUTE AND SOLVENT TRANSPORT

The preceding sections have been devoted to a basic understanding of the structure and common properties shared by epithelia. In this section we discuss epithelial polarity and why it is essential for transepithelial vectorial transport. Two models will be investigated: electrogenic Na^+ transport in tight epithelia, and solvent (water) transport in leaky epithelia.

1.4.1 Model for electrogenic Na^+ transport

The original model for transepithelial Na^+ transport was established from experiments carried out on the ventral skin of the frog, a tissue which previous studies had demonstrated was capable of net active transport of at least Na^+. The experimental procedure was to remove the skin and mount it between two solution-filled hemichambers. This chamber and all subsequent modifications have been named **Ussing chambers**, or modified Ussing chambers (Chapter 5), after the Danish scientist who designed them. The advantages of this method were that it:

- permitted control of the composition of the solutions bathing both sides of the epithelium;
- allowed measurements across the epithelium; and
- permitted determination of the electrical properties of the epithelium and the effects of experimental perturbations on these electrical properties (i.e. on the spontaneous potential and the resistance).

Using symmetrical NaCl Ringer's solutions (salt solutions with a plasma-like composition), Ussing and collaborators noted (Koefoed-Johnsen and Ussing, 1958) that the frog skin developed a spontaneous potential (V_t) which could have a value as large as −120 mV (blood side zero). The electrical resistance was about 500–1000 $\Omega \cdot cm^2$. The transepithelial voltage was reduced by anoxia, showing that it required an intact metabolism, and also by the addition of the cardiac glycoside ouabain which was known to inhibit the recently discovered Na^+,K^+-ATPase. The magnitude of V_t was reduced when the ECF side K^+ was increased. These data suggested:

- that the frog skin actively transports either a net positive charge (e.g. Na^+ or K^+) from the pond to the blood or a net negative charge (e.g. Cl^-) from the blood to the pond;
- that this transport required an operational Na^+,K^+-ATPase; and
- that the magnitude of this transport was decreased by raising blood side K^+.

The magnitude of the transepithelial net charge movement was quantitated by measuring the current that had to be passed across the epithelium to reduce V_t to zero. This current was called the **short-circuit current** (I_{sc}). If, with identical solutions on both sides, only one ion is actively transported, then the I_{sc} equals the rate of transport of that ion. If more than one ion is actively transported, then the I_{sc} is equal to the algebraic sum of all ion-transport processes (this sum must include both the direction and the valence of the transported species). Ussing and associates also determined (Koefoed-Johnsen and Ussing, 1958) that the ion species actively transported could be determined by measuring the 'net' isotopic flux of various ions across the epithelium under short-circuit conditions (V_t maintained at 0 mV) to eliminate V_t as a driving force for passive transepithelial ion transport. They found that there was a net transport of Na^+ from the pond to the blood and that this net Na^+ transport (measured using radioactive Na^+) was equal to the value of the I_{sc} (i.e. the short-circuit current). The relationship between current and ion transport is:

$$J_i = \frac{I_{sc}}{zF} \tag{1.2}$$

where J_i is the 'net' flux of ion 'i' (units of equivalents/cm^2·sec), I_{sc} is the current generated by the cells under short-circuit conditions (units of amperes = coulombs/sec), F is Faraday's constant (96 500 coulombs/equivalent) and z is the valence of the transported species (i.e. the number of charges and the polarity of its charge; $Cl^- = -1$, $Na^+ = 1$, $Ca^{2+} = 2$).

In 1958, Koefoed-Johnsen and Ussing proposed a model (Figure 1.3) to explain the mechanism by which frog skin transports Na^+ from the pond side to the blood compartment. The robustness of this model is impressive and is used to describe transport in both native and tissue-cultured epithelia (Chapter 11). Research on this epithelium (as well as on the family of Na^+-absorbing tight epithelia) has confirmed the validity of this model and has expanded it to include features such as:

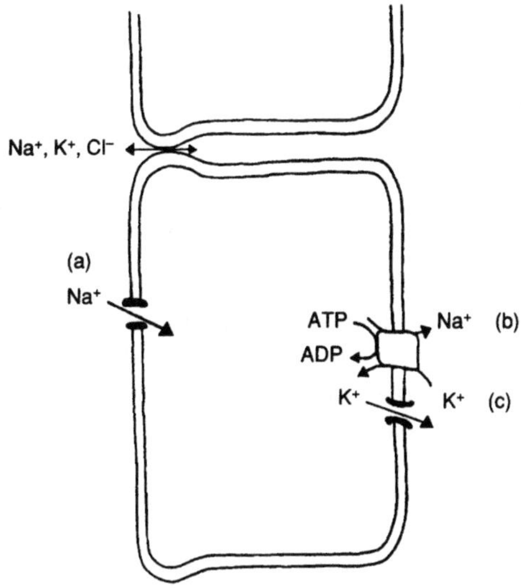

Figure 1.3 Model for Na^+ transport across frog skin epithelium. (a) Na^+ enters the cell down a net electrochemical gradient (low cell $[Na^+]$ and cell-negative voltage). (b) Once in the cell, Na^+ is extruded via the Na^+,K^+-ATPase across the basolateral membrane in exchange for K^+. (c) K^+ then exits across the basolateral membrane through a K^+ conductance in this membrane and as a consequence makes the cell interior electrically negative.

- the stoichiometry of the Na^+,K^+-ATPase;
- the magnitude of both the electrical and chemical gradients favoring Na^+ entry across the apical membrane;
- the mechanisms by which the apical membrane Na^+ conductance is regulated by hormones, neurotransmitters, intracellular factors (Ca^{2+}, pH) and second messengers (Chapter 12); and
- the role of the tight junctions in this transport system (Chapter 5).

In the *in vivo* epithelium (which is not short-circuited) the continuous flow of Na^+ alone from the pond to the ECF compartment is not possible, because of the constraints imposed by electroneutrality. Hence, concomitantly with the Na^+ flux there must be a counterflow of cations from the ECF to the pond or a flow of anions from the pond to the ECF. Either process is driven by the transepithelial voltage (open-circuit conditions) and the movement can be either transcellular or paracellular (through the tight junctions and lateral intercellular spaces). The ion species acting as counter-ion to active Na^+ transport is determined by the selectivity of the paracellular (or transcellular) pathway. For example, if the paracellular pathway is permeable only to Na^+, then the epithelium will not perform net Na^+ transport under open-circuit conditions. Conversely, if the paracellular pathway is permeable to only Cl^- then the epithelium will absorb NaCl from the pond to the ECF. The magnitude of this NaCl absorption will equal the product of the transepithelial potential and the paracellular pathway Cl^- conductance (Chapter 5).

1.4.2 Model for water transport in leaky epithelia

In the absence of hormonal stimulation, frog skin epithelium is relatively water impermeable, i.e. it is an effective ion transporter but since it has a low water permeability there is no appreciable water flux in response to imposed osmotic gradients. This is in contrast with the properties of epithelia such as renal proximal tubule and gallbladder.

Diamond (1962a,b) investigated the transport properties of fish gallbladder. Using an experimental system similar to that designed by Ussing and co-workers, he found that R_t was very low (in the order of 100 Ωcm^2) and V_t was negligible (ca. −1 mV). In the light of Ussing's model for ion transport across frog skin, a possible interpretation was that the gallbladder had been irreversibly damaged during excision or mounting in the experimental apparatus. In a simple yet elegant series of experiments, Diamond demonstrated that fish gallbladder was able to transport large volumes of water and electrolytes from the luminal compartment to the blood compartment, apparently by electroneutral NaCl transport. Of interest is that salt and water were transported in near-isosmotic proportions, i.e. there was no apparent change in the osmolality of either the luminal or blood compartments. This observation was intriguing, since the expected mechanism of water movement between two compartments separated by a barrier was osmotic flow, i.e. water flow driven by an osmotic pressure difference between the two compartments. Yet in Diamond's studies there was no measurable osmotic pressure difference between the mucosal and serosal compartments. Additional studies demonstrated that from estimates of the epithelial water permeability, one would require a substantial (and measurable) transepithelial osmotic gradient for the water flow measured. A number of researchers developed models to explain this paradox.

The essential idea is that of a three-compartment model in which the middle compartment is a restricted diffusion space separated from the adjacent compartment by barriers with different water permeability properties. The middle compartment is rendered hyperosmotic by solute transport from the mucosal compartment. This produces water flow from the mucosal compartment into the middle compartment (the barrier between these compartments is passively permeable to water but not to solute). This flow in turn elevates the hydrostatic pressure in the middle compartment, which results in the bulk flow of ions and water across the second membrane (which is permeable to both solute and water). That such a system works was demonstrated by Curran and MacIntosh (1962). It was suggested that the anatomical location of this middle compartment was the lateral intercellular spaces of the epithelium. Mathematical modeling required that salt transport across the basolateral membrane be localized to an area immediately after the tight junctions and that the lateral intercellular space be long and tortuous. Despite the theoretical appeal of a localized salt extrusion mechanism, experimental evidence did not support this requirement. These complex models were required because of the apparent low water permeabilities of the cell membranes.

Diamond suggested that perhaps the water permeability measurements were in error and the actual permeability was larger than estimated. Recent advances in light microscopy and electrophysiology (Chapter 8) have proven that Diamond's suggestion is correct, i.e. the actual water permeability is some 10 times larger than initially measured. Thus, in the amphibian gallbladder, one only needs a gradient of ca. 3 mosmol/kg (i.e. very small indeed) to generate the measured rate of water transport.

A model and the location of the osmotic gradients is shown in Figure 1.4. In brief, NaCl entry across the apical membrane is via Na^+/H^+ and Cl^-/HCO^-_3 exchangers (see Chapter 2 for more details of these transporters) and results in an increase in cytoplasmic osmolality, by about 2

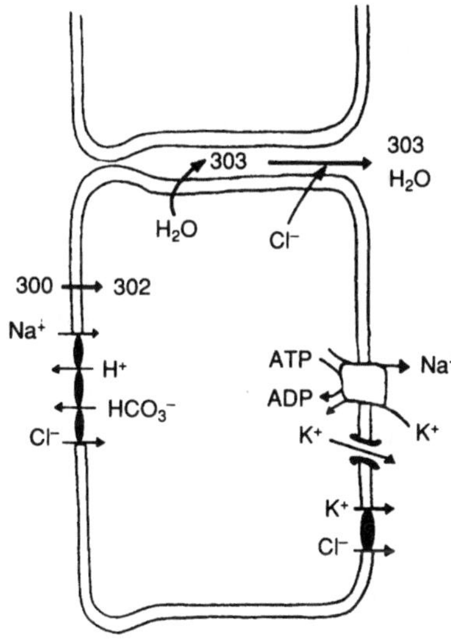

Figure 1.4 Gallbladder epithelial cell and location of osmotic gradients (and corresponding restricted diffusional spaces) required to cause net osmotic water transport coupled to NaCl absorption. It is important to note that the transporters are uniformly distributed over the entire basolateral membrane and not localized as suggested in this figure.

milliosmol/kg. This gradient is sufficient, given the water permeability of the apical membrane, to account for the net movement of water across the membrane. The exit of Na^+ and Cl^- across the basolateral membrane is via Na^+,K^+-ATPase for Na^+, and via KCl cotransport and Cl^- channels for Cl^- (see Chapter 2 for details). This results in an osmotic pressure which is about 1 milliosmol/kg greater than the intracellular osmolality. Given the water permeability of the basolateral membrane, this driving force is sufficient to account for the net water movement across this membrane.

1.5 MODULATION OF EPITHELIAL TRANSPORT

From the structure of an epithelium and its surrounding environment, it is obvious that net transepithelial transport can be modulated at five different locations:

1. the apical membrane;
2. the basolateral membrane;
3. the tight junctions;
4. the cytoplasm; and
5. the composition of either the apical or basolateral membrane bathing solutions.

Using the model for sodium transport (briefly described in section 1.4.1), we will outline some possible mechanisms of regulation and indicate the chapters where these mechanism are described in more detail.

In the model described in section 1.4.1, sodium is transported across the epithelium by electrodiffusion from the apical bathing solution across the apical membrane (through sodium-selective channels) down both electrical and chemical gradients. Once in the cell Na^+ diffuses to the basolateral membrane where it is transported via the Na^+,K^+-ATPase into either the lateral intercellular space or the basal interstitial space. For every three Na^+ extruded across the basolateral membrane, two K^+ enter the cell. These K^+ then diffuse back across the basolateral membrane (into the interstitial space) through K^+-selective channels. This active transport process creates a net current flow through the cell. The current loop is completed by the flow of ions (chloride, potassium and sodium) through the paracellular pathway.

A reduction of the $[Na^+]$ in the mucosal solution will reduce the rate of Na^+ entry across the apical membrane by two mechanisms: a decrease in the chemical driving force favoring sodium entry, and a decrease in the apical membrane Na^+ conductance (Figure 1.5). In addition, since the predominant counter-ion for net Na^+ absorption is Cl^-, a decrease in mucosal solution Cl^- will also reduce the net sodium absorption. This effect is covered in more detail in Chapter 5.

At the apical membrane itself, Na^+ transport depends on the number of

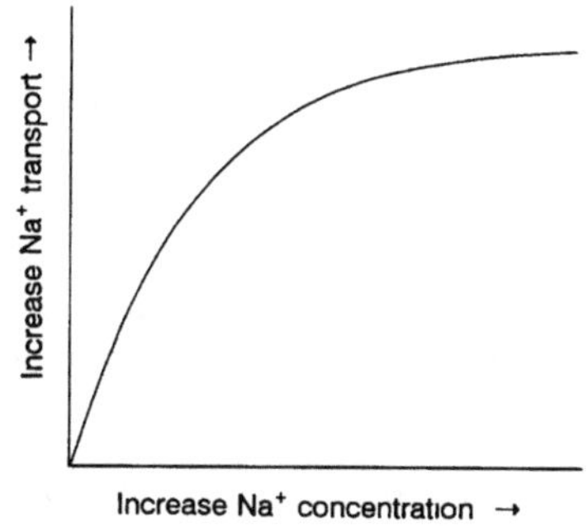

Figure 1.5 Relationship between extracellular [Na^+] and sodium transport across a sodium-transporting tight epithelium. Note that the transport saturates as the [Na^+] increases. This is due to a saturation of the conductance of Na^+ channels in the apical membrane, inhibition of Na^+ channels by external [Na^+] as well as changes in the net electrochemical gradient favoring Na^+ entry.

Na^+ channels in the apical membrane. An increase in the number of channels increases the rate of entry; conversely, a decrease in the number of Na^+ channels leads to a decrease in the rate of Na^+ entry (Figure 1.6). There are three different mechanisms by which plasma membrane channel activity can be regulated: activation, insertion (or withdrawal) and modulation. Activation is a process in which the channel present in the plasma membrane in a non-conductive state becomes active upon the appropriate stimulus. Activation is an all-or-none process. Insertion is a process by which channels in a cytoplasmic store (upon receiving the appropriate signal) are translocated from this storage site and inserted

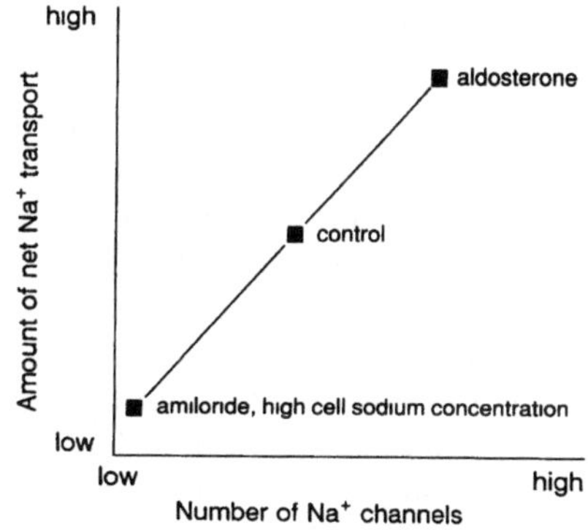

Figure 1.6 Relationship between number of Na^+ channels and rate of transepithelial Na^+ transport. The greater the number of channels, the greater is the rate of transport.

into the plasma membrane. Modulation is a process in which a channel already active (i.e. open part of the time) changes its open probability by some stimulus. Finally, channels can be altered due to mutations in the genes which code for the channel. Such alterations can include the conductance of a single channel, regulatory sites of the channel or targeting of the channel to the membrane (Chapter 14).

Changes in the number of Na^+ channels can be due to extracellular factors such as:

- the pharmacological agent amiloride, which blocks the mouth of the channel and thus inhibits the movement of sodium into the cell;
- aldosterone, which activates pre-existing but quiescent channels in the apical membrane;
- antidiuretic hormone and hypo-osmotic serosal bathing solution, which increase the number of channels; and
- hyper-osmotic bathing solutions, which decrease the number of channels.

Cytoplasmic regulation can occur by alterations in cell Na^+ activity, pH, ATP concentrations, calcium, availability of nutrients, levels of second messengers, etc. (Chapter 12). Although **most of these mechanisms have their ultimate effect at the membrane level**, i.e. altering channel activation, an increase in intracellular sodium activity will directly decrease the rate of sodium entry by reducing the favorable chemical driving force for sodium. In addition, an increase in intracellular sodium has been shown to inhibit the sodium channel.

Basolateral membrane alterations can also affect the net rate of sodium transport. An example of such a regulation is the increase or decrease in the synthesis or insertion of Na^+,K^+-ATPase into the basolateral membrane (Chapter 9). A decrease in the number of pumps will result in a rise in cell sodium which will then decrease the chemical gradient favoring sodium entry and reduce the activity of the apical membrane Na^+ channels. The converse might be true when the density of pumps is increased, i.e. increased driving force for entry and less inhibition of apical membrane sodium channels. Another mechanism for altering the pump activity is by alterations in the intracellular ATP concentration, where a decrease in the cell concentration will result in a decrease in pump activity and as a consequence an increase in cell sodium. Finally, the rate of transepithelial Na^+ transport can be changed by altering the activity of basolateral membrane K^+ channels. Since the tight junctions have a finite resistance, the measured apical membrane potential (i.e. the electrical driving force for Na^+ entry across the apical membrane) is a function of the apical membrane electromotive force (e.m.f.) and resistance, the basolateral membrane e.m.f. and resistance and the junctional resistance and its e.m.f. (Chapter 5). The highly K^+-selective basolateral membrane, in conjunction with a high intracellular [K^+] and low plasma [K^+], makes the basolateral membrane zero-current voltage (e.m.f.) cell-negative. This negative basolateral membrane e.m.f. tends to make the apical membrane potential less positive and thus more favorable for Na^+ entry. For example, if the basolateral membrane K^+ channels are blocked by a pharmacological agent such as Ba^{2+} (which depolarizes the basolateral membrane), Na^+ transport decreases because of a diminution of the electrical driving force favoring Na^+ entry. The equations which describe the interrelationship between the membrane potentials and resistances are given in Chapter 5.

Regulation can also occur at the tight junctions. It is important to remember that for every sodium ion that is actively transported from the apical membrane bathing solution to the basolateral membrane bathing solution, there must be an equivalent flow of either a cation from the basolateral membrane bathing solution to the apical solution or of an anion from the apical membrane bathing solution to the basolateral membrane bathing solution. In the model under discussion, this flux must occur through the tight junction and is therefore determined by the permeability (or conductance) of the junction to that ion (Figure 1.7). If the tight junction is only permeable to Na^+, the epithelium will not be able to perform net active Na^+ transport since for every Na^+ that is transported through the cell (apical to basolateral membrane bathing solution) another Na^+ must move from the basolateral to the apical membrane bathing solution. If only Cl^- is permeable through the tight junction, then for every Na^+ that is transported through the cell a Cl^- will flow through the tight junctions.

Although at first sight an anion-selective paracellular pathway is a

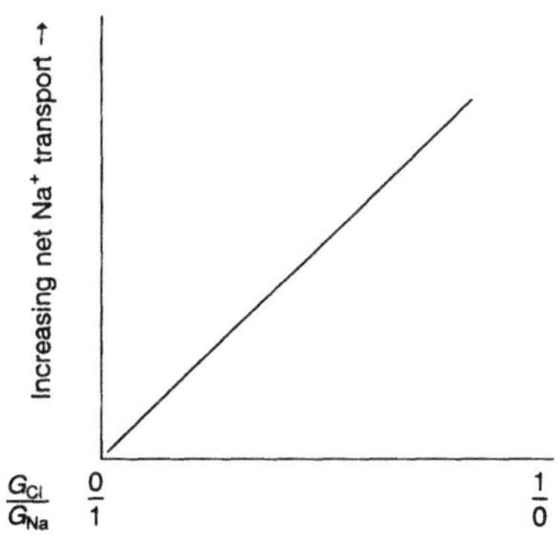

Figure 1.7 The ratio of the Cl^- conductance to the Na^+ conductance of a Na^+-transporting epithelium determines the rate of net Na^+ absorption. Note that as the junction becomes more permeable to chloride, the rate of net Na^+ absorption is increased.

desirable component in a Na^+ absortive epithelium, it has a serious limitation: a lumen-negative voltage is generated across the tight junction when the apical bathing solution [Cl^-] is decreased. This dramatically decreases the electrical driving force for Na^+ across the apical membrane. Thus, such an epithelium will only be able to absorb a limited quantity of ions from the luminal compartment before it reaches a 'static head', i.e. an apical membrane voltage that prevents Na^+ influx across this membrane. If the junction is equally permeable to sodium and chloride, then the epithelium will be able to generate a large concentration gradient; however, the system is only 50% efficient, since for every two Na^+ ions which are transported through the cell, one Na^+ will move from the basolateral to the apical bathing solution via the tight junction.

REFERENCES

Cereijido, M. (ed.) (1992) *Tight Junctions*, CRC Press, Inc., Boca Raton, Florida, USA.

Claude, P. and Goodenough, D.A. (1973) Fracture faces of zonulae occludentes from 'tight' and 'leaky' epithelia. *J. Cell Biol.* 58:390–400.

Curran, P.F. and MacIntosh, J.R. (1962) A model system for biological water transport. *Nature* 193:347–48.

Diamond, J.M. (1962a) The absorptive function of the gall-bladder. *J. Physiol.* 161:442–73.

Diamond, J.M. (1962b) The mechanism of solute transport by the gall-bladder. *J. Physiol.* 161:474–502.

Frömter, E. and Diamond, J.M. (1972) Route of passive ion permeation in epithelia. *Nature (New Biology)* 245:9–11.

Gumbiner, B. (1987) Structure, biochemistry and assembly of epithelial tight junctions. *Am. J. Physiol.* 253:C749–C758.

Koefoed-Johnsen, V. and Ussing, H.H. (1958) The nature of the frog skin potential. *Acta Physiol. Scand.* 42:298–308.

Whittembury, G. and Reuss, L. (1991) Mechanisms of coupling of solute and solvent transport in epithelia, in *The Kidney: Physiology and Pathophysiology*, (eds D.W. Seldin and G. Giebisch), Raven Press, New York, pp. 317–60.

2

Epithelial transport proteins

Luis Reuss, Nancy K. Wills and Simon A. Lewis

Epithelia separate fluid compartments within the body and transport ions, inorganic solutes and water between these compartments. Active transport occurs through the cells and is primarily dependent upon the operation of so-called ion pumps (primary active transport). These pumps create chemical or electrochemical driving forces that cause the passive movement of a single substance (passive transport via carriers or channels) or the coupled movement of two or more substances (secondary active transport via carriers). In addition to performing net transepithelial transport, membrane transport proteins can also perform housekeeping functions which regulate the volume and/or internal composition (e.g. pH or calcium concentration) of the epithelial cell.

2.1 ACTIVE AND PASSIVE TRANSPORT

Active transport is the net movement of a substance that occurs either without or against an existing electrochemical gradient. Thus active transport requires energy input. In contrast, passive transport is movement of a substance down an existing electrochemical gradient.

Epithelial Transport: A guide to methods and experimental analysis.
Edited by Nancy K. Wills, Luis Reuss and Simon A. Lewis.
Published in 1996 by Chapman & Hall, London. ISBN 978-94-010-7168-0

2.1.1 Primary active transport

The energy required for primary active transport of a substance is supplied usually from the hydrolysis of ATP by the transporter, i.e. 'pump' or 'ATPase'. All epithelia express one or more of the following ATPases: Na^+,K^+-ATPase; H^+-ATPase; H^+,K^+-ATPase; and Ca^{2+}-ATPase.

2.1.2 Secondary active transport

Secondary active transport utilizes energy stored in the electrochemical gradient for one substrate to transport another substrate or substrates. In epithelial cells, secondary active transport is in most instances coupled to Na^+ transport. Thus the energy stored in the Na^+ electrochemical gradient (produced by the Na^+ pump) is used to transport Na^+ and other substrates. This sodium-coupled transport is called **cotransport** when Na^+ moves down its electrochemical gradient and the other substrate moves in the same direction (in many instances against a net chemical or electrochemical gradient). In **countertransport**, the movement of Na^+ down its electrochemical gradient produces the movement of the substrate in the opposite direction.

2.1.3 Passive transport

Passive transport across a biological membrane can follow two pathways: through the lipid environment of the cell membrane, or via membrane transport proteins (i.e. carriers or channels). This type of transport is energetically downhill: it is driven by an existing chemical or electrochemical gradient, and involves the movement of a single substrate. Under physiological conditions (the same temperature and pressure in lumen and plasma compartments), the direction and magnitude of the driving force is dependent upon the differences in concentration of the substrate and (if the substrate is charged) the electrical potential difference across the membrane. The 'driving force' for passive transport is given by:

$$\Delta\tilde{\mu}_j = z_j V_m F + RT \ln (C_j^i / C_j^o)$$

where $\Delta\tilde{\mu}_j$ is the electrochemical potential difference (with units of joules/mole), the subscript j denotes the transported substance (in this instance an ion), z is the valence of the ion, V_m is the membrane voltage (in volts and with a polarity convention of V_i-V_o), F is the Faraday constant (96 487 coulombs/mole), R is the gas constant (8.314 joules/°K·mole; remember that joule = volt × coulomb), T is the temperature in degrees Kelvin (°K = 273.15 plus the temperature in degrees Celsius), C is concentration and the superscripts i and o refer to the two sides of the membrane (inside and outside, respectively).

2.2 PATHWAYS AND CONDITIONS FOR TRANSEPITHELIAL TRANSPORT

Substrates can move across an epithelium by either a paracellular or a transcellular route. The **paracellular pathway** is located between the epithelial cells (i.e. intercellular) and is composed of the tight junctions in series with the lateral intercellular spaces. The transport via this pathway is passive and occurs through a watery and sometimes electrically charged pathway. Although the architecture of the tight junction permeation pathway has not been determined, it is thought to be channel or pore-like in nature. Paracellular transport occurs predominantly in leaky epithelia (Chapter 1) and involves the movement (by diffusion or electrodiffusion) of small hydrophilic solutes and possibly water.

Transcellular transport is through the cells and thus it involves three steps: transport across one membrane; movement through the intracellular compartment; and transport across the opposite membrane. In general, active transepithelial transport requires a **pump** in one membrane and secondary active transport or diffusive transport ('**leak**') in the opposing membrane. In most epithelia composed of more than one cell type, net transport can be the result of substrate transport through at least two different and parallel transcellular pathways, as well as a paracellular pathway. These parallel pathways can have very different transport properties.

For highly lipid soluble substrates, passive transport occurs predominantly through the lipid bilayer, by a process usually referred to as **solubility-diffusion**. Hydrophilic solutes (which are not very soluble in lipid) such as ions and water use integral membrane proteins to facilitate their transit across the cell membrane by active transport, secondary active transport or passive transport mechanisms.

Two conditions must be met by the epithelium during steady-state transport. First, during transepithelial ion transport, the fluid compartments separated by the epithelial membranes and tight junction (two extracellular and one intracellular compartment) must remain electroneutral, i.e. there is no continuous accumulation of only a positive or a negative charge in any of the compartments (Chapter 1). As an example, net transepithelial transport of a cation must be balanced by net transport of anion(s) in the same direction and/or net transport of other cation(s) in the opposite direction. These cation and anion fluxes can occur through different transport molecules or different transport pathways. As an example, Na^+ absorption by tight epithelia occurs through a transcellular pathway and Cl^- follows passively (down the electrical gradient) through the paracellular pathway (Chapter 1). Alternatively Cl^- could follow through a transcellular route in which both the apical and basolateral membranes possess chloride conductances. Such a transcellular movement of the counter-ion has been proposed for the sweat duct and the frog skin epithelium.

In the absence of electrical or chemical gradients between the two external compartments, net transepithelial transport is a transcellular process. Thus there are continuous fluxes of solute and/or water across the epithelial cells. In the steady state (while cell volume and composition remain constant) the net amounts of solute and water entering the cell per unit time must be equal to the net amounts leaving across the opposing membrane over the same time period. Last, an increase or a decrease of net movement across one membrane must be matched by the opposing membrane so that cell volume and composition remain constant. This mechanism of matching influx with efflux has been termed **cross-talk**.

2.3 CLASSIFICATION OF EPITHELIAL MEMBRANE TRANSPORTERS

The transporters expressed in all cell membranes can belong to one of three categories: pumps, carriers and channels.

Ion pumps are ATPases which use the energy liberated by ATP hydrolysis to move the substrate against a net electrochemical gradient. The energy released during ATP hydrolysis is thought to change the conformation of the ATPase protein such that the substrate which initially associated with the protein from one compartment can only dissociate from the protein into the opposite compartment. The number of substrate molecules transported by a pump per unit time, (i.e. the 'turnover number') is typically 100 to 200 per second. For an excellent review of ion pumps, see Läuger (1991).

Carriers can perform either passive transport or secondary active transport. Carriers that transport only one subtrate are termed uniporters. Carriers that transport more than one substrate can use the electrochemical potential of one of these substrates to drive the energetically uphill movement of another substrate. The transport of two or more substrates in the same direction is performed by symporters (cotransporters), while the transport of two or more substrates in opposite directions is performed by antiporters (exchangers, countertransporters). Carriers can have multiple binding sites of different selectivities. In addition they are subject to pharmacological inhibition and saturation kinetics. Similarly to pumps, carriers must undergo a conformational change for every transport cycle. Typical turnover numbers for carriers are 10^2–10^4 molecules per second. For a review on carriers, see Stein (1990).

Channels perform only passive transport; thus they are unable to couple an energetically downhill movement of one substrate to the energetically uphill movement of another substrate. Channels have at least two states: an open state which allows the movement of the substrate from one compartment to the opposing compartment, and a closed state in which there is no movement of substrate. The transition between these two states is termed '**gating**' Like carriers, channels can display substrate selectivity, pharmacological inhibition and, in some instances, saturation kinetics. Channels differ from carriers and pumps in that the substrate can access the binding site (selectivity filter) from both solutions when the channel is 'open'. When the channel is in the open state, it allows the free movement of the substrate from one compartment to the other. Some channels rectify; i.e. in symmetric bathing solutions ions can traverse the channel more readily in one direction than the other. The typical turnover numbers for channels range from 10^6 to 10^7 ions per second. Given the differences in turnover number for pumps and channels, it is obvious that many ion pumps are necessary to match the ion flow through a channel. As an example, in the basolateral membranes of epithelial cells, it is likely that there are about 5000–10 000 Na^+ pumps per K^+ channel. For a review on channels, see Hille (1992).

The remainder of this chapter provides brief descriptions of the most common transporters found in epithelia.

2.4 PUMPS

2.4.1 H^+ pumps

Two kinds of H^+-transporting ATPases are expressed in plasma membranes of epithelial cells: vacuolar H^+-ATPases in the kidney; and H^+,K^+-ATPases in gastric mucosa, colon epithelium and renal tubule epithelial cells.

(a) H^+-ATPase

Recent review article
Gluck (1992).

Overview
The vacuolar H^+-ATPase contributes to H^+ secretion in proximal (probably a minor role) and distal nephron. H^+ transport mediated by this pump is regulated by and contributes to the maintenance of systemic acid–base homeostasis. This pump is shunted by FCCP, DCCD and moved by N-ethylmaleimide.

Distribution and regulation
The vacuolar H^+-ATPase in intercalated cells of the renal collecting duct is found in an intracellular pool of vesicles that can be inserted into the plasma membrane by exocytosis. There is evidence which suggests that this ATPase can be involved in both H^+ secretion and HCO_3^- secretion.

This apparent dual function results from pump insertion into either the apical membrane or the basolateral membrane.

Molecular characteristics

The H^+-ATPase is a multimeric protein of molecular weight ca. 580 kDa. SDS-PAGE of immunopurified enzyme reveals over ten polypeptides, of apparent molecular weights ranging from ca. 12 to ca. 70 kDa.

(b) Gastric H^+,K^+-ATPase

Recent review articles

Hersey and Sachs (1995); Klaassen and De Pont (1994).

Overview

The gastric H^+,K^+-ATPase (found in apical membrane of gastric parietal cells) is an α,β heterodimer that belongs to the P-type ion-transporting ATPases. The function of this ATPase is to secrete H^+. It performs this function by an uphill, electroneutral, obligatory exchange of a cell H^+ for a luminal K^+, coupled to ATP hydrolysis. K^+ and Cl^- enter the luminal compartment through apical membrane K^+ and Cl^- channels. Hence the net movement is HCl secretion (K^+ is recycled). This group of transporters can generate transmembrane H^+ concentration ratios greater than 10^6. The pump is inhibited by omeprazole.

Distribution and regulation

The enzyme is expressed in the tubulovesicular system and the apical membrane of the parietal cells of the gastric epithelium. Stimulation of the parietal cell with agonists results in translocation of H^+,K^+-ATPase molecules from tubulovesicular (intracellular) structures to the apical membrane domain, as well as in increased apical membrane conductance for both K^+ and Cl^-, which could result from activation and/or insertion of channels.

Molecular characteristics

The α-subunit of this ATPase, responsible for transport and ATP hydrolysis, has an apparent molecular mass of 114 kDa. Both the α and β subunits are quite homologous to the corresponding subunits of the Na^+,K^+-ATPase (for reviews see Hersey and Sachs, 1995; Klaassen and De Pont, 1994). The membrane topology of the α-subunit has not been definitively established. The β-subunit is composed of 291 amino acids and has an apparent molecular weight of 60–90 kDa. Similar to the Na^+,K^+-ATPase, this subunit has no catalytic function, but it is essential for expression and activity of the enzyme.

Colon and kidney H^+,K^+-ATPases

Recent review article
Klaassen and De Pont (1994).

Overview
In colon epithelial cells, the apical membrane expresses a Na^+-independent, K^+-stimulated ATPase activity sensitive to vanadate, omeprazole and SCH 28080. This ATPase appears to perform active H^+ secretion and is both functionally and immunologically similar to the gastric H^+,K^+-ATPase.

Distribution and regulation
The H^+,K^+-ATPase is expressed in intercalated cells of the cortical collecting duct and other distal nephron segments. The pump is located in the apical membrane and contributes to H^+ secretion, but its main physiological function appears to be K^+ reabsorption, which is stimulated in animals in low-K^+ diets.

Molecular identity
The colonic H^+,K^+-ATPase α-subunit is ca. 60% identical to the catalytic subunit of gastric H^+,K^+-ATPase and to the Na^+,K^+-ATPase isoforms (Klaassen and De Pont, 1994). The mRNA for the colonic H^+,K^+-ATPase is also expressed in kidney.

2.4.2 Ca^{2+}-ATPase

Recent review articles
Carafoli (1991, 1994).

Overview
This ATPase actively transports calcium from the cytoplasm to the interstitial fluid. For every cycle of the pump, two calcium ions are extruded. The affinity on the cytoplasmic side of the pump is high for calcium with a dissociation constant of 0.25 μM. The pump is inhibited by vanadate.

Distribution and regulation
This pump is found in the basolateral membrane of enterocytes and in renal tubule cells. It is important for the transepithelial transport of calcium. The pump is upregulated by calmodulin. In the small intestine, 1,25 dihydroxyvitamin D increases the calcium reabsorptive capacity. In vitamin D deficiency, the calcium ATPase activity is about 30% of that during maximal plasma levels of vitamin D.

Molecular characteristics

It has a molecular mass of 120–140 kDa. The intestinal isoform of the Ca^{2+}-ATPase has been partially purified and sequenced.

2.4.3 Na^+,K^+-ATPase

Recent review article

Lingrel and Kuntzweiler (1994).

Overview

This transport pump is ubiquitous in epithelia. Per cycle, it transports three Na^+ from the cytoplasm to the interstitial fluid and two potassium ions in the opposite direction. Thus, the pump is electrogenic, transporting an excess positive charge from the cytoplasm to the interstitial space. The turnover number is about 100–200 per second. The cardiac glycoside ouabain is a very specific blocker of this pump; however, the affinity of ouabain for the α-subunit is isoform dependent.

Distribution and regulation

This pump is found in all vertebrate epithelia studied to date. It is generally expressed in the basolateral membrane but in the choroid plexus and the retinal pigment epithelium the pump is in the apical membrane. One mechanism of regulation, demonstrated in the lacrimal gland acinar cell, is the rapid movement of cytoplasmic vesicles into the basolateral membrane. Another form of regulation includes an increase in basolateral pump density (many hours to days) by the hormone aldosterone, which might also require an increase in cell sodium activity.

Molecular characteristics

The pump is composed of α and β subunits in a 1:1 stoichiometry. There are three isoforms of the α-subunit (apparent molecular mass of 120 kDa) and two isoforms of the β-subunit (β1 and β2 with apparent molecular mass of 50 kDa). The α-subunit is the catalytic unit responsible for transport and ATPase activity. The β-subunit is required for correct assembly and targeting of the pump to the plasma membrane. The pump isoforms are tissue specific and can change during development.

2.5 CARRIERS

For carriers to contribute to active transepithelial transport their function must be coordinated with that of a pump. However, carriers and/or channels may be in series with a pump (e.g. apical Na^+-glucose cotransporter or amiloride-sensitive sodium channels in series with the basolateral Na^+,K^+-ATPase in some Na^+-absorptive epithelia) or in parallel with a pump (e.g. basolateral $Na^+,K^+,2Cl^-$ cotransporter, a K^+ channel and Na^+,K^+-ATPase in Cl^--secreting epithelia). Carriers and channels in epithelial cells dissipate the ion gradients generated by ion pumps. Thus, carriers and channels can be thought of as 'leak' pathways in parallel or in series with a pump. For a list of carrier systems found in epithelial cells, see Table 2.1.

2.5.1 Sodium coupled systems

In epithelia such as those of small intestine, renal proximal tubule and gallbladder, Na^+ entry across the apical membrane is carrier-mediated. We will consider four kinds of transporters that contribute to Na^+ entry: the Na^+–glucose cotransporter, the Na^+/H^+ exchanger, the NaK2Cl cotransporter and the thiazide-sensitive NaCl cotransporter.

(a) Sodium–glucose cotransporters

Recent review articles
Wright (1993); Wright *et al.* (1992); Reithmeier (1994); Thorens (1993); Hediger and Rhoads (1994).

Overview
This transporter accounts for large fractions of Na^+ entry in small intestine and renal proximal tubule. It is electrogenic and hence addition of the cotransported solute to the apical solution results in depolarization of the apical membrane. The apical-membrane Na^+/glucose cotransporters (SGLT gene family; Wright, 1993) are essential for salt and water absorption, both in intestine and proximal renal tubule. Glucose efflux across the basolateral membrane is via glucose uniporters (GLUT gene family; Thorens, 1993).

Distribution and regulation
SGLT is found in the villous cells of the small intestine and in proximal renal tubules. Two SGLT isoforms are expressed in apical membranes (SGLT1 and SGLT2; Wells *et al.*, 1992). SGLT1 has a high affinity for glucose (K_m = 0.35 mM), low capacity, a 2:1 Na^+:glucose stoichiometry and is expressed in the late proximal tubule (S2 and S3 segments); it is the only isoform found in enterocytes. SGLT2 has a lower affinity for glucose (K_m = 1.6 mM), high capacity, a 1:1 Na^+:glucose stoichiometry and is expressed in the early renal proximal tubule (S1 segment).

Molecular characteristics
SGLT1 is a 664-residue protein with significant N-linked glycosylation and 12 transmembrane domains. The amino acid sequence of SGLT1 is highly conserved in mammals.

Table 2.1 Additional ion carriers in epithelial cells

Carrier	*Isoform*	*Epithelium*	*Domain*	*Function*	*Features*
SYMPORTERS					
Na^+-amino acid (A)	SAATI	SIE,PT	A	Na^+ and amino acid uptake	660 aa, transports A,S,C,P,G
Na^+-phosphate	NPTI	PT	A	Na^+ and phosphate uptake	465 aa
Na^+-Cl^-	CCC-3	DT,WFUB	A	Na^+ and Cl^- uptake	1002, 1023 aa, respectively
K^+-Cl^-		GBE,PT,TALH	BL	K^+ and Cl^- efflux	not cloned
ANTIPORTERS					
Anion exchangers					
Cl^-/formate		PT	A	Cl^- uptake	not cloned
Cation exchangers					
Na^+/Ca^{2+}	NACA2,3	kidney	BL	Ca^{2+} reabsorption $[Ca^{2+}]_i$ regulation	cloned, 1966, 1945 aa, respectively

Abbreviations: aa = amino-acid residues; SIE = small intestine epithelium; GBE = gallbladder epithelium; PT = proximal tubule; TALH = thick

(b) Other Na^+-organic solute transporters

The Na^+ cotransporters perform secondary active transport, i.e. promote intracellular accumulation of nutrients, vitamins, bile salts and anions. The energy for uphill transport of these solutes is provided by the Na^+ electrochemical gradient across the cell membrane, generated by the operation of the Na^+,K^+ pump. This is referred to as the sodium-gradient hypothesis.

(c) Na^+/H^+ exchanger

Recent review articles
Reithmeier (1994); Pouysségur (1994).

Overview
The Na^+/H^+ exchanger (NHE) is a carrier that exchanges a Na^+ for a H^+ (in an electroneutral fashion). NHE performs two important functions: transepithelial transport (i.e. Na^+ absorption and H^+ secretion mediated by apical-membrane NHE) and a housekeeping function (cell pH and cell volume) regulation by basolateral-membrane NHE).

Distribution and regulation
NHE1 is the housekeeping exchanger, found in most cells. NHE2, NHE3 and NHE4 are expressed in epithelial cells, where they are targeted to the apical membrane. NHE2 (amiloride-sensitive) is expressed in rat small intestine and gastric mucosa, rabbit renal medulla and descending colon. NHE3 (amiloride-resistant) is expressed in proximal convoluted tubule, small intestine and gastric mucosa. NHE4 is expressed in rat gastric mucosa.

Regulation of NHE is isoform dependent. A common feature of all NHE isoforms appears to be the existence of an internal H^+-binding site that activates exchange when intracellular pH falls (ie. $[H^+]_i$ rises). For a review of regulation of NHE1, see Noël and Pouysségur (1995). The effects of cAMP can be stimulatory or inhibitory, depending on isoform and cell type.

Molecular characteristics
Four genes have been cloned: NHE1, NHE2, NHE3 and NHE4. The NHE exchangers contain from 717 to 832 amino acids, with homology ranging from 40 to 70% relative to human NHE1. The protein is predicted to have 10–12 membrane-spanning domains with intracellular N and C terminals. The isoforms exhibit large differences in sensitivity to blockers (amiloride and its analogs) but have similar substrate affinities.

(d) Na^+–K^+–$2Cl^-$ cotransporter

Recent review article
Haas (1994).

Overview
The Na^+–K^+–$2Cl^-$ cotransporter (NKCC) is a carrier protein that, although not exclusive to epithelial cells, has a major functional role in primary Cl^--transporting epithelia. NKCC performs electroneutral translocation of Cl^-, Na^+ and K^+ with the stoichiometry 2:1:1 but other stoichiometries have been proposed (Russell, 1983). Transport is inhibited by so-called 'loop' diuretics (furosemide, bumetanide and others). Kinetics analysis suggests ordered binding (Na^+, Cl^-, K^+, Cl^-). Bumetanide is thought to compete for the second Cl^- site (Haas, 1994).

Distribution and regulation
NKCC is expressed in Cl^--transporting epithelia in the membrane domain opposite to that in which the Cl^- channel is expressed. As examples, in Cl^--absorptive epithelia (thick ascending loop of Henle) the NKCC is present in the apical membrane, allowing net Na^+ and Cl^- uptake with basolateral-membrane Cl^- exit via Cl^- channels and Na^+ exit via the Na^+,K^+-ATPase. The K^+ is recycled by apical membrane K^+ channels. In the case of Cl^- secretory epithelia (e.g. airway epithelia), NKCC is present in the basolateral membrane (Cl^- uptake) in parallel with the Na^+,K^+-ATPase and the Cl^- channel is in the apical membrane (Cl^- flux from the cell interior to the luminal solution). The Na^+ is recycled across the basolateral membrane. Electroneutrality for transepithelial transport is maintained by Na^+ secretion through the tight junctions. In both instances, the NKCC performs secondary-active Cl^- transport, which raises intracellular $[Cl^-]$ to levels above those predicted for electrochemical equilibrium. Thus, Cl^- can then exit the cell by electrodiffusion via channels across the opposite membrane.

Regulation of NKCC (Haas, 1994) appears to involve changes in the density of active transporters in the membrane. Increases in cell cAMP levels, as well as cell shrinkage, increase the NKCC activity. Intracellular $[Cl^-]$ might play a role in the activation of NKCC, thus coupling the activity of the apical-membrane Cl^- channels to the cotransporter located in the opposing membrane. The sequence of events upon agonist-mediated stimulation of transport would be as follows: activation of Cl^- channels, fall of $[Cl^-]_i$, stimulation of the cotransporter expressed in the opposite membrane domain, and stimulation of Cl^- entry. This is one of a number of examples of membrane cross-talk.

Molecular characteristics
NKCC is a 195 kDa glycoprotein in shark rectal gland. A homologous 150 kDa protein has been identified in other species. The Na^+–K^+–$2Cl^-$

cotransporter from shark rectal gland (NKCC-1) has been cloned, sequenced and functionally expressed in human cells. A second isoform (NKCC-2, 61% amino-acid identity with NKCC-1) was found in rabbit kidney. Another Na^+–K^+–$2Cl^-$ cotransporter (rBSC) was cloned from rat kidney medulla. The NKCC-1 is 1191 amino acids long and has 12 predicted transmembrane (TM) helices; TM7 and TM8 are separated by a large extracellular loop, and both C and N terminals are in the cytoplasmic domain.

The NKCC appears to be related to the thiazide-sensitive cotransporter (TSC) in flounder urinary bladder and mammalian distal tubule. Haas (1994) has proposed the name cation–chloride cotransporters (CCC), with the following grouping: CCC-1 (shark rectal gland NKCC, NKCC-1), CCC-2 (mammalian kidney NKCCs: NKCC-2, rBSC), CCC-3 (winter-flounder urinary bladder and mammalian distal tubule thiazide-sensitive NaCl cotransporter).

2.5.2 Anion transporters

(a) Anion exchangers

Recent review article
Reithmeier (1993).

Overview
The anion exchangers (AE family) carry out electroneutral exchange of Cl^- and HCO_3^-. Similar to the NHEs, the AEs perform a housekeeping function (maintenance and regulation of pH_i, $[Cl^-]_i$ and cell volume) as well as transepithelial Cl^- transport. The prototype is the red cell band 3 (AE1). The fluxes are dictated by the chemical gradients for Cl^- and HCO_3^-. Disulfonic stilbene derivatives inhibit anion exchange. However, these agents also inhibit Na^+-HCO_3^- cotransport and block some anion channels.

Distribution and regulation
AE1b (a truncated version of AE1) is expressed in the renal collecting duct (basolateral membrane of intercalated cells). Na^+-independent Cl^-/HCO_3^- exchange has also been demonstrated in basolateral membranes of rabbit late proximal tubule and rabbit medullary collecting duct, but the molecular identity of these transporters is unknown.

Molecular characteristics
The AE family includes the related genes AE2 and AE3. AE1 consists of 911–929 amino acid residues, while AE1b consists of 850 amino acid residues. Its predicted membrane topology includes intracellular N and C termini and 12 transmembrane domains.

(b) Na^+–HCO_3^- cotransporter, Na^+-dependent Cl^-/HCO_3^- exchanger

Recent review article
Boron and Boulpaep (1989).

Overview
The experimental evidence for the existence of the Na^+–HCO_3^- cotransporter is convincing, and biophysical studies suggest Na^+:HCO_3^- stoichiometries of 1:2 or 1:3, i.e. electrogenic transport. Under normal physiological conditions, base extrusion takes place when the membrane voltage (cell interior negative) is rather large. The transport is Cl^- independent and highly sensitive to disulfonic stilbene derivatives.

The Na^+-dependent Cl^-/HCO_3^- exchanger is electroneutral, with an apparent stoichiometry of $(Na^+(HCO_3^-)_2/Cl^-)$.

Distribution and regulation
The Na^+–HCO_3^- cotransporter was first described in proximal tubule and accounts for > 90% of the HCO_3^- flux across the basolateral membrane of this epithelium. Other epithelia in which this cotransporter is expressed are the thick ascending limb of Henle's loop, monkey kidney cells in culture and gastric oxyntic cells. The cotransporter is subject to acute and chronic regulation. Acutely, it appears to respond to changes in pH_i by an effect on a modifier site; chronically, activity is increased by systemic acidosis.

The Na^+-dependent Cl^-/HCO_3^- exchanger has been demonstrated in basolateral membranes of proximal tubule cells. The available experimental data are consistent with electroneutral exchange involving $Na^+(HCO_3^-)_2/Cl^-$ or an equivalent process.

In proximal tubule, the contribution of this transporter to HCO_3^- transport across the basolateral membrane appears to be small in comparison with that of the Na^+–HCO_3^- cotransporter. Expression of $Na^+(HCO_3^-)_2/Cl^-$ in epithelial cells appears to be restricted to the proximal renal tubule.

Molecular characteristics
Neither of these transporters has been identified at the molecular level.

2.6 CHANNELS

Given the enormous literature on epithelial channels and space limitations, we have chosen only a few representative channels for the following summary.

2.6.1 Sodium channels

Recent review articles
Garty (1994); Palmer (1995).

Overview
Classification is as follows.

I. High Na^+ selectivity: $P_{Na}/P_K \geq 10$; low conductance (ca. 5 pS); long open and closed times (0.5–5.0 s); found in native tissues and cultured cells.
II. Moderate selectivity: $P_{Na}/P_K = 3$–4; higher conductance (7–15 pS); much shorter open and closed times (≤ 50 ms); found in cultured cells.
III. Non-selective: $P_{Na}/P_K \leq 1.5$; either high (23–28 pS) or low (≥ 3 pS) conductance; found in cultured cells.

Channels in all three groups are amiloride-sensitive and the apparent affinity for amiloride is high ($K_i \leq 0.5$ μM). The pattern of blockage by various amiloride analogs is different from that for the Na^+/H^+ exchanger or Na^+/Ca^{2+} exchanger. The blockage of the channel by amiloride appears to be a plug-type mechanism that senses about 12% of the membrane electric field.

Distribution and regulation
Highly selective Na^+ channels are expressed in the apical membranes of 'tight' (high electrical resistance) epithelia such as the descending colon and urinary bladder, respiratory tract, salivary and sweat ducts, taste buds and renal distal tubule and cortical collecting ducts. Aldosterone-sensitive and vasopressin-sensitive epithelia demonstrate increases in the open probability and/or number of conducting sodium channels in response to hormonal stimulation. Channel activity is decreased by elevations of intracellular sodium levels, probably via activation of protein kinase C.

Moderately selective Na^+ channels were reported to coexist with highly selective channels in cultured epithelia grown on impermeable supports. They are regulated by G proteins and actin filaments. The existence of moderately selective channels in native tissues is questionable. It is possible that they represent degraded highly selective channels (see 2.6.5).

Non-selective cation channels have been described in the apical membranes of primary cultures of inner medullary collecting ducts (IMCDs), endothelial cells, thyroid cells and cultures of nasal mucosa from cystic fibrosis patients. In IMCD cells, channel activity is downregulated by increases in cyclic GMP which phosphorylates the channel and decreases the P_o.

Molecular characteristics

Purification and reconstitution of amiloride-blockable channels have proven difficult. Early protein purification studies indicated that epithelial sodium channels consisted of hetero-oligomers with molecular weights of ca. 600–700 kDa consisting of four to five subunits. One difficulty is the existence of unrelated epithelial proteins that bind amiloride with a high affinity.

Recently a heteromultimeric sodium channel membrane protein consisting of three subunits was demonstrated by expression cloning of rat colonic epithelium (rENaC; Figure 2.1). The epithelial sodium channel represents a novel class of molecules that is distinct from the Na^+ channels of excitable membranes.

The three subunits (called α, β and γ rENaC) are highly homologous. Each has two transmembrane domains (called M1 and M2), short intracellular N and C termini (9–10 kDa), and large extracellular loops (ca. 50 kDa) with multiple N-linked glycosylation sites. The three subunits have a considerable degree of homology with mechanosensitive channels and with proteins involved in neurodegenerative disorders. Other properties of the subunits are as follows:

- αrENaC (78 kDa) contains the sodium-selective conducting pore that is blocked by amiloride.
- βrENaC and γrENaC (72–79 kDa) are necessary for augmented channel expression. A defect of the β subunit (a premature stop codon within the cytoplasmic carboxyl terminus) is associated with Liddle's syndrome, an inherited rare form of hypertension in which the channels are constitutively activated and sodium absorption is abnormally increased.

2.6.2 Potassium channels

Recent review articles

Taglialatela and Brown (1994); Wang *et al.* (1992).

Overview

Potassium channels in epithelia serve three important functions:

1. They generate a cell interior negative potential as a consequence of the high intracellular [K^+] compared with the extracellular fluid [K^+]. This function contributes to the electrical driving force for transport of other ions.
2. They provide a route for transepithelial K^+ absorption or secretion. The net direction of this transport will depend on the net electrochemical driving force for this ion (Chapter 7).

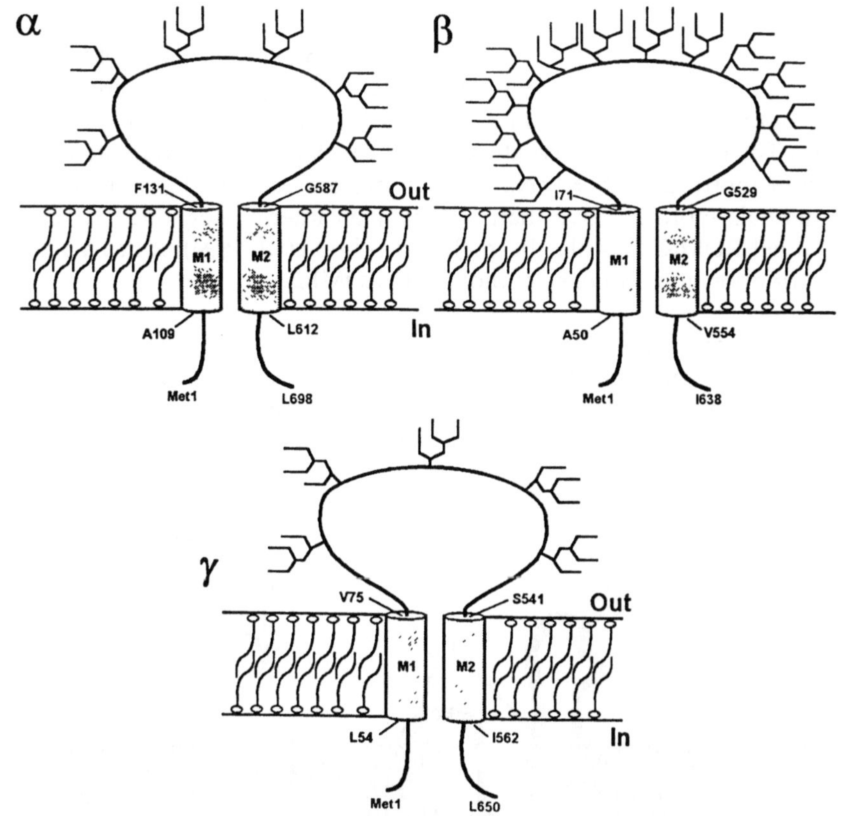

Figure 2.1 Subunits of the mammalian amiloride-sensitive Na^+ channel. (Adapted with permission from Canessa *et al.*, 1994.)

3. They are activated during cell swelling and are thought to play an important role in cell volume regulation.

Given the large number of potassium channels and their ubiquitous distribution, we will focus on a few more extensively studied channels, primarily from absorptive epithelia.

Although epithelia are not electrically excitable, many possess potassium channels that are voltage-gated (e.g. the maxi-K^+ channel). A few epithelia contain delayed rectifiers (DRK), or transient potassium current channels (TK) (e.g. lung and cornea). Many epithelia possess moderate conductance (10–60 pS) potassium channels and calcium gated maxi-K^+ channels (80–200 pS).

(a) ROMK1

An inwardly-rectifying ATP sensitive K^+ channel, ROMK1 (ROM = renal outer medulla) has been cloned (Ho *et al.*, 1993; Figure 2.2) and the properties of this channel match those from native renal ATP-sensitive K^+ channels. The channel has a conductance of ca. 39 pS (with moderate inward rectification), a high K^+/Na^+ selectivity, and a high P_o at membrane potentials more positive than –60 mV (cell interior negative). A related channel ROMK2 has also been cloned (Zhou *et al.* 1994); it has a similar distribution and sequence homology to ROMK1.

Molecular characteristics
The ROMK1 protein has a molecular weight of ca. 45 kDa. There are two potential membrane-spanning segments (M1 and M2) that flank a region that has a 59% similarity to the 17-amino-acid pore-forming segment of

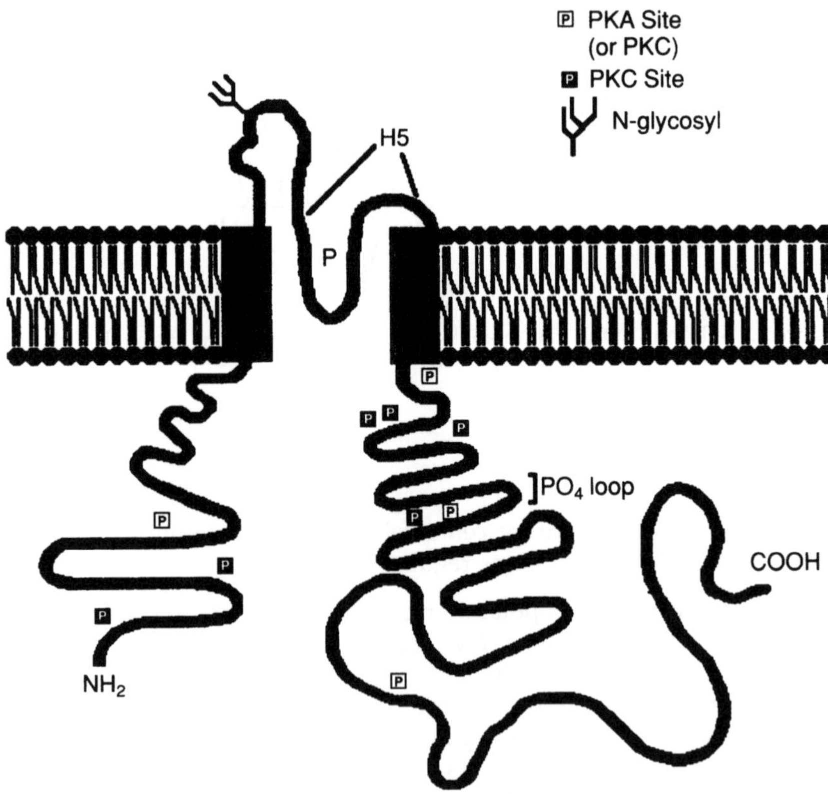

Figure 2.2 Topography of the ROMK1 potassium channel. (Adapted with permission from *Nature*, Ho *et al.*, 1993. Copyright 1993 MacMillan Magazines Ltd.)

voltage-gated K^+ channels (IRK1). There is no distinct homology to the putative voltage sensor region (S4) of voltage-gated channels. A 27-amino-acid segment following M1 contains an ATP-binding motif, and potential sites for PKA- and PKC-mediated phosphorylation. ROMK2 has almost complete homology and appears to be a splice variant that is missing the first 19 amino acids of ROMK1.

Distribution of channels and regulation
ROMK1 expression in renal cortex and outer medulla was detected in Northern blot analyses. No expression was found in inner medulla (Ho *et al.*, 1993). Data concerning the localization of ROMK1 in individual cells and membrane domains have not yet been published.

In conclusion, the functional characteristics and intrarenal distribution of ROMK1 indicate that this channel may be the ATP-sensitive K^+ channel found in mammalian kidney. Interesting features of this channel are its small size and relatively simple membrane topology, which should facilitate structure–function studies.

(b) Other potassium channels

A number of potassium channels have been characterized from patch clamp or reconstitution studies and many await further molecular characterization.

Inwardly rectifying apical potassium channels have been described for several regions of the renal tubule. These channels, referred to as SK or secretory potassium channels, have an inward conductance of ca. 20–40 pS and a high P_o that is slightly voltage dependent. Except for strong rectification (which may reflect a difference in renal cytoplasmic and intracellular oocyte Mg^{2+} levels), their properties are very similar to those of ROMK2. Partial clones of an aldosterone-regulated Shaker-type K^+ channel have also been reported and maxi-K^+ channels have been found in the apical membranes of renal proximal tubule, diluting segment and cortical collecting duct.

Rectifying basolateral potassium channels that have low or intermediate conductances and are activated by hyperpolarization have been reported in renal tubules and frog skin. A second type of inwardly rectifying, intermediate conductance channel has also been reported for renal tubules. The P_o of these channels is increased by depolarization, and decreased by internal acidification and by millimolar concentrations of ATP on the cytoplasmic side.

A number of basolateral channels have been identified in enterocytes. However, differences between the results of these studies have hindered definitive identification of the channels responsible for the basolateral K^+ conductance of these cells.

Recently I_sK, a ca. 130-amino-acid protein with a single putative transmembrane segment, was found in the apical membrane of epithelial cells of the proximal tubule and submandibular gland (Takumi, 1993). It has not been definitively established whether I_sK is a channel or channel modulator (Attall *et al.*, 1993; see also discussion of Cl^- channel modulators in section 2.6.3).

2.6.3 Chloride channels

Recent review articles
Guggino (1994); Fong and Jentsch (1995).

Overview
Chloride channels play an essential role in a number of epithelial functions, including Cl^- secretion, NaCl absorption and cell volume regulation. A review of the classification schemes for chloride channels can be found in Frizzell and Cliff (1992). Recent work has identified a number of epithelial Cl^- channels which vary with respect to primary sequences, functional characteristics and mechanisms of regulation. Major chloride channels include cystic fibrosis transmembrane conductance regulator (CFTR), outwardly-rectifying channels, Ca^{2+}-sensitive channels and swelling-activated channels.

(a) CFTR

CFTR, originally referred to as cystic fibrosis transmembrane conductance regulator, is an ATP-dependent, cAMP-regulated channel with a linear (ohmic) conductance of 7–10 pS (in symmetrical Cl^- solutions). Hallmarks of CFTR in human epithelia are a selectivity sequence of $Br^- > Cl^- > I^-$, blocking by 5-nitro-2-(3-phenylpropylamino) benzoic acid (NPPB) and glibenclamide, and insensitivity to 4,4′-diisothiocyanostilbene-2,2′disulfonic acid (DIDS). CFTR is the only putative anion channel whose function is supported by purification and functional reconstitution (Bear *et al.*, 1992).

Distribution and regulation
CFTR is found in the apical membranes of epithelia from most exocrine glands. A crucial feature of CFTR is its opening by intracellular cAMP, a regulation which is absent in individuals with cystic fibrosis (CF).

Molecular structure
CFTR has been cloned and sequenced, and shown to be similar to previously sequenced ATP-binding cassette (ABC) proteins. The predicted topology of CFTR is discussed and illustrated in Chapter 14 (Figure 14.6). In brief, CFTR consists of two regions, each containing six transmembrane

domains, some of which appear to form the pore, and two nucleotide-binding domain (NBD) consensus sequences. Between these two regions is a large hydrophilic moiety, the regulatory domain (R-domain), which requires phosphorylation for channel opening to occur.

CFTR is a multi-ion pore and replacement of a single arginine by glutamine abolishes this behavior. The results of studies using site-directed mutagenesis approaches and incorporation of partial constructs in artificial lipid bilayer membranes indicate that transmembrane segments 2 and 6 are required to form the pore, although the R-domain alone has been reported to cause anion channel activity when inserted in artificial lipid bilayer membranes.

Several laboratories have presented evidence that CFTR may also regulate the activation of other ion channels.

(b) Other chloride channels

Electrophysiological studies have identified several other epithelial Cl^- channels but they have not yet been characterized at the molecular level (for review, see Anderson *et al.*, 1992). For this reason, the properties of these channels are briefly summarized together:

- Ca^{2+}-activated Cl^- channel(s): 7–15 pS linear (or slightly outwardly rectifying) channels which are expressed in apical membranes of gastric, airway and colonic epithelial cells. They are more permeable to I^- than to Cl^- and are blocked by the disulfonic stilbene derivatives DIDS and DNDS, and by NPPB.
- Swelling-activated Cl^- channels: At least two types of swelling activated chloride channel have been reported in epithelia:
  1. a 35–50 pS linear channel with $P_I > P_{Cl}$, found in airway and intestinal epithelia and which is blocked by DIDS;
  2. a 350 pS linear channel found in cultured cells of cortical collecting duct and inner medullary cells of rabbit with the permeability sequence $P_{Cl} > P_{HCO_3} > P_{gluconate} > P_{Na}$ and which is blocked by diphenylamine-2-carboxylic acid (DPC), DIDS and NPPB.
- Outwardly rectifying Cl^- channel: This channel is expressed in the apical membranes of several epithelia and the basolateral membrane of the thick ascending limb of the loop of Henle. The channel is activated by depolarization and has a conductance of 40–50 pS, $P_I > P_{Cl}$, and is blocked by DIDS. In airway epithelia, it has been proposed that this channel is activated by external ATP and is also abnormally regulated in cystic fibrosis (for review, see Schwiebert *et al.*, 1994).

Molecular characteristics

Aside from CFTR, the Cl^- channels that have been most definitively characterized at the molecular level are those of the ClC family. The molecular masses of ClC gene products ranges from ca. 75 to ca. 110 kDa and hydropathy analyses suggest about 12 transmembrane domains. However, the topology is not precisely known. The halide selectivity sequence of ClC channels can be $Cl^- > Br^- >$ I- or $Br^- > Cl^- > I^-$.

(c) ClC-2 and ClC-K

ClC-2, a swelling-activated channel that is a member of the ClC family, is expressed in many cells, including several epithelia. Homologous proteins, specifically ClC-K1, ClC-K2, ClC-K2L and ClC-K2S, are expressed exclusively in the rat kidney. Two related isoforms have also been identified in human kidney.

The conductances of ClC-2 and ClC-K channels have not been published. Expression of ClC-K2L and CLC-K2s in *Xenopus laevis* oocytes yielded outwardly rectifying anion currents which were blocked by DIDS and DPC. An important feature of ClC-K1 is its increased transcription in dehydrated animals, indicating a possible function for the ClC-K1 group of channels in volume regulation. A third channel type, ClC-3, is found in kidney and other organs, but is less than 30% identical to other ClC proteins. The function of this channel has not been established.

Recently, a chloride channel has been cloned from apical membranes of acid-secreting cells of gastric mucosa (Malinowska *et al.*, 1995). This channel has 93% homology to ClC-2 and when expressed in oocytes was active at pH 3. The channel is linear and has a selectivity sequence similar to that of apical membrane chloride channels reconstituted in planar lipid bilayers (i.e. ca. 7 pS in 150 mM CsClm and $I^- > Cl^- > NO_3^-$).

(d) Proteins that regulate Cl^- channels

Several other proteins have been suggested to be Cl^- channels. This group includes phospholemman, pI_{Cln} and P-glycoprotein (a protein that shows increased expression in multidrug-resistant cancer cells). However, controversy exists as to whether these proteins function as chloride channels or regulators of Cl^- channel activity.

2.6.4 Calcium channels

Recent review articles
Miller (1992); Sather *et al.* (1994); Yu (1994); Friedman and Gesek (1993).

Overview
Calcium channels play an important role in Ca^{2+} absorption or reabsorption by intestinal and renal epithelia as well as an important function in

intracellular signalling pathways. Ca^{2+} channels also play a role in the increase in Ca^{2+} permeability observed after intracellular calcium depletion. Although a diverse array of plasma membrane and intracellular Ca^{2+} channels is found in cells, only a few studies have identified calcium channels in epithelial membranes. Epithelial calcium channel types include:

(a) Voltage-dependent Ca^{2+} channels

These channels are activated by depolarizing membrane potentials, inhibited by Co^{2+}, and blocked by phenylalkylamines, dihydropyridines and benzothiazepines. L-type channels are among the best studied Ca^{2+} channels and are defined as high-voltage activated channels that are sensitive to 1,4,-dihydropyridine compounds. Channels functionally similar to L-type Ca^{2+} channels have been observed in patch clamp studies of apical membranes from the rabbit connecting tubule and from cultured renal distal convoluted tubule cells. These findings are consistent with previous physiological investigations which indicated that renal Ca^{2+} reabsorption, a process regulated by parathyroid hormone, is mediated by a dihydropyridine-sensitive pathway.

Molecular characteristics and distribution
Voltage-sensitive Ca^{2+} channel cDNAs consist of five subunits (α_1, α_2, β, δ and γ). The pore-forming subunit (α) includes homologous repeats containing six membrane spanning regions. The voltage-sensing domain is thought to lie within the S4 segment of each repeat.

The diversity of calcium channel transcripts is a result of the expression of multiple genes and alternative splicing. For example, at least four genes are thought to encode α_1 subunits. Cloning studies identified a family of transcripts that encoded four putative α_1 Ca^{2+} channel subunits, all of which were expressed in the kidney (Yu, 1994). One clone (CaCh4) resembled the DHP-insensitive (i.e. P-type) Ca^{2+} channel. The role of these channels in kidney epithelial function remains to be established.

(b) Other Ca^{2+} channels and regulation

Cell swelling is thought to activate Ca^{2+} channels in a number of epithelia (Chapter 12). In addition, Ca^{2+} is known to be permeable through at least some stretch-activated cation channels (section 2.6.5).

2.6.5 Non-selective cationic channels

Recent review articles
Biel *et al.* (1994); Ahmad (1992); Marunaka *et al.* (1991).

Overview

Non-selective cation channels are widely distributed in epithelial cells, yet their function is poorly understood. Non-selective cation channels regulated by cGMP are important in sensory epithelia such as taste buds and hair cells of the inner ear. As discussed above, cGMP regulated cation channels are also found in renal inner medullary collecting duct cells. Many of these channels are blocked by amiloride. An amiloride-insensitive non-selective cationic channel from rabbit urinary bladder appears to be a degradation product of amiloride-sensitive channels (Lewis and Clausen, 1991).

A low conductance non-selective cation channel was reported from patch clamp studies of the apical membrane of cultured renal distal tubule cells. The channel had a conductance of 1 pS and was activated by nitroprusside and cGMP.

Some stretch-activated and mechanosensitive channels are cation non-selective and blockable by amiloride (see Chapter 12 for a discussion of mechanosensitive channels and their putative functions).

Molecular characteristics

Cyclic nucleotide gated (CNG) cation channels have been cloned from epithelial cells. Two proteins are expressed in the bovine kidney cortex and medulla: CNG-1, which is identical to the retinal CNG but expressed at relatively low levels, and CNG-3, which has an amino acid sequence 60% and 62% homologous to CNG from bovine outer rod segment and olfactory epithelium, respectively. A partial sequence of a 34 pS CNG channel was also reported for the M-1 cortical collecting duct cell line.

2.6.6 Water channels

Recent review articles

Sabolic and Brown (1995); Engel *et al.* (1994). An excellent source on water transport is Finkelstein (1987).

(a) Aquaporin: a family of water channel proteins

Since most lipid membranes have appreciable osmotic water permeabilities (P_f), osmotic water flow across epithelial cells does not necessarily require the existence of water channels or pores. Nonetheless, epithelia with high osmotic water permeability, such as leaky epithelia and ADH-stimulated tight epithelia, do contain such pores. The molecular features of these water channels, called aquaporins, have recently started to be identified.

The first member of this family to be discovered was AQP1 (CHIP28), a 28 kDa channel-forming integral protein (CHIP) initially found in the red blood cell membrane. Injection of *in vitro* transcribed CHIP RNA in *Xenopus* oocytes elicits an increase in P_f that is inhibited by antisense oligonucleotides. Reconstitution of this protein in liposomes also conferred an osmotic water permeability that is inhibitable by mercurial sulfhydryl reagents and has the biophysical characteristics of pore-mediated water flow.

Molecular characteristics
AQP1 is a membrane-bound glycoprotein that exists in both nonglycosylated (CHIP28) and N-glycosylated (glyCHIP; 40–60 kDa) forms. AQP1 is a tetramer in which each subunit may act as a pore. The selectivity of the pores for water is very high, i.e. there is virtually no permeation of ions or small nonelectrolytes.

Distribution and regulation
AQP1 is expressed in a number of epithelial cells. In the kidney, it is found in the apical and basolateral membranes of renal proximal tubule cells and in the descending thin limb of the loop of Henle. Other epithelia include: choroid plexus; ciliary body; lens; biliary tract; eccrine sweat gland; male reproductive tract; salivary glands; pancreatic acinar cells; and colonic crypt cells.

(b) Other water channels

Complementary DNAs encoding several proteins similar to AQP1 have been identified, including the following proteins that are important for water transport:

- the apical membrane ADH-regulated water channel of the renal collecting tubule, also called WCH-CD, AQP-CD and AQP2;
- the basolateral-membrane water channel of renal collecting tubule and intestine, also called AQP3 or BLIP;
- the mercury-insensitive water channel of the renal vasa recta, and the cells that line the subarachnoid space and ventricles in the central nervous system and conjunctiva.

In addition to conferring constitutive osmotic water permeability, aquaporins can be regulated. For example, AQP-CD, the water channel underlying the vasopressin-sensitive osmotic water permeability of apical membrane of renal collecting duct cells, is regulated by the hormone via its effect on V_2 receptors. This interaction leads to the elevation of intracellular cAMP levels and possibly other messengers mediating the fusion of vesicles containing these water pores.

REFERENCES

Ahmad, I., Kornmacher, C., Segal, A.S. *et al.* (1992) Mouse cortical collecting duct cells show nonselective cation channel activity and express a gene related to the gGMP-gated rod photoreceptor channel. *Proc. Nat. Acad. Sci. USA,* **89**:10262–6.

Anderson, M.P., Sheppard, D.N., Berger, H.A. and Welsh, M.J. (1992) Chloride channels in the apical membrane of normal and cystic fibrosis airway and intestinal epithelial. *Am. J. Physiol.* **263**:L1–L14.

Attall, B., Guillemare, E., Lesage, F., *et al.* (1993) The protein I_{sk} is a dual activator of K^+ and Cl^- channels. *Nature* **365**:850–2.

Bear, C.E., Li, C., Kartner, N. *et al.* (1992) Purification and functional reconstitution of the cystic fibrosis transmembrane conductance regulator (CFTR). *Cell* **68**:809-812.

Biel, M., Zong, X., Distler, M. *et al.* (1994) Another member of the cylic nucleotide-gated channel family, expressed in testis, kidney and heart. *Proc. Nat. Acad. Sci. USA,* **91**:3505–09.

Boron, W.F. and Boulpaep, E.L. (1989) The electrogenic Na/HCO_3 cotransporter. *Kidney Int.* **36**:392–403.

Canessa, C.M., Merillat, A.M. and Rossier, B.C. (1994) Membrane topology of the epithelial sodium channel in intact cells. *Am. J. Physiol.* **267**:C1682–C1690.

Carafoli, E. (1991) Calcium pump of the plasma membrane. *Physiol. Rev.* **71**:129–53.

Carafoli, E. (1994) Biogenesis: plasma membrane calcium ATPase: 15 years of work on the purified enzyme. *FASEB J.* **8**:993–1002.

Engel, A., Walz, T. and Agre, P. (1994) The aquaporin family of membrane water channels. *Curr. Opin. Struct. Biol.* **4**:545–53.

Finkelstein, A. (1987) Water Movements Through Lipid Bilayers, *Pores and Plasma Membranes: Theory and Reality.* Wiley, New York.

Fong, P. and Jentsch, T. (1995) Molecular basis of epithelial Cl- channels. *J. Memb. Biol.,* **144**:189–97.

Friedman, P.A. and Gesek, F.A. (1993) Calcium transport in renal epithelial cells. *Am. J. Physiol.* **264**:F181–F198.

Frizzell, R.A. and Cliff, W.H. (1992) Chloride channels. No common motif. *Curr. Biol.* **2**:285–7.

Garty, H. (1994) Molecular properties of epithelial, amiloride-blockable Na^+ channels. *FASEB J.* **8**:522–8.

Gluck, S.L. (1992) The structure and biochemistry of the vacuolar H^+ ATPase in proximal and distal urinary acidification. *J. Bioenerg. Biomembr.* **24**:351–8.

Guggino, W.B. (1994) Chloride Channels. *Current Topics in Membrane and Transport,* Vol. 42, Academic Press, New York, 487 pp.

Haas, M. (1994) The Na–K–Cl cotransporters. *Am. J. Physiol.,* **267**:C869–C885.

Hasegawa, H., Ma, T., Skach, W. *et al.* (1994) Molecular cloning of a mercurial-insensitive water channel expressed in selected water-transporting tissues. *J. Biol. Chem.,* **269**:5497–5500.

Hayashi, M., Sasaki, S., Tsuganezawa, H., *et al.* (1994) Expression and distribution of aquaporin of collecting duct are regulated by vasopressin V2 receptor in rat kidney. *J. Clin. Invest.* **94**:1778–83.

Hediger, M.A. and Rhoads, D.B. (1994) Molecular physiology of sodium-glucose cotransporters. *Physiol. Rev.* **74**:993–1026.

Hersey, S. and Sachs, G. (1995) Gastric acid secretion. *Physiol. Rev.* **75**:155–89.

Hille, B. (1992) *Ionic Channels of Excitable Membranes,* 2nd edn, Sinauer Associates, Inc., Sunderland, MA.

Ho, K., Nichols, C.G., Lederer, W.J. *et al.* (1993) Cloning and expression of an inwardly rectifying ATP-regulated potassium channel. *Nature* (Lond.) **362**:31–8.
Klaassen, C.H.W. and De Pont, J.J.H.H.M. (1994) Gastric H^+/K^+-ATPase. *Cell Physiol. Biochem.* **4**:115–34.
Läuger, P. (1991) Electrogenic ion pumps, in *Distinguished Lecture Series of the Society of General Physiologists,* Vol. 5, Sinauer Associates, Inc., Sunderland, MA, pp. 1–313.
Lewis, S.A. and Clausen, C. (1991) Urinary proteases degrade epithelial sodium channels. *J. Membr. Biol.* **122**:77–88.
Lingrel, J.B. and Kuntzweiler, T. (1994) Na^+,K^+-ATPase. *J. Biol. Chem.* **269**:19659–62.
Malinowska, D.H., Kupert, E.Y., Bahiniski, A. *et al.* (1995) Cloning, functional expression, and characterization of a PKA-activated gastric Cl- channel. *Am. J. Physiol.* **268**:C191–C200.
Marunaka, Y., Ohara, A., Matsumoto, P. and Eaton, D.C. (1991) Cyclic GMP-activated channel activity in renal epithelial cells (A6). *Biochim. Biophys. Acta BioMembr.* **1070**:152–6.
Miller, R.J. (1992) Voltage-sensitive calcium channels. *J. Biol. Chem.* **267**:1403–6.
Noël, J. and Pouysségur, J. (1995) Hormonal regulation, pharmacology, and membrane sorting of vertebrate Na^+/H^+ exchanger isoforms. *Am. J. Physiol.* **268**:C283–C296.
Palmer, L.G. (1995) Epithelial Na channels and their kin. *NIPS* **10**:61–7.
Pouysségur, J. (1994) Molecular biology and hormonal regulation of vertebrate Na^+/H^+ exchanger isoforms. *Renal Physiol. Biochem.* **17**:190–3.
Pusch, M. and Jentsch, T.J. (1994) Molecular physiology of voltage-gated chloride channels. *Physiol. Rev.* **74**:813–28.
Reithmeier, R.A.F. (1993) The erthrocyte anion transporter (Band 3). *Curr. Opin. Struct. Biol.* **3**:515–23.
Reithmeier, R.A.F. (1994) Mammalian exchangers and co-transporters. *Curr. Opin. Cell Biol.* **6**:583–594.
Russell, J.M. (1983) Cation-coupled chloride influx in squid axon. Role of potassium and stoichiometry of the transport process. *J. Gen. Physiol.* **81**:909–26.
Sabolic, I. and Brown. D. (1995) Water channel in renal and nonrenal tissues. *NIPS,* **10**:12–17.
Sather, W.A., Yang, J. and Tsien, R.W. (1994) Structural basis of ion channel permeation and selectivity. *Curr. Opin. Neurobiol.* **4**:313–23.
Schwiebert, E.M., Lopes, A. and Guggino, W.B. (1994) Chloride channels along the nephron, in *Chloride Channels* (ed. W.B. Guggino), Academic Press, New York, pp. 265–316.
Stein, W. D. (1990) *Channels Carriers, and Pumps. An Introduction to Membrane Transport,* Academic Press, Inc., San Diego, CA, 326 pp.
Taglialatela, M. and Brown, A.M. (1994) Structural correlates of K^+ channel function. *NIPS* **9**:169–73.
Takumi, T. (1993) A protein with a single transmembrane domain forms an ion channel. *NIPS* **8**:175–7.
Thorens, B. (1993) Facilitated glucose transporters in epithelial cells. *Ann. Rev. Physiol.* **55**:591–608.
Wang, W., Sackin, H. and Giebisch, G. (1992) Renal potassium channels and their regulation. *Ann. Rev. Physiol.* **54**:81–96.
Wells, R.G., Pajor, A.M., Kanai, Y. *et al.* (1992) Cloning of a human kidney cDNA with similarity to the sodium–glucose cotransporter. *Am. J. Physiol,* **32**:F459–F465.

Wright, E.M. (1993) The intestinal Na^+/glucose cotransporter. *Ann. Rev. Physiol.* **55**:575–89.
Wright, E.M., Hager, K.M. and Turk, E. (1992) Sodium cotransport proteins. *Curr. Opin. Cell Biol.* **4**:696–702.
Yu, A.S. (1994) Calcium channels. *Curr. Opin. in Nephrology & Hypertension* **3**(5):497–503.
Zhou, H., Tate, S.S. and Palmer, L.G. (1994) Primary structure and functional properties of an epithelial K channel. *Am. J. Physiol.* **266**:C809-C824.

3

Epithelial polarity

M. Cereijido, R.G. Contreras, M.R. García-Villegas, L. González-Mariscal and J. Valdés

The exchange of substances between the organism and the environment requires that fluxes across epithelia be highly vectorial, i.e. the amount of a given substance that crosses in one direction (lumen to blood, or blood to lumen) is usually orders of magnitude larger than in the opposite one. This vectoriality is due to a sharp structural, biochemical and physiological asymmetry of epithelial cells (polarity): the apical pole, facing the lumen, has translocating mechanisms (pores, channels, carriers, pumps, etc.) that are essentially different from those present on the basolateral side facing the interstitium (Cereijido and Rotunno, 1970).

Yet polarity is not the exclusive domain of epithelial cells, as it is also a fundamental feature of neurons, and may be even observed in single cells, such as spermatozoa, macrophages and fibroblasts that may show a tail or move in a given direction when triggered by specific stimuli (Figure 3.1). Finally, polarity is not reduced to the plasma membrane, because the nucleus of epithelial cells is frequently displaced towards the basal pole, the Golgi apparatus is interposed between the nucleus and the apical

Epithelial Transport: A guide to methods and experimental analysis.
Edited by Nancy K. Wills, Luis Reuss and Simon A. Lewis.
Published in 1996 by Chapman & Hall, London. ISBN 978-94-010-7168-0

domain, and microtubules and microfilaments are oriented and polarized even at the molecular level (Chapter 13). Therefore our understanding of the mechanisms for achieving and maintaining polarity in epithelial cells stems in fact from studies on practically all cell types, from bacteria to mammals. This view was stressed in recent years by the observation that the mechanisms involved in polarization, from yeasts to humans, appear to be controlled by homologous genes.

However, a given molecular species (e.g. $Na^+K^+ATPase$) may be distributed at random in the membrane of erythrocytes, basolaterally in cells of the intestinal mucosa, and apically in the choroid plexus. Some molecules reverse their polarity (e.g. IgG receptors) depending on whether they are occupied by the ligand, phosphorylated, etc. Finally, some epithelial cells in culture (Chapter 11) can reverse the position of the entire apical and basolateral domains, in response to changes in the composition of the bathing medium.

3.1 POLARITY OF EPITHELIAL CELLS

In most epithelial cells the tight junction (TJ), itself a polarizedly distributed structure (Cereijido, 1991; Cereijido *et al.*, 1989a,b), divides the plasma membrane into two major domains, the **apical** and the **basolateral**, although in some instances it is useful to subdivide the second one into basal and lateral.

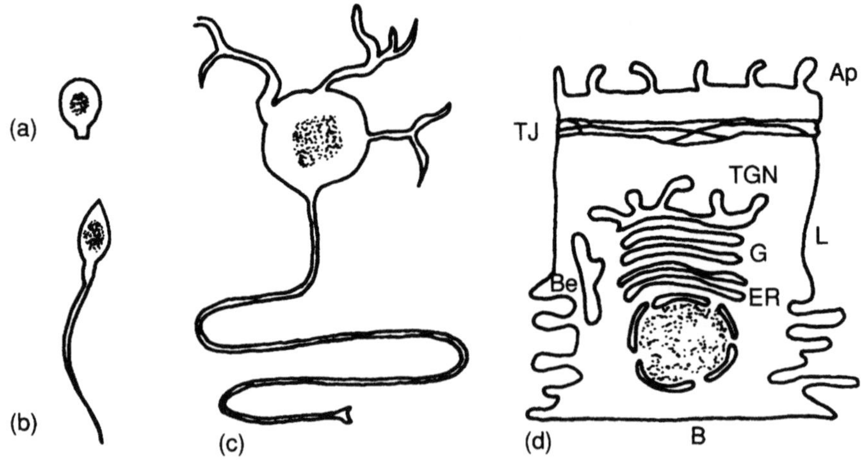

Figure 3.1 Polarity is a widely distributed property: (a) yeast cell; (b) spermatozoid; (c) neuron; (d) epithelial cell (not to scale). Epithelial cells have an apical domain (Ap) limited by the tight junction (TJ), a lateral (L) and a basal (B) domain. The distribution of internal organelles may be also polarized: endoplasmic reticulum (ER), Golgi apparatus (G), *trans*Golgi network (TGN) and basal endosome (Be).

3.1.1 The apical domain

This region is in contact with the lumen; it has microvilli, glycolipids, hydrolases, H^+-ATPase, Na^+-dependent cotransporters for sugars and amino acids, ion channels of a given type (e.g. the amiloride-sensitive Na^+ one, some types of Cl^- channels, some types of K^+ channels), receptors for IgG as in the intestinal mucosa of lactating animals, proteins anchored to the membrane by glycosylphosphatidylinositol, etc.

The main functions of the apical domain are the uptake of nutrients, vitamins, ions and water, regulated secretion towards the lumen, and protection.

3.1.2 The lateral domain

This is the domain where epithelial cells contact each other. It contains the zonula occludens, zonula adherens, desmosomes (Farquhar and Palade, 1963), gap junctions, cell adhesion molecules such as uvomorulin, and interdigitations. Pumps, ion channels and cotransporters on the lateral membrane are different from the ones present on the apical domain (Figure 3.2). The lateral membrane shares with the basal domain several types of receptors such as those for antidiuretic hormone, growth factor and neurotransmitters. Accordingly, it also has the molecular machinery for transducing signals from these receptors, such as phospholipase C (PLC), protein kinase C (PKC), adenylate cyclase, etc. (Chapter 12). Therefore this region is usually involved in the generation of ion gradients, secretion, signal reception and transduction.

3.1.3 The basal domain

This region shares with the lateral domain many molecular species, albeit at a different density; it has hemidesmosomes and specific receptors, and its contact with the basement membrane involves laminin, collagen type IV and proteoglycans. Freeze-fracture replicas of MDCK cells (epithelial of renal origin) show that the basal and the lateral domain have a much higher density of intramembrane particles in the P face than the apical one (Cereijido *et al.*, 1980).

3.2 APPROACHES TO THE STUDY OF CELL POLARITY

3.2.1 Microscopy

The earliest methods for studying polarity combined light microscopy with specific staining. Today the use of antibodies containing fluorescent tags (Figure 3.2) or gold particles, followed by optical sectioning with

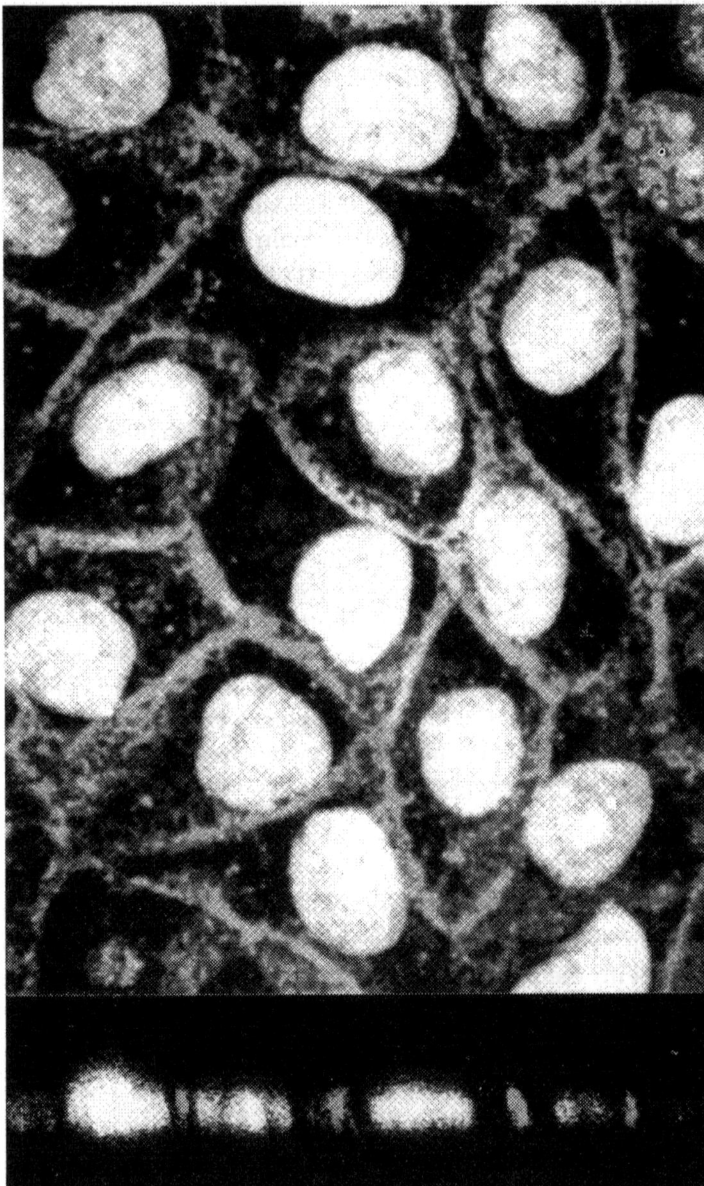

Figure 3.2 Monolayer of epithelial MDCK cells (from the dog kidney) viewed from above (up) and in a lateral optical section (down). Nuclei are stained with propidium iodine and the β-subunit of the Na^+,K^+-ATPase is stained with a monoclonal antibody using FITC-labeled secondary antibody. Note that the enzyme only occupies the lateral membrane.

confocal or electron microscopy, allows a precise mapping of the different regions of the cell membrane.

3.2.2 Biochemical

Biochemical studies depend on the isolation and purification of apical and basolateral membranes, guided by specific markers for each domain (Mircheff, 1989; and Chapter 9). Alternatively, the marker is identified through its union to a very specific inhibitor, such as labeled derivatives of ouabain, amiloride, phloridzin, etc. (Rabito, 1991; Rabito and Ausiello, 1980; Rabito and Tchao, 1980).

Most studies combine several methods. This is illustrated in Figure 3.3, showing a monolayer of epithelial cells cultured on a permeable support (chapter 11). Membrane proteins on the apical side are tagged with sulfo-NHS-biotin added to the upper bathing solution. Upon extraction with triton X-114, proteins are subjected to temperature-induced phase separation. The detergent phase is enriched in membrane proteins. Treatment with PI-PLC cleaves GPI-anchored proteins between the phosphate group and the diacylglycerol. After a second phase separation GPI-anchored proteins can be found in the aqueous phase. Biotinylated proteins are detected after SDS-PAGE transfer to nitrocellulose by [^{125}I)-streptavidin blotting and autoradiography. When sulfo-NHS-biotin is instead added only to the basolateral side, it labels a set of protein species which is different from the apical ones (Lisanti *et al.*, 1988, 1989; Sargiacomo *et al.*, 1989).

3.2.3 Physiological and pharmacological techniques

These methods consist of mounting the epithelium as a flat sheet between two Lucite chambers containing saline solution and then gauging vectorial secretion, studying unidirectional fluxes by adding tracers to one chamber and sampling the contralateral one, and studying the uptake or wash-out of tracers from the apical or from the basolateral side (Cereijido *et al.*, 1964).

Polarity can also be studied through the sidedness of the effect of inhibitors such as amiloride, ouabain and phloridzin. However, administration of a given substance to one side does not necessarily mean that it elicits its effect on that particular pole of the cell. Thus antidiuretic hormone only acts when added to the basolateral side, but the effect is elicited mainly on the apical one (Cereijido and Rotunno, 1971).

3.2.4 Electrical methods

One can also study the response of the spontaneous electrical potential to

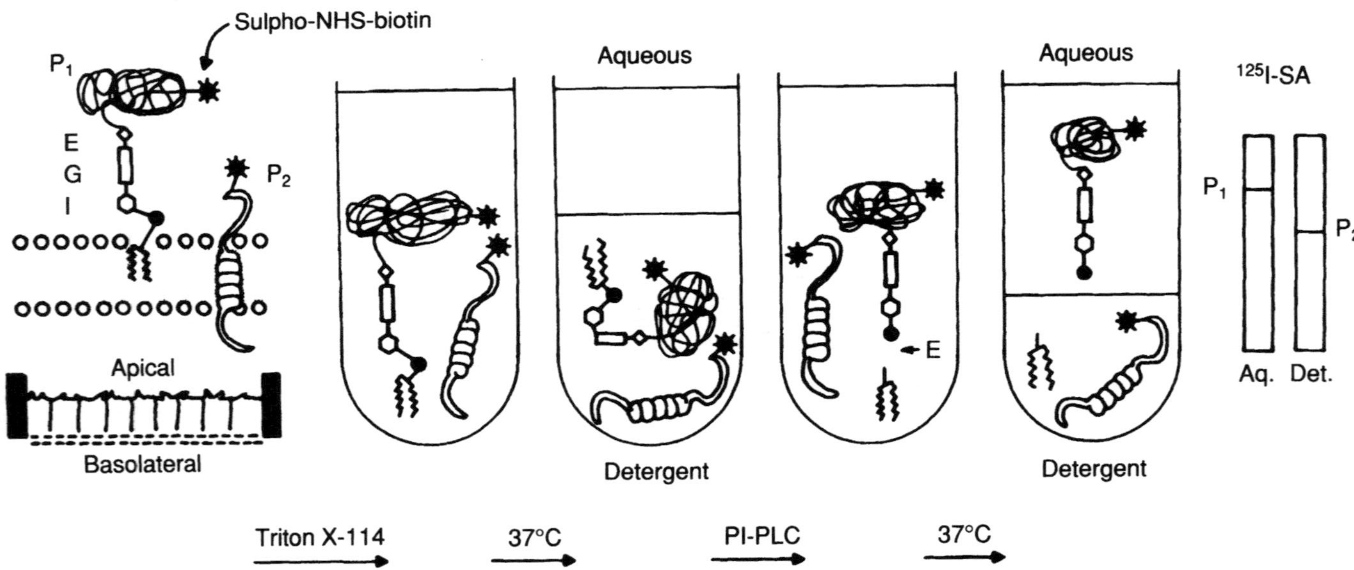

Figure 3.3 The study of the polarized distribution of membrane proteins usually involves several procedures. In this case a monolayer of epithelial cells, resembling an epithelium, is cultured on a polycarbonate filter (lower left). Sulpho-NHS-biotin is added only to the apical side. Because of the tight junctions, it cannot reach basolateral proteins and therefore can only label apical proteins (P_1 and P_2). P_1 is anchored to a phosphatidic acid of the plasma membrane, via an inositol (I), a glycan (G) and an ethanolamine (E). P_2 is a transmembrane protein. Proteins are then extracted with Triton X-114. Upon switching the temperature to 37 °C, the mixture separates into an aqueous and a detergent phase containing P_1 and P_2. Treatment of detergent phase with phospholipase C (E) splits P_1 at the phosphate level, thus releasing the hydrophilic moiety containing biotin. A new phase separation sorts the two proteins: P_1 goes to the aqueous phase and P_2 remains in the detergent one. P_1 and P_2 can be individualized through a SDS-PAGE/transfer to nitrocellulose, $[^{125}I]$-streptavidin blotting and autoradiography. The same procedure, but performed with biotin added to the basolateral side (not shown), would label a different set of proteins.

changes in the concentration of specific ions in one or the other bathing solution (Koefoed-Johnson and Ussing, 1958; Chapter 5).

Detection and analysis of ion channels with patch clamp is a particularly sensitive method for studying polarity of ion channels (Ponce *et al.*, 1991a,b; Ponce and Cereijido, 1991; Chapter 10). Thus, while identification of a given molecule such as Na^+,K^+-ATPase through the binding of a labeled marker (e.g. [^{3}H]-ouabain) results from the subtraction of millions of cold-competed sites per cell, patch clamp allows detection and characterization of a single channel at a time from millions of total sites labeled by the tagged drug. This method is particularly useful to study monolayers that can be patch clamped from either side (Figure 3.4). For further discussion of patch clamp methods, see Chapter 10.

3.3 APPROACHES TO STUDY THE MECHANISM OF POLARIZATION

Some tissues present their cells ordered topographically in different stages of differentiation. Thus, in the intestinal mucosa, as cells progress from crypt to apex of the villi, membrane markers are displaced from one region of the plasma membrane to another, providing information on the pathways they follow to achieve a polarized distribution. Another preparation amenable for dynamic studies, is the trophectoderm in the preimplantation mammalian embryo (Rodríguez-Boulán and Nelson, 1989).

Another valuable tool is offered by cell lines that retain the epithelial phenotype, and that can be cultured *in vitro* as monolayers on permeable supports, resembling natural epithelia in many respects (Misfeldt *et al.*, 1976; Cereijido, 1991; Cereijido *et al.*, 1978a,b, 1981a,b; Taub, 1985). Harvesting these cells with trypsin-EDTA makes them lose TJs and polarity – features that are quickly regained upon replating, in particular if replating is done at confluence so that cells do not have to spend time in proliferation to fill the available area.

Epithelial cells cultured in suspension show no sign of polarity or TJs, but in the presence of Ca^{2+} they attach to each other, gradually develop microvilli directed towards the bathing medium, and establish TJs at the outermost end of the intercellular space. TJs, in turn, constitute an effective fence between apical and basolateral domains (Mauchamp *et al.*, 1979; Nitsch *et al.*, 1985). Interestingly, as cells forming the suspended clump proliferate, they begin to generate a lumen at the center of the cluster; TJs are displaced to the opposite side of the intercellular space; and the apical and the basolateral domain, with their typical markers, are transposed to the opposite sides, indicating that polarity is a highly dynamic steady-state situation.

Since polarization depends on a host of intracellular mechanisms, one way to study these mechanisms is to use cell-free (Balch *et al.*, 1984) or

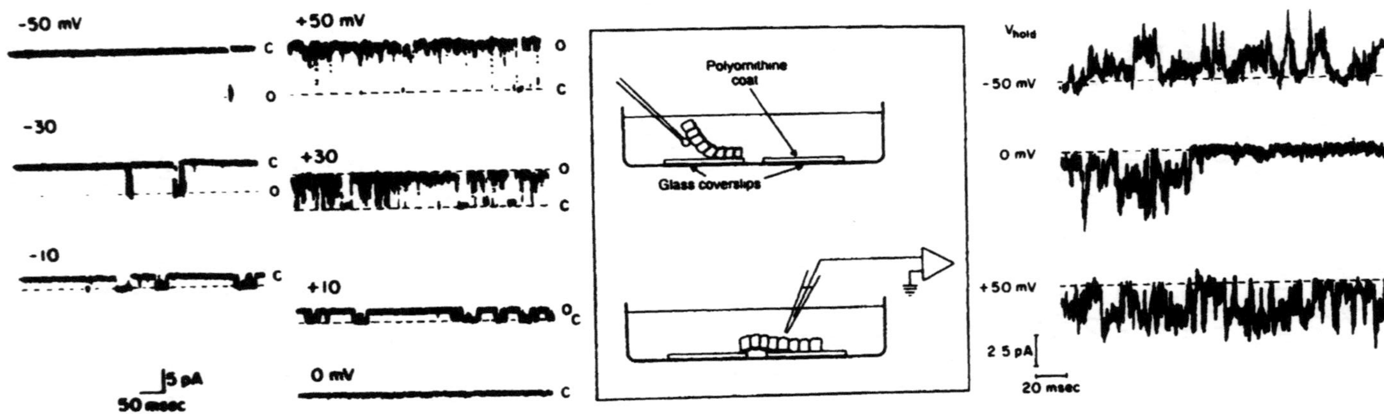

Figure 3.4 Exploration of polarity using patch clamp. A monolayer of MDCK cells is patch clamped from the apical side (left) and detects certain types of channels (in this example a Maxi-K^+). The monolayer is then carefully detached with glass rods, and bent over a polyornithine coated coverslip (center). Patch clamping of the basolateral area demonstrates that the types of channels found on the apical side are absent from the basolateral, and that this domain contains other types of channel (in this example a typically noisy Cl^- channel).

permeabilized cell systems (Ahnert-Hilger *et al.*, 1989; Schulz, 1990). Cells can be opened by 'filter stripping', scraping, high voltage discharge (Schulz, 1990), or by chemical permeabilization with either mild non-ionic detergents, such as saponin and digitonin (Perrin *et al.*, 1987) or bacterial toxins such as α-toxin and streptolysin-O (Gravotta *et al.*, 1990). Requirements for polarization can then be studied by substituting ATP with analogues, adding GTP, Mg^{2+}, antibodies, non-permeable inhibitors, diverse fractions of cytosol, etc. (Gravotta *et al.*, 1990; Pimplikar and Simons, 1993).

The process of directional addressing of proteins to the apical or to the basolateral domain can be also followed by infecting epithelial cells with enveloped RNA viruses (Rodriguez-Boulán and Sabatini, 1978; Rodriguez-Boulán and Pendergast, 1980). Certain envelope viral glycoproteins (e.g. hemagglutinin of the influenza virus) are targeted to the apical surface, while others (e.g. the envelope glycoprotein of vesicular stomatitis virus) are addressed to the basolateral domain. Infection shuts off the biosynthesis of endogenous proteins, and the cell uses its own machinery to synthesize envelope proteins of the replicated viruses and to recognize addressing signals, as well as to target, transport and insert them into the plasma membrane. The fate of viral components can be then followed with antibodies and several types of microscopy. The use of mutants and chimeric proteins helps to search for the sequence of amino acids constituting the sorting signals.

Yeasts are an important source of knowledge on polarity. Thus, *Saccharomyces cerevisiae*, a unicellular eukaryote, recruits cytoplasmic vesicles to a restricted area of the surface during mating and cell division (budding) and this results in polarized cell growth. The advantage offered by yeasts stems from the possibility of preparing mutants that die at 37 °C, but not at 25 °C, and identifying genes involved in the different stages of surface expansion.

Neurons transmit signals vectorially, a function that, of course, depends on the polarization of the plasma membrane into an axon and a cell body full of dendrites (Figure 3.1). The axon is in several respects analogous to the apical membrane of epithelial cells, and the cell body posseses mechanisms analogous to the basolateral domain (De Camilli and Jahn, 1990). A series of *unc*-mutants of the nematode *Caenorhabditis elegans*, which are deficient in neurotransmission, are helpful in the characterization of the steps leading to secretion and polarized insertion of membrane proteins (Hata *et al.*, 1993; García *et al.*, 1994).

3.4 ACHIEVEMENT AND MAINTENANCE OF POLARITY

Cells have a continuous flow of membrane vesicles, from the endoplasmic reticulum (ER) to the cisternae of the Golgi apparatus (G), to the trans-

Golgi network (TGN) and to the plasma membrane (Figure 3.1). Vesicles addressed to the plasma membrane carry basically two types of proteins: those contained in the lumen of the vesicles, that upon exocytic fusion become secreted to the extracellular medium, and those anchored to the membrane of the vesicle, that upon exocytic fusion become residents of the plasma membrane (Figure 3.5, P_1 and P_2). In this respect, the realm of polarity overlaps with the field of protein synthesis, with the study of membrane traffic from the ER to the G, to the TGN, to lyzosomes, to several types of cytoplasmal vesicles, and to the apical or basolateral domain of the plasma membrane, and with the field of secretion.

Furthermore, the study of polarity affords examples of almost all the possibilities that might have been conceived *a priori*:

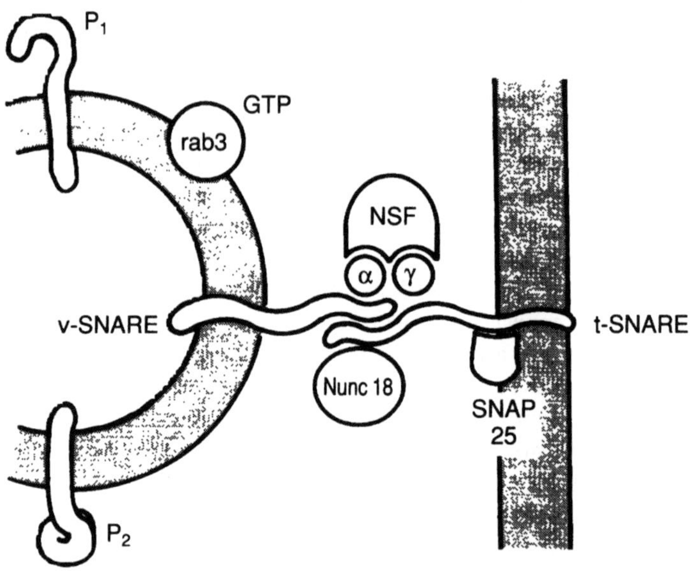

Figure 3.5 Apical/basolateral polarity depends on the addressing and incorporation of proteins into a specific membrane domain. These proteins are carried in vesicles (arched shade) split from the TGN, that fuse with a specific domain of the plasma membrane (vertical shade). SNAP proteins α and γ are recognized by specific receptors (SNAREs): v-SNARE is a resident of the vesicle membrane and t-SNARE is a resident of the target membrane. NSF, Nunc18 and SNAP25 are proteins required for recognition, attachment, fusion and retrieval of membrane; Rab3 is a small GTP-binding protein known to take part in the mechanism. Fusion incorporates proteins P_1 and P_2 into the target membrane. This mechanism is analogous to the one used by the cell to maintain a highly specific traffic of proteins between its different membrane compartments. Specificity is provided presumably by the nature of the SNAREs; other participants seem to be ubiquitous.

1. The simplest is that, once synthesized, proteins are sorted in the TGN into two different sets of vesicles, one addressed to the apical and the other to the basolateral domain (Rindler *et al.*, 1984; Wandinger-Ness *et al.*, 1990).
2. A second possibility is that the TGN delivers non-sorted proteins to the whole plasma membrane, but only those pertaining to the domain they happened to reach would be retained; the rest would keep recycling, until eventually they contacted their correct destination (Hammerton *et al.*, 1991; Matter *et al.*, 1990a,b).
3. A third possibility is that all protein species are targeted to a single domain, where only those belonging to that domain would be retained (Bartles *et al.*, 1987).
4. A fourth possibility would be a default delivery of all proteins to a particular domain, unless they have a signal that addresses them to a different target.
5. To these scenarios one has to add the delivery of proteins to intracellular organelles, either as a final destination such as the lyzosome (Nabi *et al.*, 1991), or as an intermediate step in the recycling of the plasma membrane, such as the basal endosome (Rodríguez-Boulán and Powell, 1992).
6. The possibility also exists that proteins would not leave the TGN via vesicles, but through ephemeral tubules, transiently connecting this compartment with adjacent intracellular compartments or with the plasma membrane (Cooper *et al.*, 1990).

3.5 MEMBRANE TRAFFIC AND POLARITY

The key process in the traffic routes mentioned above is the budding of vesicles from one membrane and fusion to another. So it is pertinent to pursue our study of polarity by reviewing briefly what is known about this traffic (Figure 3.5).

3.5.1 SNAPs and SNAREs

Vesicles trafficking from one membrane to the other have resident proteins called v-SNAREs, attached to their exterior by a single carboxy-terminal. Correspondingly, target membranes belonging either to a cisterna of the Golgi or to the plasma membrane have t-SNAREs, attached in the same manner (Söllner *et al.*, 1993; Ferro-Novick and Jahn, 1994). In nerve terminals v-SNAREs are called synaptobrevins and t-SNAREs are called syntaxins. On the other hand, there is a homotetramer known as NSF (N-ethylmaleimide-sensitive factor) that interacts with attachment proteins called α- and γ-SNAPs. When NSF binds to SNAPs, it develops an affinity for SNAREs. In the presence of Mg^{2+}, NSF cleaves ATP and triggers fusion of the vesicle with the target membrane. The fact that v-SNAREs do not accumulate on the target membrane suggests that they are recycled after fusion. While NSF and SNAPs are common, there probably exists a variety of SNAREs that would be responsible for selective matching of a given vesicle with its specific target.

A cohort of 'assistant' proteins participate in these processes, helping in v-SNARE/t-SNARE recognition, fusion and recycling. Tetanus and botulinum toxins are homologous heterodimers, with a light chain that penetrates the cell membrane of certain neurons and acts as Zn^{2+}-dependent protease that cleaves synaptobrevin and blocks exocytic membrane fusion. However, in spite of their structural similarity, the substrates of the two neurotoxins are mutually exclusive, i.e. they do not act on the same synaptobrevin species (Schiavo *et al.*, 1992; Link *et al.*, 1992).

3.5.2 Small GTP-binding proteins

In addition to the proteins mentioned above, there is a family of small GTP-binding proteins (called Ras in mammalian cells, and Ypt and Sec4 in yeast) that may act as 'switches' regulating the assembly of docking/fusion complexes, but do not seem to play a role in membrane recognition (Brennwald and Novick, 1993; Dunn *et al.*, 1993). In recent years, a number of other proteins involved in these process have been detected but the review of this topic is beyond the scope of this chapter (see Bennett and Scheller, 1993; Ferro-Novick and Novick, 1993).

3.6 SIGNALS

Phosphorylated mannose side chains on lysosomal hydrolases are recognized in the Golgi complex by specific receptors that target the hydrolases to lysosomes (Sly, 1982). Yet N-linked carbohydrates do not appear to act as signals in proteins addressed to the plasma membrane, as inhibition of glycosylation with tunicamycin does not alter the polarized distribution of glycoproteins (Green *et al.*, 1981). It has recently been observed (Scheillele *et al.*, 1995) that N-glycans act as apical sorting signals in secretory proteins. Chimeras prepared with viral proteins addressed to the apical and to the basolateral membrane suggest that signals reside in the ectodomain of the protein (Roth *et al.*, 1987). However, the transmembrane and cytoplasmic domains do not lack targeting signals, as these portions are responsible for the basolateral addressing of the envelope protein of VSV virus, and are also able to address to this domain chimeric peptides containing secretory proteins, such as human chorionic gonadotropin (Brown and Rose, 1988). This suggests that there is a hierarchy of signals, and that some of them only become apparent when the others are absent. Furthermore, there is a transmission of information between different segments of a peptide. Thus, while information for the apical sorting of the poly-Ig receptor seems to reside in the amino acid sequence of the ectodomain (Mostov *et al.* 1980), phosphorylation of the cytoplasmic side modifies its apical and basolateral expression (Casanova *et al.*, 1990).

There is a special family of plasma membrane proteins anchored through the glycophospholipid glycosylphosphatidylinositol (GPI). This anchoring has been correlated strictly with the apical localization of certain proteins in a variety of epithelial cells. Exogenous GPI-anchored proteins expressed from their transfected cDNA are also expressed to the apical domain. Recombinant DNA replacing the transmembrane and cytoplasmic domains of otherwise basolaterally targeted viral glycoproteins by GPI results in their rerouting to the apical domain. Therefore GPI anchoring constitutes a distinct mechanism for achieving polarity (Lisanti *et al.*, 1990, 1991; Lisanti and Rodríguez-Boulán, 1990).

Lipid molecules in each leaflet of a membrane tend to pack by accommodating to each other according to van der Waals attraction forces between hydrocarbon chains and attraction/repulsion forces between charged polar groups. These interactions can lead to segregation of a given lipid species in a patch of membrane. Likewise, a proteolipid in the TGN might be incorporated in those patches, by virtue of its lipid moiety. Then, in principle, if some of the molecules in the lipid patch are sorted to the apical or basolateral membranes, other molecular species occupying the patch may become (secondarily) polarized (Simons and Wandinger-Ness, 1990; Rodríguez-Boulán and Powell, 1992).

3.7 THE ROLE OF CALCIUM

Harvesting with trypsin-EDTA opens cell–cell junctions and provokes hydrolysis of membrane proteins exposed to the extracellular protease, endocytosis of membrane and randomization of the distribution of proteins remaining in the plasma membrane. Replating seems to suffice to promote resynthesis of the destroyed proteins, yet newly synthesized peptides are not addressed or distributed in the membrane unless Ca^{2+} is present in the bathing solution (González-Mariscal *et al.*, 1985a,b; Salas *et al.*, 1985; Rindler *et al.*, 1984). The crucial role played by this ion is studied using the Ca-switch protocol depicted in Figure 3.6. MDCK cells incubated overnight in 'Ca-free' medium have a very low cytoplasmic concentration (ca. 20 nM) and a very high permeability to this ion (González-Mariscal *et al.*, 1990). Upon switching to 1.8 mM Ca^{2+} the intracellular concentration shows a sharp peak (Figure 3.7), TJs form, polarity develops and calcium permeability drops sharply and quickly (González-Mariscal *et al.*, 1990). Blocking the penetration of Ca^{2+} into the cytoplasm with La^{3+} or verapamil keeps cytoplasmic Ca^{2+} at very low levels but does not inhibit its effect, suggesting that this ion acts on an extracellular site. Although La^{3+} can

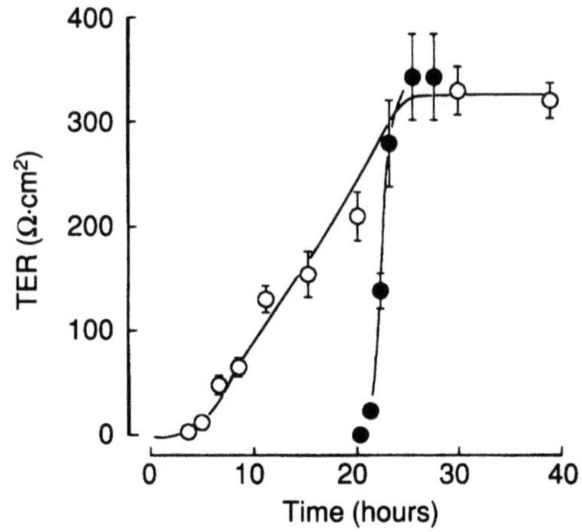

Figure 3.6 Ca^{2+} is a crucial factor participating in the last steps of polarization and TJ formation, which can be studied in a Ca-switch protocol. MDCK cells plated at confluence contact each other and establish tight junctions. As a consequence, a transepithelial electrical resistance (TER) develops (open circles). In the absence of Ca^{2+}, junctions do not form. Added at the 20th hour, Ca^{2+} triggers a cascade of intracellular reactions, as a result of which cells establish tight junctions and TER develops (filled circles). Junction formation is a particular case of polarization that proceeds with the simultaneous insertion of other apical and basolateral markers.

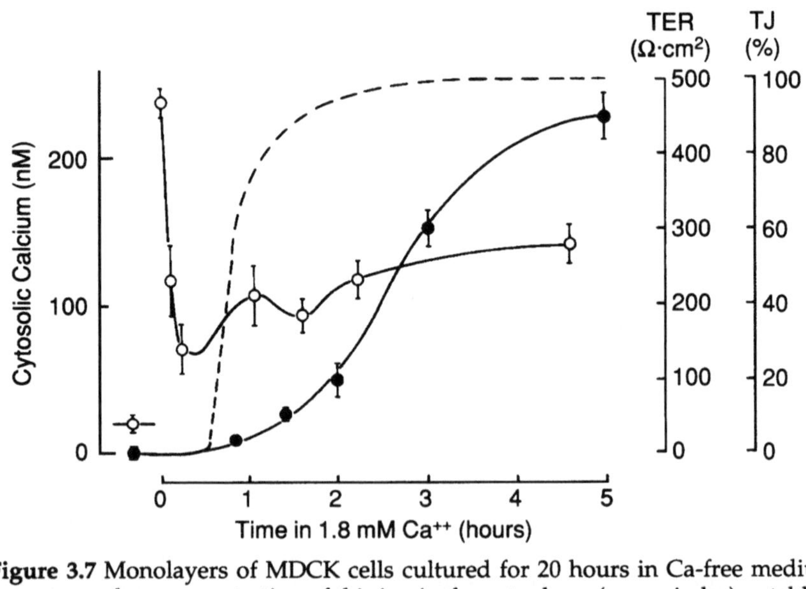

Figure 3.7 Monolayers of MDCK cells cultured for 20 hours in Ca-free medium have a very low concentration of this ion in the cytoplasm (open circles), establish no tight junctions (dashed curve) and show no transepithelial electrical resistance (filled circles). Transfer to media with 1.8 mM Ca^{2+} (time zero) produces a sharp peak of intracellular Ca^{2+}, followed by a slower rise to a steady state concentration.

block the penetration of Ca^{2+}, it cannot replace this ion in the triggering of the differentiated phenotype. Instead, Cd^{2+} can block penetration of Ca^{2+} and also suppress its ability to trigger junction formation and polarization, indicating that the extracellular sites involved have a high selectivity for Cd^{2+}. Nevertheless this selectivity is not enough for Cd^{2+} to substitute Ca^{2+} (Contreras *et al.*, 1992a,b). One of the extracellular sites where Ca^{2+} acts is uvomorulin, a molecule that in the presence of this ion interacts with homologous ones in the plasma membrane of neighboring cells (Gumbiner and Simons, 1986).

Binding of Ca^{2+} to the extracellular site triggers a series of cellular events, such as exocytic fusion (González-Mariscal *et al.* 1990), rearrangement of the cytoskeleton (Meza *et al.*, 1980, 1982; Nelson and Veshnock, 1987), incorporation to the plasma membrane of Na^+,K^+-ATPase (Contreras *et al.*, 1989), insertion of ion channels (Talavera *et al.*, 1995) and the establishment of several types of cell–cell contact (Musil, 1994; Grunwald, 1993). The fact that Ca^{2+} acts on extracellular sites, but these processes occur on the cytoplasmic side indicates that there must be a mechanism to transduce the information across the plasma membrane. Balda *et al.* (1991) have shown that this transduction involves at least two G-proteins, phospholipase C (PLC), protein kinase C (PKC) and calmodulin (Figure 3.8) (see also González-Mariscal *et al.*, 1994).

3.8 THE ROLE OF THE CYTOSKELETON

Microtubules are oriented along the apical/basolateral axis, with the minus end toward the apical surface (Bacallao *et al.*, 1989), suggesting that besides their role in the distribution of intracellular organelles (Kelly, 1990), microtubules also play a role in membrane polarity. However, studies with drugs that depolymerize microtubules show controversial results, a situation that may reflect an indirect participation of these structures (Salas *et al.*, 1985, 1986; Rindler *et al.*, 1987; Matter *et al.*, 1990b).

Newly plated MDCK cells lack TJs and have their membrane proteins distributed at random, a situation maintained if, as mentioned above, monolayers are kept overnight in Ca-free media. Contreras *et al.* (1989) have shown that, after the switch to Ca^{2+}, TJs form so rapidly that a fraction of Na^+,K^+-ATPase is trapped in the newly formed apical position. The cell then inserts new Na^+,K^+-ATPase in the correct basolateral location (Caplan *et al.*, 1986) and simultaneously removes it from the apical domain (Contreras *et al.*, 1989). The reason Na^+,K^+-ATPase is not retained in the apical (wrong) position, but is anchored in the lateral (correct) side, is that in the latter position it forms an insoluble complex with fodrin/ankyrin (Hammerton *et al.*, 1991). In turn, the reason this complex is stabilized on the basolateral side is the formation of the cytoskeleton of actin filaments (McNeill *et al.*, 1990), suggesting that the polarized distrib-

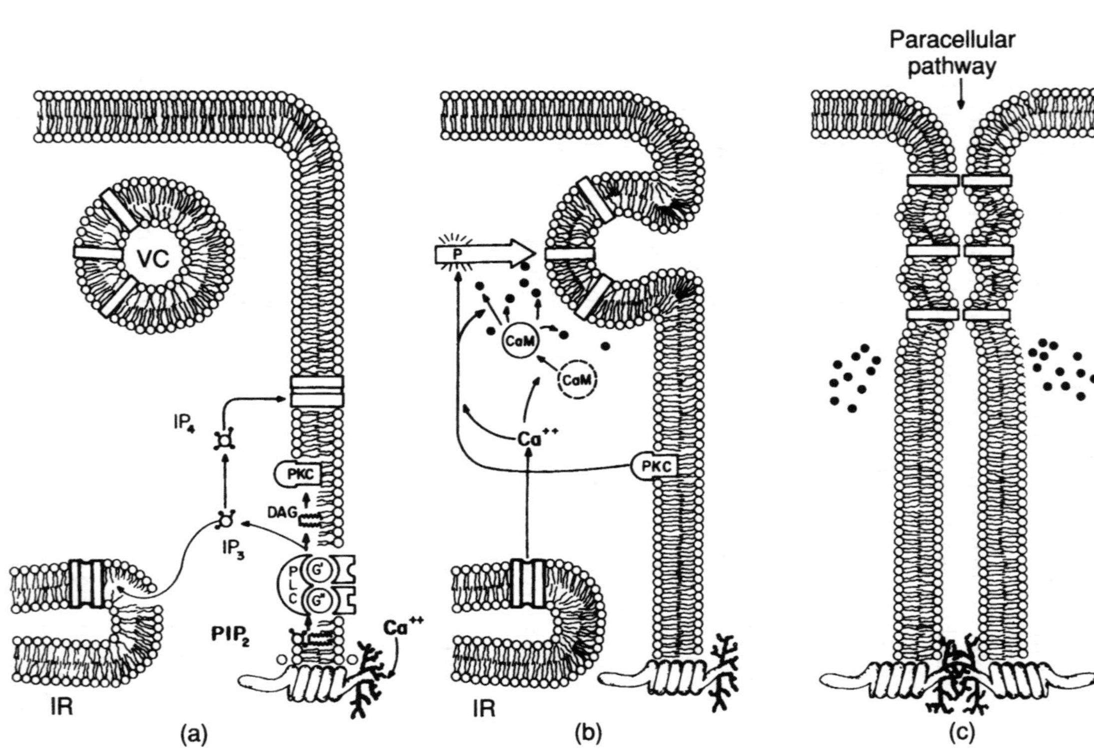

Figure 3.8 Schematic view of a corner of an epithelial cell, to illustrate the effect of Ca^{2+} in polarization and tight junction formation. (a) Cells incubated for 20 hours in Ca-free medium have several populations of cytoplasmic vesicular compartments like VC, containing molecules targeted to the apical, the basal, the lateral or the TJ region. Extracellular Ca^{2+} activates glycosylated uvomorulin molecules and permits the attachment to neighboring cells. This contact stimulates phospholipase C (PLC) that is connected to membrane receptors via two G-proteins and converts phosphatidyl inositol (PIP_2) into inositol triphosphate (IP_3) and diacylglycerol (DAG). IP_3 mobilizes Ca^{2+} from an internal reservoir (IR), D-*myo*-inositol 4-monophosphate (IP_4) decreases Ca^{2+} permeability, and DAG activates protein kinase C (PKC). (b) This cascade of reactions provokes phosphorylation (P) of still undeterminated peptides, and incorporation of membrane components through exocytic fusion of the VCs (Figure 3.5) and activation of calmodulin (CaM). (c) CaM is activated and actin filaments organize into a continuous ring that encircles the cell (represented by a group of filled circles in each neighboring cell).

ution of Na^+,K^+-ATPase arises as a consequence of the peculiar positioning of the actin cytoskeleton. (For further discussion of cytoskeleton, see Chapter 13.)

REFERENCES

Ahnert-Hilger, G., Mach, W., Fohr, K.J. and Gratzl, M. (1989) Poration by alpha-toxin and streptolysin-O: an approach to analyze intracellular processes. *Methods Cell Biol.* **31**:63–91.

Bacallao, R., Antony, C., Dotti, C. *et al.* (1989) The subcellular organization of Madin–Darby Canine Kidney cells during the formation of a polarized epithelium. *J. Cell Biol.* **109**:2817–32.

Balch, W.E., Dunphy, W.G., Braell, W.A. and Rothman, J.E. (1984) Reconstitution of the transport of protein between successive compartments of the Golgi measured by the coupled incorporation of N-acetylglucosamine. *Cell* **39**: 405–16.

Balda, M.S., González-Mariscal, L., Contreras, R.G., and Cereijido, M. (1991) Assembly and sealing of tight junctions: possible participation of G-proteins, phospholipase C, protein kinase C and calmodulin. *J. Membr. Biol.* **122**:193–202.

Bartles, J.R., Feracci, H.M., Stieger, B. and Hubbard, A.L. (1987) Biogenesis of the rat hepatocyte plasma membrane in vivo: Comparison of the pathways taken by apical and basolateral proteins using subcellular fractionation. *J. Cell Biol.* **105**:1241–51.

Bennett, M.K. and Scheller, R.H. (1993) The molecular machinery for secretion is conserved from yeast to neurons. *Proc. Nat. Acad. Sci.* USA **90**:2559–63.

Brennwald, P. and Novick, P. (1993) Interactions of three domains distinguishing the Ras-related GTP-binding proteins Ypt1 and Sec4. *Nature* **362**:560–3.

Brown, D.A. and Rose, J.K. (1988) Polarized expression of hybrid proteins in MDCK cells. *J. Cell Biol.* **107**:782abs.

Caplan, M.J. Anderson, H.C., Palade, G.E. and Jamieson, J.D. (1986) Intracellular sorting and polarized cell surface delivery of (Na^+,K^+) ATPase, an endogenous component of MDCK cell basolateral plasma membranes. *Cell* **46**:623–31.

Casanova, J.E., Breitfeld, P.P., Ross, S.A. and Mostov, K.E. (1990) Phosphorylation of the polymeric immunoglobulin receptor required for its efficient transcytosis. *Science* **248**:742–45.

Cereijido, M. (1991) Evolution of ideas on the tight junction, in *Tight Junctions*, (ed. M. Cereijido), CRC Press, Boca Raton.

Cereijido, M. and Rotunno, C.A. (1970) *Introduction to the Study of Biological Membranes*, Gordon & Breach, London.

Cereijido, M. and Rotunno, C.A. (1971) The effect of antidiuretic hormone on Na^+ movement across frog skin. *J. Physiol.* **213**:199–233.

Cereijido, M., Herrera, F., Curran, P.F. and Flanigan, W. (1964) The influence of Na^+ concentration on Na^+ transport across frog skin. *J. Gen. Physiol.* **47**:879–93.

Cereijido, M., Rotunno, C.A., Robbins, E.S. and Sabatini, D.D. (1978a) Polarized epithelial membranes produced *in vitro*, in *Membrane Transport Processes*, (ed. J.F. Hoffman), Raven Press, New York, pp. 433–61.

Cereijido, M., Robbins, E.S., Dolan W.J. *et al.* (1978b) Polarized monolayers formed by epithelial cells on a permeable and translucent support. *J. Cell Biol.* 77:853–80.

Cereijido, M., Ehrenfeld, J., Meza, I. and Martínez-Palomo, A. (1980) Structural and functional membrane polarity in cultured monolayers of MDCK cells. *J. Membr. Biol.* **52**:147–59.

Cereijido, M. Ehrenfeld, J., Fernández-Castelo, S. and Meza, I. (1981a) Fluxes, junctions and blisters in cultured monolayers of epithelioid cells (MDCK). *Ann. N.Y. Acad. Sci.* **372**:422–41.
Cereijido, M., Meza, I., and Martínez-Palomo, A. (1981b) Occluding junctions in cultured epithelial monolayers. *Am J. Physiol.* **24**:C96–C102.
Cereijido, M., González-Mariscal, L. and Contreras, R.G. (1989a) Tight junction: barrier between higher organism and environment. *News in Physiological Sciences*, **4**:72–5.
Cereijido, M., Ponce, A. and González-Mariscal, L. (1989b) Tight junctions and apical/basolateral polarity. *J. Membrane Biol.* **100**:1–9.
Contreras,R.G., Avila, G., Gutirrez, J. *et al.* (1989) Repolarization of Na^+–K^+ pumps during establishment of epithelial monolayers. *Am. J. Physiol.* **257** (*Cell Physiol.* **26**):C896–C905.
Contreras, R.G., González-Mariscal, L., Balda, M.S. *et al.* (1992a) The role of calcium in the making of a transporting epithelium. *NIPS* **7**:105–8.
Contreras, R.G., Miller, J.H., Zamora, M. *et al.* (1992b) Interaction of calcium with plasma membrane of epithelial (MDCK) cells during junction formation. *Am. J. Physiol.* **263** (*Cell Physiol.*, **32**):C313–C318.
Cooper, M.S., Cornell-Bell, A.H., Chernjavsky, A. *et al.* (1990) Tubulovesicular processes emerge from trans-Golgi cisternae, extend along microtubules, and interlink adjacent trans-Golgi elements into a reticulum. *Cell* **61**:135–45.
De Camilli, P. and Jahn R. (1990) Pathways to regulated exocytosis in neurons. *Ann. Rev. Physiol.* **52**:625–45.
Dunn, B., Streams, T. and Botstein, D. (1993) Specificity domains distinguish the Ras-related GTPases Ypt1 and Sec4. *Nature* **362**:560–5.
Farquhar, M.G. and Palade, G.E. (1963) Junctional complexes in various epithelia. *J. Cell Biol.* **17**:375–412.
Ferro-Novick, S. and Jahn, R. (1994) Vesicle fusion from yeast to man. *Nature*, **370**:191–3.
Ferro-Novick, S. and Novick, P. (1993) The role of GTP-binding proteins along the exocytic pathway. *Ann. Rev. Cell Biol.* **9**:575–99.
García, E.P., Gatti, E., Butler, M. *et al.* (1994) A rat brain Sec1 homologue related to Rop and UNC18 interacts with syntaxin. *Proc. Natl. Acad. Sci.* USA **91**:2003–7.
González-Mariscal, L., Borboa, L., López-Vancell, R. *et al.* (1985a) Electrical properties of epithelial cells, in *Tissue Culture of Epithelial Cells*, (ed. M. Taub), Plenum Press, New York, 25–36.
González-Mariscal, L., Chávez de Ramírez, B. and Cereijido, M. (1985b) Tight junction formation in cultured epithelial cells (MDCK) *J. Membrane Biol.* **86**:113–25.
González-Mariscal, L., Contreras, R.G., Bolívar, J.J. *et al.* (1990) Role of calcium in tight junction formation between epithelial cells. *Am. J. Physiol.* **259** (*Cell Physiol.* **28**):C978–C986.
González-Mariscal, L., Contreras, R.G., Valds, J. *et al.* (1994) Extracellular and intracellular regulation of junction assembly in epithelial cells, in *Molecular Mechanisms of Epithelial Cell Junctions*, (ed. S. Citi), R.G. Landes, Austin.
Gravotta, D., Adesnik, M. and Sabatini, D.D. (1990) Transport of influenza HA from the trans-Golgi network to the apical surface of MDCK cells permeabilized in their basolateral membranes: energy dependence and involvement of GTP-binding proteins. *J. Cell Biol.* **111**:2893–908.
Green, R., Meiss, H.K. and Rodríguez-Boulán, E. (1981) Glycosylation does not determine segregation of viral envelope proteins in the plasma membrane of epithelial cells. *J. Cell Biol.* **89**:230–9.

Grunwald, G.B. (1993) The structural and functional analysis of cadherin calcium-dependent cell adhesion molecules. *Curr. Opin. Cell Biol.* **5**:797–805.

Gumbiner, B. and Simons, K. (1986) A functional assay for proteins involved in establishing an epithelial occluding barrier: identification of an uvomorulin-like peptide. *J. Cell Biol.* **102**:457–68.

Hammerton, R.W., Krzeminski, K.A., Mays, R.W. *et al.* (1991) Mechanism for regulating cell surface distribution of Na^+,K^+-ATPase in polarized epithelial cells. *Science* **254**:847–50.

Hata, Y., Slaughter, C.A. and Südhof, T.C. (1993) Synaptic vesicle fusion complex contains unc-18 homologue bound to syntaxin. *Nature* **366**:347–51.

Kelly, R.B. (1990) Microtubules, membrane traffic, and cell organization. *Cell* **61**:5–7.

Koefoed-Johnson, V. and Ussing, H.H. (1958) Nature of the frog skin potential. *Acta Physiol. Scand.* **42**:289.

Link, E., Edelmann, L., Chou, J.H. *et al.* (1992) Tetanus toxin action: inhibition of neurotransmitter release linked to synaptobrevin proteolysis. *Biochem. Biophys. Res. Commun.* **189**:1017–23.

Lisanti, M.P. and Rodríguez-Boulán, E. (1990) Glycophospholipid membrane anchoring provides clues to the mechanism of protein sorting in polarized epithelial cells. *TIBS (Trends Biochem. Sci.)* **15**:113–18.

Lisanti, M.P., Sargiacomo, M., Graeve, L. *et al.* (1988) Polarized apical distribution of glycosyl-phosphatidylinositol-anchored proteins in a renal epithelial cell lines. *Proc. Natl. Acad. Sci. USA* **85**:9557–61.

Lisanti, M., Caras, I.P., Davitz, M.A. and Rodríguez-Boulán, E. (1989) A glycolipid membrane anchor acts as an apical targeting signal in polarized epithelial cells. *J. Cell Biol.* **109**:2145–56.

Lisanti, M.P., Rodríguez-Boulán, E. and Saltiel, A.R. (1990) Emerging functional roles for the glycosyl-phosphatidylinositol membrane protein anchor. *J. Memb. Biol.* **117**:1–10.

Lisanti, M.P., Caras, I.W. and Rodríguez-Boulán, E. (1991) Fusion proteins containing a minimal GPI-attachment signal are apically expressed in transfected MDCK cells. *J. Cell Sci.* **99**:637–40.

Matter, K., Brauchbar, M. Bucher, K. and Hauri, H.P. (1990a) Sorting of endogenous plasma membrane proteins occurs from two sites in cultured human instestinal epithelial cells (Caco-2). *Cell* **60**:429–37.

Matter, K., Bucher, K. and Hauri, H.P. (1990b) Microtubule perturbation retards both the direct and the indirect apical pathway but does not affect sorting of plasma membrane proteins in intestinal epithelial cells (Caco-2). *EMBO J.* **9**:3163–70.

Mauchamp, J., Margotat, A., Chambard, M. *et al.* (1979) Polarity of three-dimensional structures derived from isolated hog thyroid cells in primary culture. *Cell Tissue Res.* **204**:417–30.

McNeill, H., Ozawa, M., Kemler, R. and Nelson, W.J. (1990) Novel function of the cell adhesion molecule uvomorulin as an inducer of cell surface polarity. *Cell* **26**:309–16.

Meza, I., Ibarra, G., Sabanero, M. *et al.* (1980) Occluding junctions and cytoskeletal components in a cultured transporting epithelium. *J. Cell Biol.* **87**:746–54.

Meza, I., Sabanero, M., Stefani, E. and Cereijido, M. (1982) Occluding junctions in MDCK cells: modulation of transepithelial permeability by the cytoskeleton. *J. Cell Biochem.* **18**: 407–21.

Mircheff, A.K. (1989) Isolation of plasma membranes from polar cells and tissues: apical/basolateral. *Meth. Enzymol.* **172**:18–34.

Misfeldt, D.S., Hamamoto, S.T. and Pitelka, D.R. (1976) Transepithelial transport in cell culture. *Proc. Natl. Acad. Sci. USA*, **73**:1212–16.

Mostov, K.E., Kraehenbuhl, J.P. and Blobel, G. (1980) Receptor-mediated transcellular transport of immunoglobulin: synthesis of secretory component as multiple and larger transmembrane forms. *Proc. Natl. Acad. Sci. USA* **77**:7257–61.

Musil, L.S. (1994) Structure and assembly of gap junctions, in *Molecular Mechanisms of Epithelial Cell Junctions*, (ed. S. Citi), R.G. Landes, Austin.

Nabi, I.R., Le Bivic, A., Fambrough, D. and Rodríguez-Boulán, E. (1991) An endogenous MDCK lysosomal membrane glycoprotein is targeted basolaterally before delivery to lysosomes. *J. Cell Biol.* **115**:1573–84.

Nelson, W.J. and Veshnock, P.J. (1987) Ankyrin binding to (Na^+ + K^+) ATPase and implications for the organization of membrane domains in polarized cells. *Nature* **328**:533–6.

Nitsch, L., Tramontano, D., Ambesi-Impiombato, F.S. *et al.* (1985) Morphological and functional polarity in an epithelial thyroid cell line. *Eur. J. Cell Biol.* **38**:57–66.

Perrin, D., Langley, O.K. and Aunis, D. (1987) Anti-alpha-fodrin inhibits secretion from permeabilized chromaffin cells. *Nature*, **326**:498–501.

Pimplikar, S.W. and Simons, K. (1993) Apical transport in epithelial cells is regulated by a Gs class of heterotrimeric G protein. *Nature* **362**:456–8.

Ponce, A. and Cereijido, M. (1991) Polarized distribution of cation channels in epithelial cells. *Cell Physiol. Biochem.*, **1**:13–23.

Ponce, A., Bolivar, J.J., Vega, J. and Cereijido, M. (1991a) Synthesis of plasma membrane and potassium channels in epithelial (MDCK) cells. *Cell Physiol. Biochem.* **1**:195–204

Ponce, A., Contreras, R.G. and Cereijido, M. (1991b) Polarized distribution of chloride channels in epithelial cells. *Cell Physiol. Biochem.* **1**:160–9.

Rabito, C.A. (1991) Tight junction and apical/basolateral polarity, in *Tight Junctions*, (ed. M. Cereijido), CRC Press, Boca Raton.

Rabito, C.A. and Ausiello, D.A. (1980) Na^+-dependent sugar transport in a cultured epithelial cell line from pig kidney. *J. Membrane Biol.* **54**:31–8.

Rabito, C.A. and Tchao, R. (1980) [^{3}H] Ouabain binding during the monolayer organization and cell cycle in MDCK cells. *Am. J. Physiol.* **238**:C43–C48.

Rindler, M.J., Ivanov, I.E., Plesken, H. *et al.* (1984) Viral glycoproteins destined for apical or basolateral plasma membrane domains traverse the same Golgi apparatus during their intracellular transport in doubly infected Madin–Darby canine kidney cells. *J. Cell Biol.* **98**:1304–19.

Rindler, M.J., Ivanov, I.E. and Sabatini, D.D. (1987) Microtubule-acting drugs lead to the nonpolarized delivery of the influenza hemagglutinin to the cell surface of the polarized. Madin–Darby canine kidney cells. *J. Cell Biol.* **104**:231–41.

Rodríguez-Boulán, E. and Nelson, W.J. (1989) Morphogenesis of the polarized epithelial cell phenotype. *Science* **245**:718–25.

Rodríguez-Boulán, E. and Pendergast, M. (1980) Polarized distribution of viral envelope proteins in the plasma membrane of infected epithelial cells. *Cell* **20**:45–54.

Rodríguez-Boulán, E. and Powell, S.K. (1992) Polarity of epithelial and neuronal cells. *Ann. Rev. Cell Biol.* **8**:395–427.

Rodríguez-Boulán, E. and Sabatini, D.D. (1978) Asymmetric budding of viruses in epithelial monolayers: a model system for study of epithelial polarity. *Proc. Natl. Acad. Sci. USA* **75**:5071–75.

Roth, M.G., Gundersen, D., Patil, N. and Rodríguez-Boulán, E. (1987) The large external domain is sufficient for the correct sorting of secreted or chimeric influenza virus hemagglutinins of polarized monkey kidney cells. *J. Cell Biol.* **104**:769–82.

Salas, P.J., Vega-Salas, D., Misek, D. and Rodríguez-Boulán, E. (1985) Intracellular routes of apical and basolateral plasma membrane proteins to the surface of epithelial cells. *Pflügers Arch.* **405**:152–7.

Salas, P.J., Misek, D. Vega-Salas *et al.* (1986) Microtubules and actin filaments are not critically involved in the biogenesis of epithelial cell polarity. *J. Cell Biol.* **102**:1853–7.

Sargiacomo, M., Lisanti, M., Graeve, L. *et al.* (1989) Integral and peripheral proteins compositions of the apical and basolateral membrane domains in MDCK cells. *J. Membr. Biol.* **107**:277–86.

Schiellele, P., Perähen, J. and Simons, K. (1995) N-glycans as apical sorting signals in epithelial cells. *Nature* **378**:96–8.

Schiavo, G., Benfenati, F., Poulain, B. *et al.* (1992) Tetanus and botulinum-B neurotoxins block neurotransmitter release by proteolytic cleavage of synaptobrevin. *Nature* **359**:832–5.

Schulz, I. (1990) Permeabilizing cells: some methods and applications for the study of intracellular processes. *Methods Enzymol.* **192**:280–300.

Simons, K. and Wandinger-Ness, A. (1990) Polarized sorting in epithelia. *Cell* **62**:207–10.

Sly, W. (1982) The uptake and transport of lysosomal enzimes. Section 1, in *The Glycoconjugates*, (ed. M.I. Horowitz), Academic Press, New York, **4**:3–25.

Söllner, T., Whiteheart, S.W., Brunner, M. *et al.* (1993) SNAP receptors implicated in vesicle targeting and fusion. *Nature* **362**:318–24.

Talavera, D., Ponce, A., Fiorentino, R. *et al.* (1995) Expression of potassium channels in epithelial cells depends on calcium activated cell–cell contacts. *J. Membrane Biol.*, in press.

Taub, M. (ed) (1985) *Tissue Culture of Epithelial Cells*, Plenum Press, New York.

Wandinger-Ness, A., Bennett, M.K., Antony, C. and Simons, K. (1990) Distinct transport vesicles mediate the delivery of plasma membrane proteins to the apical and basolateral domains of MDCK cells. *J. Cell. Biol.* **111**:987–1000.

4

Characterization of epithelial ion transport

Calvin U. Cotton and Luis Reuss

Epithelia are barriers between the body and the external world, or between extracellular fluid compartments within the body. They also function as asymmetric transporters of salt, non-electrolytes and water, thereby influencing the volume and composition of the compartments they define. The main purpose of transepithelial transport in tissues such as the kidney is homeostatic, i.e. to maintain proper water and electrolyte balance in the body. In other tissues, such as the gastric epithelium, the asymmetric transfer of ions (HCl secretion) generates an ionic environment suitable for a highly specialized function, namely digestion.

The architecture of a simple, homogenous (i.e. comprised of a single cell type), epithelium is illustrated in Figure 4.1. This model epithelium consists of two membrane domains (apical and basolateral) that define three compartments (apical, basolateral and cellular). The movement of a substance across such a barrier is limited to two pathways: paracellular (across the junctional complex and lateral intercellular space) and transcellular (across the apical and basolateral cell membranes).

Epithelial Transport: A guide to methods and experimental analysis.
Edited by Nancy K. Wills, Luis Reuss and Simon A. Lewis.
Published in 1996 by Chapman & Hall, London. ISBN 978-94-010-7168-0

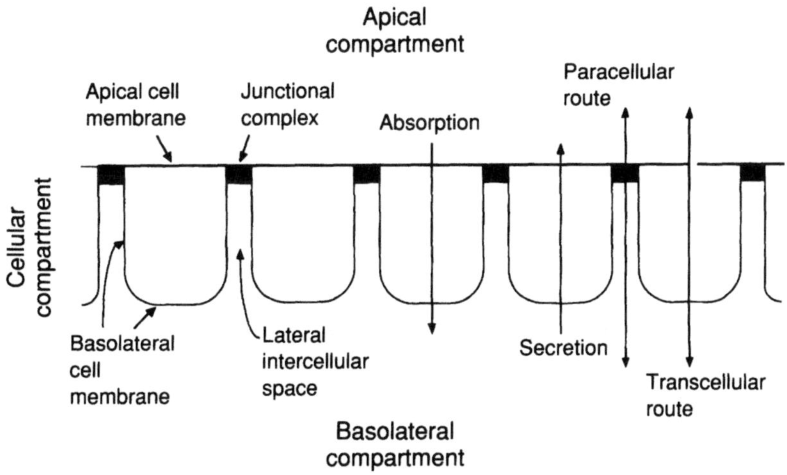

Figure 4.1 Simple cuboidal epithelium, consisting of a cellular compartment that separates the apical and basolateral compartments. Synonyms for apical include mucosal, luminal and outer. Synonyms for basolateral include serosal, submucosal, blood and inner. The junctional complex is located near the apical aspect of the cell and separates the apical and basolateral cell membranes. Transepithelial flow can occur via the cellular or paracellular route. Net flux is characterized as either absorption or secretion.

One of the primary goals of studies of epithelial transport is to establish the direction (absorption or secretion) and magnitude of the net transfer of solutes and water and to identify mechanisms of control and regulation of the transport processes. A crucial step in understanding the transport function of any epithelium is to identify, characterize and localize specific transport proteins responsible for the overall properties of the tissue. The choice of experimental methods used to accomplish these goals depends in part upon the structure and properties of the epithelium of interest. For example, the detailed methods used to study a high-resistance renal tubule segment could be quite different from those selected for the study of a low-resistance small-intestine epithelium.

In spite of the differences between the many types of epithelia and their transport properties, there are general approaches that may be applied to most epithelia. This chapter outlines those approaches most frequently used to study 'planar' epithelia, which include sheets or large tubes and sacs that can be opened to yield flat sheets (e.g. small and large intestine, stomach, gallbladder, urinary bladder, skin, trachea, etc.). Other epithelia that form small tubes, ducts or acini (e.g. renal tubule, pancreatic acinar and ductal cells, salivary acinar and ductal cells, sweat gland, etc.) are not amenable to study as flat sheets. However, cells from many of these epithelial tissues can be grown in culture as confluent monolayers

(Chapter 11) and studied as planar epithelia. Highly specialized experimental techniques such as *in vitro* renal tubule perfusion are beyond the scope of this chapter and will not be discussed. Although a number of elegant experimental approaches have been developed to study ion transport *in vivo*, this chapter focuses on the more commonly used techniques applied to isolated *in vitro* epithelial preparations.

Three approaches that are widely used to characterize epithelial transport systems are discussed: transepithelial fluid transport; transepithelial electrical responses to experimental perturbations; and transepithelial radioisotope tracer fluxes. The three approaches are not mutually exclusive and in many instances they are complementary and should be combined whenever appropriate. Specific examples will be described to illustrate the experimental approach rather than reviewing the results of many such experiments. Particular emphasis will be placed on what the experiments do and do not allow the investigator to conclude.

4.1 TRANSEPITHELIAL FLUID TRANSPORT

Net movement of water across animal cell membranes is coupled osmotically to the movement of osmolytes, primarily salt (NaCl and KCl) (House, 1974; Finkelstein, 1987). Many epithelia effect net absorption or secretion of salt, and if the water permeability of the epithelium is high, a volume of water is transferred per unit time (J_w) in proportion to the amount of solute moved per unit time (J_{solute}) (see Chapter 8 for a discussion of transmembrane water flux). Thus, in epithelia that exhibit near-isosmotic fluid transport, the rate of fluid transfer (J_v) is a good measure of net ion transport (Reuss and Cotton, 1987). Measurements of fluid transport across 'planar' epithelia generally fall into two categories: gravimetric and non-gravimetric. Gravimetric methods are simple and widely applicable, yet suffer from poor temporal resolution. Non-gravimetric methods are technically more difficult and usually require specialized equipment, but the temporal resolution and sensitivity are superior.

4.1.1 Gravimetric measurements of transepithelial J_v

These methods are commonly used to measure fluid transport in tissues that can be handled as a sac (bladder) or tube (intestine). The principle of the method is that the weight of a fluid-filled epithelial sac or tube is increased or decreased by net transepithelial fluid flow, and that the rate of weight gain or loss is equal to the rate of net fluid secretion or absorption. In this section we describe as an example studies of mouse small intestine.

The small intestine is removed from the animal and placed in a fluid-filled bath. A flared polyethylene tube (PE-190) is inserted in the lumen and tied in place with silk thread. The tube is connected to a fluid-filled reservoir and the reservoir is elevated to initiate perfusion of the intestine. Lumen contents are washed out and the distal end of the intestine is closed with a tie (#2 silk suture). Additional double-ties are made along the length of the partially inflated intestine. The tissue is cut between adjacent double-ties to yield several (usually four to five) isolated segments (ca. 1.5 cm in length) each with a closed lumen fluid compartment. The partially inflated segments are transferred to vials filled with the appropriate bathing solution and placed in a heated water bath. At fixed times thereafter (usually 5 min intervals) the tissue is removed from the bath, blotted in a standardized way on moist filter paper to remove adherent fluid, weighed and returned to the bath. At the end of the experiment the segment is opened, the lumen fluid is drained and the wet weight of the tissue is determined in order to normalize transport rates among different tissues. The slope of the plot (weight vs. time) is a measure of the rate of net fluid transport (Figure 4.2).

The effects of experimental perturbations expected to alter the rate of ion transport may be evaluated from changes in the rate of fluid transport. In the case of small intestine, elevation of intracellular cAMP by exposure to forskolin and theophylline causes a reversal from net fluid absorption to net fluid secretion. The response is due to stimulation of Cl^- secretion in the crypts and partial inhibition of NaCl absorption (inhibition of parallel Na^+/H^+ and Cl^-/HCO_3^- exchangers) in the villi (Powell, 1987). When the intestine is pretreated with bumetanide to inhibit Cl^- secretion (O'Grady *et al.*, 1987), cAMP causes a decrease in fluid absorption (i.e. NaCl absorption in the villi is inhibited) but net fluid secretion is not observed. As expected, the response to elevation of cAMP in an intestinal segment from a transgenic cystic fibrosis (CF) mouse (Snouwaert *et al.*, 1992; Clarke *et al.*, 1992), which lacks the Cl^--channel activity necessary for Cl^- secretion, is similar to that obtained by pharmacologic inhibition of Cl^- secretion in the non-CF tissue. Thus, the measurements of fluid absorption reveal that the cAMP-induced inhibition of NaCl absorption is preserved in the CF intestine.

Additional experiments could be done to define more clearly contributions of specific transport pathways thought to be responsible for absorption and secretion in the normal and CF mouse intestine. For example, replacement of lumen and bath solution HCO_3^- with Cl^- or replacement of lumen solution Na^+ with tetramethylammonium (a cation that is not transported by the Na^+/H^+ exchanger) would be expected to reduce net NaCl and fluid absorption in intestinal segments from both CF and normal mice. Similarly, exposure of the tissues to pharmacologic inhibitors of Cl^-/HCO_3^- or Na^+/H^+ exchangers (see Chapter 2) should also reduce NaCl and fluid absorption. Since the cAMP-induced down-regulation of NaCl and fluid absorption is not complete, then the maneuvers

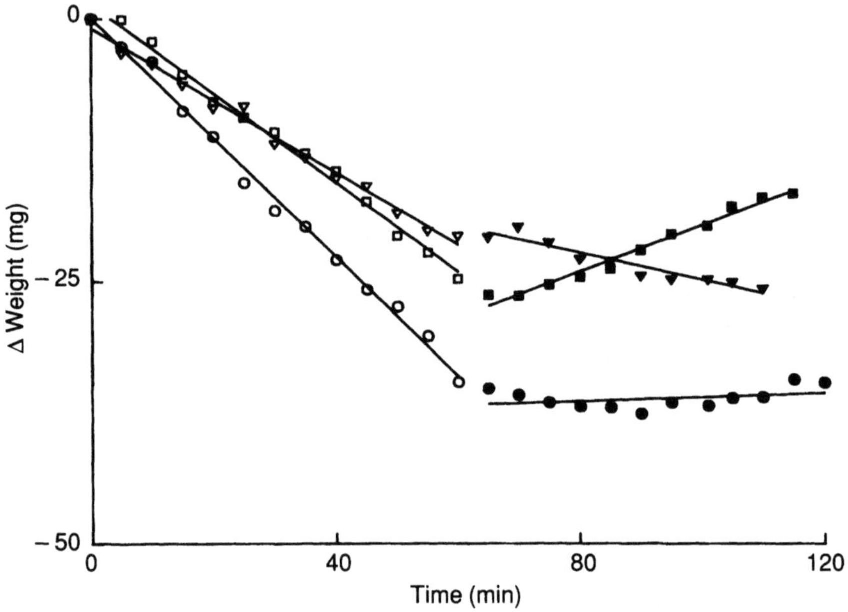

Figure 4.2 Measurements of transepithelial fluid transport across mouse intestine. Isolated ileal segments from normal and cystic fibrosis mice were incubated in a bicarbonate-buffered salt solution. At 5-min intervals, the tissues were removed, blotted, weighed and returned to the bath. Circles and squares represent tissues from a normal mouse; triangles represent a tissue from a CF mouse. One of the normal-mouse tissues (circles) was incubated with bumetanide to inhibit Cl^- secretion. All tissues were exposed to forskolin and theophylline (to elevate intracellular cAMP) during the times indicated by the filled symbols (60–115 min). Transport rates were calculated from the slopes of the lines before and after addition of forskolin and theophylline.

described above to reduce exchanger activity might be expected to increase further the rate of fluid secretion in normal but not CF tissues.

The major disadvantages of the gravimetric method for measuring fluid transport are the inability to make other types of measurements concomitantly and the relatively poor temporal resolution. Poor temporal resolution is not a serious limitation if transport rate and alterations in transport rate are sustained. There is one other potentially serious technical problem that should be considered when using gravimetric methods to measure transepithelial fluid transport. It is implicitly assumed that once fluid crosses the epithelial cell layer it enters the bulk fluid and no longer contributes to the weight of the tissue. However, the transported fluid retained in the interstitial space will not be detected as transported. If this problem is suspected then another method (i.e. volume marker method) should be used to confirm the validity of the measurements (Bello-Reuss *et al.*, 1981). The major advantages of the gravimetric method are its technical simplicity and the modest equipment requirements (standard laboratory analytical balance and water bath).

4.1.2. Non-gravimetric measurements of transepithelial J_v

Two types of non-gravimetric methods have been applied to fluid transport in epithelia mounted as flat preparations: volume displacement and changes in the concentration of a volume marker. Volume displacement methods rely on measurement of changes in the volume of either the apical or basolateral, or both, fluid compartments. Generally what is measured is a change in the position of the fluid/air or fluid/oil interface in the chamber.

In early experiments, movement of the meniscus in a horizontal capillary was monitored (Diamond, 1968; Reuss *et al.*, 1979); subsequently more sensitive electrical methods were developed (Reuss, 1984). One apparatus that deserves special mention is that developed by van Os and coworkers (Wiedner, 1976; van Os *et al.*, 1979). The detectors in their system were probes positioned above the air/solution interfaces in the apical and basolateral compartments. The electrical capacitance is an inverse function of the distance between the probe and the surface of the conductive fluid. Thus, net transepithelial fluid transport alters the volume of the compartments and the positions of the air/fluid interfaces which are sensed as changes in capacitance. Later versions of the apparatus included motorized syringes coupled to the output of the capacitance probes so that fluid was injected into or withdrawn from the compartments to maintain the air/fluid interface positions constant. If tissue-independent changes in volume (i.e. those due to evaporation and temperature changes) are taken into account, then the volume of fluid injected or withdrawn represents the net transepithelial fluid flux.

There are several advantages to this method and the associated instrumentation. Small rates of fluid transport are detected with great accuracy (± 1 nl/min) and the temporal resolution is good. The chamber geometry is suitable for simultaneous measurements of transepithelial electrical properties (voltage, resistance and short-circuit current) and fluid flow (Hughes *et al.*, 1984; Jiang *et al.*, 1993). The major disadvantage is the demanding requirement for specially designed chambers, which must be built from hydrophobic material, and additional equipment. The approach and the types of experimental perturbation that one can impose are essentially the same as those described for gravimetric measurements and similar information is obtained. However, the ability simultaneously to measure fluid transport and to monitor transepithelial electrical properties is a distinct advantage in epithelia in which electrogenic transporters play a dominant role in net transport.

The other non-gravimetric method relies upon measurement of changes in the concentration of a substance proven to be a water-volume marker added to the bathing solution (Reuss *et al.*, 1979; Cotton *et al.*, 1989). The approach is based upon the indicator-dilution principle, i.e. the volume of a solution (V) can be determined if the amount (Q_i) and concentration (C_i) of an indicator (i) are known, as stated in equation (4.1):

$$V = Q_i/C_i \tag{4.1}$$

Thus, changes in the volume of the fluid will cause changes in the concentration of the indicator if the amount of indicator remains constant. To apply this principle to measurements of transepithelial fluid transport, several properties of the indicator must hold: homogenous distribution; impermeability (or low permeability relative to water permeability – see Jovov *et al.*, 1991); chemical stability; inertness; and the existence of a method to measure its concentration accurately.

The transepithelial water permeability of *Necturus* gallbladder epithelium was measured using this method (Cotton *et al.*, 1989). Net water flux was driven by an imposed transepithelial osmotic difference and the concentration of indicator (tetramethylammonium) was measured with an ion-selective microelectrode. The major disadvantage of this approach is that it is practical for the measurements of transepithelial J_v only if the volume of the compartment that contains the indicator is small, so that net transport of fluid by the tissue generates a measurable change in the concentration of the indicator. This approach has not been widely used but it appears well suited for studies of cultured epithelial cell monolayers.

Each of these three methods provides the same type of information, namely the rate of net fluid transport. Without additional experiments, the investigator obtains no information about the ion transport mechanism(s) responsible for net fluid flux. For example, in the case of mouse intestine, elevation of cAMP caused a switch from net fluid absorption to net fluid secretion in the non-CF mouse. Without prior knowledge of the ion transport properties of the unstimulated epithelium and the effect of cAMP upon ion transport, it would not be possible to conclude from this result alone that cAMP inhibits villus-cell NaCl absorption and stimulates crypt-cell Cl^- secretion. The usefulness of these techniques is that they provide a measure of the algebraic sum of net ion transport by the tissue. If the ion transport properties are well known, fluid transport measurements offer a simple assay to screen the effects of experimental perturbations and to make predictions that can be tested with other types of measurement. Fluid transport measurements are especially useful in leaky epithelia that often exhibit high rates of isosmotic fluid transport, usually by electroneutral transport pathways that cannot be monitored by conventional transepithelial electrical measurements.

4.2 TRANSEPITHELIAL ELECTRICAL MEASUREMENTS

The principles and details of transepithelial electrical measurements are presented in Chapter 5 and will not be discussed here. We will instead illustrate how specific experimental perturbations that alter the electrical properties of an epithelium can be used to identify and characterize epithelial transport pathways. The transepithelial electrical parameters that we are concerned with are:

- the voltage between the apical and basolateral bathing solutions (V_t);
- the current that is required to reduce V_t to zero (short-circuit current, I_{sc});
- the electrical resistance of the epithelium (R_t).

The short-circuit current technique is illustrated in Figure 4.3. Across different types of epithelia, the range of values for each of these parameters can be large, e.g. R_t ranges from < 10 Ωcm^2 in mammalian proximal tubule to > 10 000 Ωcm^2 in mammalian urinary bladder (Frömter and Diamond, 1972; Lewis and Diamond, 1976). Furthermore, the polarity (V_t)

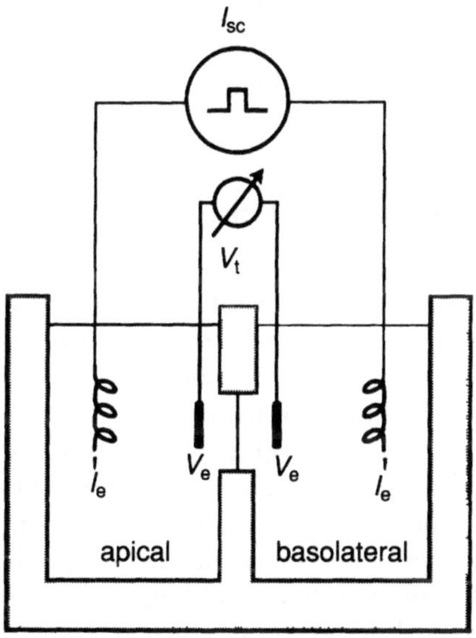

Figure 4.3 Short-circuit current technique. The epithelium is mounted in the middle partition of an Ussing chamber, separating solutions of identical composition, at the same pressure. Current is applied across the epithelium with an external circuit (I_e = current electrode) and the transepithelial voltage (V_t) is measured (V_e = voltage electrode). The short-circuit current (I_{sc}) is the current flow needed to make $V_t = 0$. This current, in the conditions defined, is equivalent to the sum of active electrogenic ion fluxes across the epithelium.

and numerical values of these parameters alone offer little insight into the underlying transport mechanisms. For instance, a lumen-negative transepithelial potential difference may arise from the net absorption of cation (or cations), the net secretion of anion (or anions), or some combination of transport processes. Thus, simple transepithelial electrical measurements alone can neither establish the nature of the transported ion(s), nor identify the mechanism(s) responsible for their transport.

The other major shortcoming inherent in transepithelial electrical measurements is the inability to monitor electroneutral transport processes. Tissues such as gallbladder epithelium transport salt and water at high rates, yet do not normally generate an appreciable V_t or I_{sc}. Therefore, transepithelial electrical measurements are of little value for studies of this type of tissue.

The premise of this section is that there are, however, experimental maneuvers with predictable outcomes that may provide insight into the transport pathways that are operative in many epithelia. This approach is especially useful when faced with a new preparation in which one seeks a preliminary characterization of the properties of the tissue or when the properties of the tissue are well established and a new experimental perturbation is to be examined. In the paragraphs that follow, we describe some experimental details relevant to the study of planar epithelia and discuss maneuvers that are frequently used to characterize transport systems.

4.2.1 Tissues, chambers and solutions

The choice of tissue will obviously depend upon the goals of the investigator but certain generalizations apply. In most cases, excised tissue is washed with chilled physiologic salt solution. If necessary, subepithelial tissue is removed by dissection. The subepithelial tissue will effectively establish a minimum thickness for the basolateral-solution unstirred fluid layer (Barry and Diamond, 1984). Electrically, the subepithelial connective tissue space may be considered as an equivalent thickness of bathing solution. However, the time course of changes in solution composition at the epithelial surface in experiments in which the bath solution is replaced, or in which pharmacologic agents are added, will be limited by diffusion through the unstirred fluid layer. Practical considerations for the time required to attain a steady-state change in concentration at the cell membrane may limit the use of extremely thick tissues with large unstirred layers.

In addition to the problems associated with slow response to experimental perturbations, there are several other reasons why subepithelial tissue may pose a problem. The underlying tissue may contain nerve endings capable of modulating the properties of the epithelium.

Tetrodotoxin is sometimes added to the basolateral bathing solution to block neuronal Na^+ channels and thus limit neuronal discharge and muscle contraction (Cooke *et al.*, 1983). Other cells (macrophages, neutrophils, etc.) may reside in the subepithelial space and release mediators that can affect epithelial function. Although the epithelium may be well oxygenated via the luminal bath, a thick subepithelial layer will not be well oxygenated and tissue damage resulting in the release of cellular mediators can occur. On the other hand, the subepithelial tissue supports the epithelium and may protect it from mechanical damage caused by handling. Furthermore, even fastidious dissection is unlikely to eliminate all subepithelial tissue and thus the problem of mediator release may not be entirely solved.

After the epithelium is removed and cleaned, it is opened to form a sheet. Care must be taken to avoid touching the portion of the sheet that will form the diaphragm that separates the bathing solutions. Many tissues can be stretched without apparent damage to the epithelium. This is especially true of epithelia with macroscopic folds that normally inflate and deflate *in vivo* (e.g. urinary bladder). Tissues are generally held in place in the chamber by pins but more elaborate systems have been devised to accommodate different tissues.

In most cases the tissue is compressed between lucite surfaces to form a seal, which may result in some degree of edge damage (Helman and Miller, 1971). The magnitude and importance of the damage to the epithelium is variable and should be minimized. In general, it is advantageous to use the largest piece of tissue available, since the ratio of surface area to perimeter increases as the diameter of the chamber aperture increases. In addition, edge damage may be reduced or prevented by the use of silicone compressible gaskets (Lewis and Diamond, 1976), or a Sylgard disk (Reuss and Finn, 1975), coated with silicone grease. In theory, the I_{sc} is not affected by edge damage since the damaged tissue is in fact a non-selective leak in parallel with the undamaged tissue. Damage to the epithelium might be expected to alter the apparent ion selectivity of the epithelial barrier, its conductance (and resistance) and its spontaneous transepithelial voltage.

Once the epithelium is mounted in a suitable chamber, the preparation is bathed on both sides by well-mixed solutions maintained at the appropriate temperature. This is usually done by attaching water-jacketed fluid reservoirs to the chamber. The solutions are circulated by gas lifts that serve the dual function of insuring the correct gas tension (O_2, CO_2, etc.) and mixing the solutions. Most Ussing flux chambers and fluid reservoirs are custom made but there are at least two commercial vendors (World Precision Instruments, Sarasota, FL and Precision Instrument Design, Los Altos, CA).

The choice of bathing solutions will again depend upon the tissue to be studied. In most cases phosphate- or bicarbonate-buffered physiologic salt solutions are used. Since the experiments are frequently of long duration (several hours) most investigators include glucose in the bathing solutions to maintain glycolysis. Calcium deserves special mention since it plays a particularly important role in epithelial integrity (González-Mariscal *et al.*, 1990; Contreras *et al.*, 1992; Jovov *et al.*, 1994). Most epithelia tolerate removal of calcium from the luminal solution but removal from the basolateral bathing solution usually elicits a rapid decrease in transepithelial resistance, due to disruption of the junctional complexes. In some cases the resistance decrease is reversible, depending upon the period of exposure to low calcium.

4.2.2 Ion replacements

If one suspects that a particular ion-transport process is responsible for the generation of a transepithelial potential difference (or short-circuit current) then removal of that ion from both sides of the tissue should reduce V_t and I_{sc}. Figure 4.4a depicts a cellular transport model for a Na^+-absorbing epithelium such as frog skin (Koefoed-Johnsen and Ussing, 1958). Sodium enters the cell via an apical Na^+ channel and is pumped across the basolateral cell membrane by the Na^+,K^+-ATPase, in exchange for K^+. Potassium ions recycle across the basolateral cell membrane via K^+

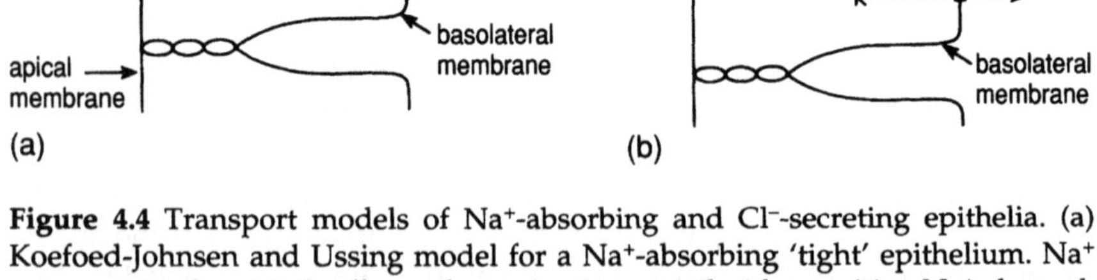

Figure 4.4 Transport models of Na^+-absorbing and Cl^--secreting epithelia. (a) Koefoed-Johnsen and Ussing model for a Na^+-absorbing 'tight' epithelium. Na^+ entry across the apical cell membrane is via an amiloride-sensitive Na^+ channel. Extrusion of Na^+ across the basolateral cell membrane is mediated by a ouabain-sensitive Na^+,K^+-ATPase. K^+ recycles across the basolateral cell membrane via a K^+-channel. (b) Transport model for a Cl^--secreting epithelium. Intracellular $[Cl^-]$ is maintained above electrochemical equilibrium by electroneutral influx with Na^+ and K^+ across the basolateral cell membrane. Na^+ and K^+ are recycled across the basolateral cell membrane by the Na^+,K^+-ATPase and a K^+ channel, respectively. (Reproduced from Reuss and Cotton (1994), with permission.)

channels. This model predicts a lumen-negative transepithelial voltage and a short-circuit current equal to net Na^+ absorption. Replacement of Na^+ on both sides of the tissue with a non-transported cation such as tetramethylammonium would be expected to reduce V_t and I_{sc} to zero.

It is necessary to exercise caution in interpreting this kind of experiment, because replacement of an ion on one or both sides of a tissue may cause changes in ion transport indirectly. Consider, for example, the transport model for Cl^- secretion (Silva *et al.*, 1977) illustrated in Figure 4.4b. In this case, Cl^- enters the cell across the basolateral cell membrane coupled to Na^+ and K^+. The cations recycle across the basolateral membrane via the Na^+,K^+-ATPase and K^+ channels whereas Cl^- exits across the apical cell membrane. This model also predicts a lumen-negative transepithelial voltage but in this case the I_{sc} is due to net Cl^- secretion. Replacement of Na^+ with a non-transported cation such as tetramethylammonium (TMA^+) would also reduce V_t and I_{sc} to zero since Cl^- entry across the basolateral cell membrane is Na^+ dependent. It is therefore not possible from this type of experiment to distinguish between two epithelia with different transport properties, namely Na^+ absorption and Cl^- secretion. It is possible, however, to design ion replacement experiments that would distinguish between the two transport models. Replacement of Cl^- with a non-transported anion such as gluconate would be expected to reduce V_t and I_{sc} to zero in the Cl^- secretory epithelium but not in the Na^+-absorbing tissue. In contrast, replacement of Cl^- with a less permeant anion (gluconate) actually increases V_t with little or no effect on I_{sc} in tissues such as frog skin. The reason for the increase is that the Cl^- permeability of the paracellular pathway exceeds the gluconate permeability, i.e. V_t is 'shunted' to a lesser extent by gluconate than by chloride in the open circuit condition. This analysis is complicated if one or both cell membranes of a Na^+-transporting epithelium have a Cl^--conductive pathway. In this case, the I_{sc} may change because of the change in membrane voltage elicited by Cl^- removal.

In the example cited above, we illustrated how bilateral replacement of Na^+ would reduce V_t and I_{sc} directly in Na^+-absorbing and indirectly in Cl^--secreting epithelia. In the Na^+-absorbing tissue, apical-solution Na^+ is essential, whereas in the Cl^--secreting tissue it is basolateral-solution Na^+ that is required. Thus, it would appear that unilateral replacement of Na^+ with a non-transported cation (TMA^+) would elicit different responses in the two types of epithelia. This is not necessarily true since the response to a unilateral solution replacement may be complicated by transepithelial diffusion (bi-ionic) potentials (Barry and Diamond, 1970; Barry *et al.*, 1971). The magnitude and polarity of the bi-ionic potential (V_{bi}, apical–basolateral) that would arise in the example cited above can be approximated by the following simplified form of the Henderson equation (see Barry, 1989):

$$V_{bi} = -2.3\frac{RT}{F}\log\frac{u_{TMA^+} + u_{Cl^-}}{u_{Na^+} + u_{Cl^-}} \tag{4.2}$$

where u is mobility, and R, T and F have their usual meanings. This equation is written for TMACl in the apical solution and NaCl in the basolateral solution, at equal concentrations, neglecting all other ions; 2.3 RT/F is ca. 58 mV at 23 °C. If the paracellular pathway has a high conductance and is cation selective, then replacement of apical solution Na^+ with a less permeant cation (TMA^+) will reduce (or even reverse) the lumen negative V_t in both Cl^--secreting and Na^+-absorbing tissues by an amount that depends upon the relative permeabilities of Na^+, TMA^+ and Cl^-. In the former case, this will occur because of the bi-ionic potential, and in the latter because of both the bi-ionic potential and the contribution from removal of the transported ion. The response to replacement of basolateral solution Na^+ also fails to distinguish unambiguously between the two types of epithelia. In the Na^+-absorbing tissue, V_t would increase (due to the bi-ionic potential); in the Cl^--secreting tissue, V_t would either increase (due to the bi-ionic potential) or decrease (due to inhibition of Cl^- secretion). It should be clear from the examples cited above that replacement of bathing solution ions is fraught with complexity. However, the results obtained from carefully designed experiments may assist in the characterization of epithelial transport systems.

4.2.3 Single-membrane approaches

The interpretation of transepithelial electrical measurements during experiments designed to localize a particular transport pathway to either the apical or the basolateral cell membrane is complicated by the existence of a paracellular route of finite conductance. The effect of the paracellular pathway is to couple electrically the two membranes, so that changes in electrical properties elicited at one membrane affect the opposite one. The traditional approach to this problem is to use intracellular microelectrodes to measure individual membrane voltages and determine membrane resistances (Chapter 5). It is possible from such measurements to identify the site of the primary effect of an experimental perturbation (Altenberg *et al.*, 1990). However, intracellular microelectrode techniques are difficult and not suitable for all epithelial preparations. Alternative methods have been devised to examine single-membrane (apical or basolateral) electrical properties independently of the opposite membrane.

Early experiments designed to study the electrical properties of the apical cell membrane of frog skin epithelium were done by replacing the Na^+ in the basolateral bathing solution with K^+ (Fuchs *et al.*, 1977). Although this maneuver reduces the voltage and the resistance of the basolateral cell membrane, it does not allow the investigator to control the composition of the solution on the cytosolic side of the apical cell membrane, i.e. the intracellular compartment. More recently, detergents and pore-forming molecules such as digitonin, nystatin, amphotericin B and *Staphylococcus aureus* α-toxin have been used to permeabilize selectively either the apical or the basolateral cell membranes of epithelia (Lewis *et al.*, 1977; Anderson and Welsh, 1991; Ostedgaard *et al.*, 1992). A gradient for the ion(s) of interest is established across the tissue, and the resulting current is monitored. This approach is particularly useful for studies of electrodiffusive pathways (i.e. channels) and electrogenic pumps (Lewis *et al.*, 1977). There are four major disadvantages of this approach:

- It is difficult to assure that changes in intracellular compartment composition do not affect the transporter of interest.
- Electroneutral transporters do not generate a current.
- The paracellular pathway remains in parallel with the unmodified membrane and thus changes in its properties could affect the transepithelial measurements.
- It is difficult to prove that only the target membrane is affected by the permeabilizing agent.

In spite of these limitations, the approach is technically simple and in some cases provides an acceptable alternative to the use of other techniques to measure single-membrane electrical properties.

4.2.4 Hormones and second messengers

Many epithelial cells are characterized by acute and/or chronic changes in ion transport rates, elicited by hormones or neurotransmitters and mediated by second-messenger systems. Thus, transepithelial ion transport provides a sensitive and convenient system to identify and characterize signal transduction pathways in the intact epithelial cell. Even though changes in net transport mediated by electrogenic transport pathways are readily detected by transepithelial electrical measurements, the investigator is faced with the problem that a change in V_t, R_t or I_{sc} may arise from many combinations of events. Therefore, it is necessary to identify the alterations in net transport by other means in conjunction with electrical measurements (e.g. tracer fluxes). In tissues in which the transport properties are well established or can be defined pharmacologically it may be possible to infer regulatory mechanisms from changes in electrical parameters alone. For example, the entire short-circuit current across frog skin is due to amiloride-sensitive Na^+ absorption (Ussing and Zerahn, 1951; Dörge and Nagel, 1970). Exposure of the frog skin to aldosterone increases the short-circuit current, but it remains fully inhibitable by amiloride. These results suggest that aldosterone stimulates net Na^+ absorption across frog skin, a conclusion supported by direct measurements of Na^+ flux.

4.2.5 Transport inhibitors

A relatively simple and powerful approach to study epithelial ion transport systems based upon transepithelial electrical measurements has evolved with the discovery and characterization of a vast array of ion transport inhibitors. The main impetus for discovery and development of pharmacologic agents active against epithelial transport proteins is the interest in modulating salt and water transport in the kidney.

The major concerns with the use of pharmacologic agents to characterize transport systems are specificity and secondary effects. With the possible exception of ouabain, drugs within the major classes of ion transport inhibitors have been demonstrated to affect at least two different types of transport system. The potential complications attributable to secondary effects of transport inhibitors may pose an even more severe limitation for their use. In spite of these problems, careful experimental design may allow one to characterize a transport system based upon the responses to inhibitors.

Amiloride is a diuretic that is known to inhibit both epithelial Na^+ channels and Na^+/H^+ exchangers (among other transporters). It is relatively easy to distinguish experimentally between inhibition of these two transport pathways based on the time course of the response and inhibitor sensitivity (Benos, 1982). The K_i for epithelial Na^+ channels is ca. 100 nM and inhibition causes rapid decreases in V_t and I_{sc} in tissues that express apical Na^+ channels. In contrast, the K_i for inhibition of Na^+/H^+ exchange is much higher (10 μM to 1.0 mM) and does not directly cause changes in electrical properties since the Na^+/H^+ exchanger is electroneutral. Secondary changes in electrogenic transport as a consequence of changes in intracellular pH are likely to accompany inhibition of Na^+/H^+ exchanger activity, but they develop slowly.

In addition to the problem of inhibitor specificity, indirect effects due to changes in driving forces may complicate attempts to characterize epithelial transport systems based upon electrical responses to transport inhibitors. Changes in electrical and/or chemical driving forces resulting from inhibition of a particular transport pathway may induce secondary effects at the same membrane, at the opposite membrane or both. For example, if we include an apical membrane Na^+ channel in the model depicted in Figure 4.4b, the tissue should be capable of both Na^+ absorption and Cl^- secretion. Let us assume that Na^+ absorption and Cl^- secretion contribute equally to both the lumen-negative V_t and the I_{sc} at the steady state. Then, inhibition of the Na^+ channel by amiloride would be expected to abolish Na^+ absorption and presumably reduce V_t and I_{sc} by

50%. However, this result is not observed in tissues with these properties because inhibition of apical membrane Na^+ channels causes a hyperpolarization of membrane voltage, and therefore increases Cl^- secretion (Boucher and Gatzy, 1983).

Finally, inhibition of one type of transporter may modulate the activity of another transporter independent of changes in driving forces. For instance, pharmacologic inhibition of the electroneutral Cl^-/HCO_3^- exchanger would be expected to increase intracellular pH. Since some classes of potassium channel expressed in epithelial cells are known to be sensitive to changes in intracellular pH (e.g. activated at alkaline pH), it is likely that intracellular alkalinization would increase K^+ permeability (P_K) of the cell membrane, and thereby influence V_t and I_{sc}.

Changes in intracellular Ca^{2+} activity also exert profound regulatory effects on transport pathways in many epithelial cells. These changes are normally elicited by and studied in the context of receptor-mediated signaling. However, it was recently demonstrated that exposure of cultured intestinal T84 cells to high concentrations of the anion transport inhibitor, 4,4′-diisothiocyanostilbene-2,2′-disulfonic acid (DIDS) induces a transient increase in intracellular Ca^{2+} activity by an undefined mechanism (Brayden *et al.*, 1993). Thus, instead of the expected decrease in I_{sc} due to inhibition of apical membrane Cl^- channels, DIDS stimulated electrogenic Cl^- secretion by activation of Ca^{2+}-sensitive Cl^- channels.

These examples are meant to illustrate the complexity and interdependency of transport pathways in epithelia and to serve as a reminder that transepithelial electrical measurements alone may not be sufficient to interpret correctly the results of even simple experimental perturbations.

4.3 TRANSEPITHELIAL RADIOISOTOPE TRACER TECHNIQUES

The two major limitations of the methods for study of epithelia outlined above are inability to identify directly the transported ions, and inability to monitor electroneutral transport with transepithelial electrical measurements. A powerful method for characterizing transepithelial ion transport, which does not have the above limitations, utilizes measurements of unidirectional transepithelial radioisotope tracer fluxes. The major advantage of using this approach is that one can actually identify individual ionic species that are transported by the tissue regardless of the mechanism of transport. The fundamental assumption required for the use of radioisotope tracers to measure unidirectional fluxes is that low-abundance, unstable isotopes of relevant ions (such as sodium, chloride, calcium and potassium) interact with biological systems in a manner identical to their stable congeners. In other words, transport proteins such as the Na^+,K^+-ATPase interact with $^{22}Na^+$ or $^{24}Na^+$ in the same manner as the more common, naturally occurring $^{23}Na^+$. Inasmuch as this assumption is valid, tracer amounts of the radioisotopes may be used to monitor unidirectional fluxes of the ion of interest. In the paragraphs that follow we describe the principles and experimental details underlying the use of radioisotopes to study transepithelial ion transport. We assume that the reader is familiar with the basic electrical parameters of epithelia (transepithelial potential difference, resistance and short-circuit current) mentioned above and described in detail in Chapter 5.

The use of radioisotopes to study ion transport was developed and applied to epithelia by Ussing and Zerahn (1951) as part of their studies of the transport potential generated by frog skin. The goal of these experiments was to measure unidirectional fluxes of Na^+ in the absence of all external driving forces, i.e. transepithelial differences in chemical or electrical potential and pressure (osmotic or hydrostatic). The frog skin was mounted as a diaphragm between solutions of identical composition and a transepithelial current was passed to clamp V_t to zero (i.e. the tissue was short-circuited). The unidirectional fluxes of Na^+ (inner-to-outer and outer-to-inner) were measured using tracer amounts of $^{24}Na^+$. They found that the absorptive flux (outer-to-inner) exceeded the secretory flux (inner-to-outer) by ca. 50-fold and that the difference between unidirectional fluxes was equal to the short-circuit current. They concluded that the so-called transport potential generated by frog skin results from active absorption of Na^+. Since the initial experiments, this approach has been widely used to characterize transepithelial ion transport across many types of epithelia. The application of this technique is described in the next sections.

4.3.1 Radioisotopes

Radioisotopes of most ions of biological importance are readily available (NEN, Amersham, ICN). These include $^{22}Na^+$, $^{36}Cl^-$, $^{42}K^+$, $^{125}I^-$, $^{35}SO_4^{2-}$ and $^{45}Ca^{2+}$. The use of radioisotopes to measure H^+ and HCO_3^- fluxes presents a special problem and will not be discussed. The specific properties of radionuclides (such as half-life and emission type and energy) and details on detector systems (e.g. liquid scintillation counters and gamma counters) for measuring the radioactivity of samples can be found in reference books (Slater, 1990) and in literature available from radioisotope vendors.

4.3.2 Experimental protocols, calculations and interpretations

The goal of these measurements is to determine the net flux of ions across the tissue of interest and to obtain some insight into the mechanisms responsible for transport. The net fluxes of ions across an epithelial barrier are calculated from the differences in oppositely directed unidirectional fluxes. Multiple tissues from the same animal are usually grouped as pairs, matching the magnitude of the transepithelial resistance. The rationale behind the decision to pair tissues according to transepithelial resistance derives from the expectation that a significant fraction of the transepithelial conductance ($G_t = 1/R_t$) is paracellular. Inasmuch as this is true, unidirectional ion fluxes via this route are expected to be symmetric when the bathing solutions are identical and $V_t = 0$, and thus if the tissues are properly paired will not contribute to the calculated net flux. The tissues are mounted in suitable chambers (see above and Chapter 5), bathed on both sides by identical solutions, and short-circuited. A small amount of one or more radiotracers (0.1–1 μCi/ml) is added to the bathing solution on either the apical or the basolateral side of the tissue (Cotton *et al.*, 1983, 1988). An equal amount of tracer is added to the bathing solution on the opposite side of the other tissue of the pair. The bathing solution to which the isotope is added is referred to as the 'hot or source' side, and the solution on the other side of the tissue is called the 'cold or sink' side. At timed intervals thereafter, samples (usually 1 ml) are removed from the 'cold or sink' side of each tissue. An equal volume of isotope-free bathing solution is immediately added back, to maintain the volume constant. Throughout the course of the experiment several small samples (usually 10–20 μl) are removed from the 'hot side'. The amount of radiotracer present in each sample is then determined by the appropriate detector (e.g. gamma counter, liquid scintillation counter). The unidirectional permeability coefficient (P_j, $\text{cm}\cdot\text{sec}^{-1}$) and flux rate ($J_j$, $\mu\text{Eq}\cdot\text{cm}^{-2}\cdot\text{hr}^{-1}$) are calculated from equations 4.3 and 4.4, respectively, with appropriate corrections for additions of solution to replace the samples:

$$P_j = \frac{\text{steady-state rate of tracer appearance in sink (cpm}\cdot\text{sec}^{-1})}{\text{source radioactivity (cpm}\cdot\text{cm}^{-1}) \times \text{area (cm}^2)} \tag{4.3}$$

$$J_j = \frac{\text{steady-state rate of tracer appearance in sink (cpm}\cdot\text{hr}^{-1})}{\text{specific activity of source (cpm}\cdot\mu\text{Eq}^{-1}) \times \text{area (cm}^2)} \tag{4.4}$$

The difference between the unidirectional fluxes ($J_j^{ab} - J_j^{ba}$) is equal to the net flux for the ion (J_j^{net}); where J_j^{ab} is the unidirectional apical-to-basolateral flux, and J_j^{ba} is the unidirectional basolateral-to-apical flux.

There are several possible outcomes from experiments of this type. If the unidirectional fluxes for an ion are of equal magnitude, and therefore the net flux is zero, then the most likely conclusion is that transepithelial flux of that ion is passive. This conclusion can be checked at different transepithelial voltages using the flux-ratio equation (equation (4.5) below). In contrast, if the fluxes are asymmetric, then it is likely that flux in at least one direction is 'active'. In this context, the designation of 'active' carries no mechanistic connotation, only that simple electrodiffusive flows cannot account for both unidirectional fluxes. If only one ion is actively transported, then there should be equivalence between the short-circuit current and the net flux for that ion. The most frequent scenario is that more than one ion is actively transported. Since the short-circuit current is the algebraic sum of all net ion fluxes, the isotopically determined net flux of an ion will represent only part of the short-circuit current. In order to account fully for the short-circuit current it may be necessary to measure net fluxes of several ions. Alternatively, in many tissues there is net ion transport which greatly exceeds the measured short-circuit current. In this case it is likely that the transport pathways are electroneutral, and thus generate no short-circuit current.

One potential problem that arises from measurements of tracer fluxes under short-circuit conditions is that the geometry of some epithelial tissues precludes uniform voltage clamping. For example, a number of epithelial tissues consist of a relatively flat layer of surface epithelial cells in parallel with crypts (colon) or subepithelial glands (upper airways) connected to the apical compartment via narrow openings. Such an arrangement prevents uniform short circuiting due to non-uniform current density across the entire epithelium. In other words, the surface epithelial cells are probably 'over-clamped' and the crypt or gland cells are 'under-clamped'. This problem may severely compromise measurements of tracer fluxes since uncontrolled transepithelial voltage differences will exist throughout the epithelial barrier.

4.3.3 Tracer fluxes measured when V_t is not zero

If ion fluxes are measured in a uniformly short-circuited tissue ($V_t = 0$), then the unidirectional fluxes of an ion that crosses the epithelium exclusively via passive electrodiffusive pathways should be of equal magnitude. Ussing (1949) derived a general expression that provides an excellent test for passive transepithelial ion transport (Ussing, 1949). This relationship is the flux-ratio equation, which establishes that:

$$J_j^{ab}/J_j^{ba} = (C^a/C^b)\exp(zV_tF/RT) \tag{4.5}$$

where J_j denotes unidirectional ion flux (superscripts *a* and *b* refer to apical and basolateral compartments), *C* is concentration, V_t is transepithelial voltage, and *z*, *R*, *T* and *F* have their usual meanings. For this type of analysis, there can exist both chemical and electrical potential differences across the epithelium. If the experimentally determined ratio of unidirectional fluxes does not fit the prediction, then active transport must be suspected. This is not a definitive test however, since exchange diffusion and single-file diffusion can cause deviations from the predicted ratio (see Schultz, 1980). If the ratio of unidirectional fluxes does fit the prediction, then it is likely that the fluxes are passive. Experiments of this type provide data from fluxes measured under the more physiologic open-circuit condition.

4.4 LIMITATIONS OF TRANSEPITHELIAL STUDIES

Measurements of unidirectional ion fluxes help in the characterization of the transport properties of an epithelium, and in some cases may provide insight into the mechanisms responsible for transport. However, caution should be exercised when suggesting specific transport mechanisms based upon transepithelial measurements alone. The studies of Frizzell *et al.* (1975) and Cremaschi and Hénin (1975) demonstrated that under steady-state conditions the transepithelial absorption of Na^+ and Cl^- by rabbit gallbladder epithelium requires the presence of the counter-ion in the apical bathing solution. From these studies they concluded that the entry mechanism for Na^+ and Cl^- at the apical cell membrane is via carrier-mediated cotransport of Na^+ and Cl^- as a ternary complex. The results of a number of studies in this and other tissues (Duffey *et al.*, 1978; Spring and Kimura, 1978; García-Díaz and Armstrong, 1980) supported the conclusion. Subsequent experiments, performed under non-steady state conditions, revealed that other entry mechanisms, namely Na^+/H^+ and Cl^-/HCO_3^- exchange, are probably more important (Baerentsen *et al.*, 1983; Reuss, 1984; Reuss and Stoddard, 1987). In this model, the coupling between Na^+ and Cl^- entry is not due to the existence of a NaCl cotransporter, but instead is due to indirect coupling mediated by changes in intracellular pH. Thus, at the steady state the exchangers operate independently, but at the same rate (Reuss, 1984).

This example is meant to illustrate that transepithelial measurements may provide a rather complete description and characterization of the transport properties of an epithelial tissue when viewed as a 'black box', yet may lead to ambiguous or incorrect conclusions about the molecular basis of a particular transport step. Thus, it is apparent that measurements other than those obtained by traditional transepithelial methods are required in order to look inside the 'black box' and to understand the mechanisms and the regulation of those mechanisms which enable epithelia to accomplish their many and varied transport functions.

REFERENCES

Altenberg, G.A., Copello, J., Cotton, C. *et al.* (1990) Electrophysiological methods for studying ion and water transport, in *Methods in Enzymology*, Vol. 192. (eds S. Fleischer and B. Fleischer), Academic Press, Inc., Orlando, FL, pp. 650–83.

Anderson, M.P. and Welsh, M.J. (1991) Calcium and cAMP activate different chloride channels in the apical membrane of normal and cystic fibrosis epithelia. *Proceedings of the National Academy of Sciences, USA* **88**:6003–7.

Baerentsen, H.J., Giraldez F. and Zeuthen T. (1983) Influx mechanisms for Na^+ and Cl^- across the brush border membrane of leaky epithelia: a model and microelectrode study. *Journal of Membrane Biology* **75**:205–18.

Barry, P.H. (1989) Ionic permeation mechanisms in epithelia: bi-ionic potentials, dilution potentials, conductances, and streaming potentials. *Methods in Enzymology* **171**:679–714.

Barry, P.H. and Diamond, J.M. (1970) Junction potentials, electrode standard potentials, and other problems in interpreting electrical properties of membranes. *Journal of Membrane Biology* **3**:93–122.

Barry, P.H. and Diamond, J.M. (1984) Effects of unstirred layers on membrane phenomena. *Physiological Reviews* **64**:763–872.

Barry, P.H, Diamond, J.M. and Wright, E.M. (1971) The mechanism of cation permeation in rabbit gallbladder. Dilution potentials and bi-ionic potentials. *Journal of Membrane Biology* **4**:358–94.

Bello-Reuss, E., Grady, T.P. and Reuss L. (1981) Mechanism of the effect of cyanide on cell membrane potentials in *Necturus* gallbladder epithelium. *Journal of Physiology* (London) **314**:343–57.

Benos, D.J. (1982) Amiloride: a molecular probe of sodium transport in tissues and cells. *American Journal of Physiology* **242**:C131–C145.

Boucher, R.C. and Gatzy, J.T. (1983) Characteristics of sodium transport by excised rabbit trachea. *Journal of Applied Physiology: Respiratory Environmental and Exercise Physiology* **55**:1877–83.

Brayden, D.J., Krouse, M.E., Law, T. *et al.* (1993) Stilbenes stimulate T84 Cl^- secretion by elevating Ca^{2+}. *American Journal of Physiology* **264**:G325–G333.

Clarke, L.L., Grubb, B.R., Gabriel, S.E. *et al.* (1992) Defective epithelial chloride transport in a gene-targeted mouse model of cystic fibrosis. *Science* **257**:1125–8.

Clarke, L.L., Grubb, B.R., Yankaskas, J.R. *et al.* (1994) Relationship of a non-cystic fibrosis transmembrane conductance regulator-mediated chloride conductance to organ-level disease in CFTR (-/-) mice. *Proceedings of the National Academy of Sciences* **91**:479–83.

Contreras, R.G., Miller, J.H., Zamora, M. *et al.* (1992) Interaction of calcium with plasma membrane of epithelial cells during junction formation. *American Journal of Physiology* **263**:C313–C318.

Cooke, H.E., Shonnard, K., Highison, G. and Wood, J.D. (1983) Effects of neurotransmitter release on mucosal transport in guinea pig ileum. *American Journal of Physiology* **245**:G745–G750.

Cotton, C.U., Lawson, E.E., Boucher, R.C. *et al.* (1983) Bioelectric properties and ion transport of airways excised from adult and fetal sheep. *Journal of Applied Physiology: Respiratory Environmental and Exercise Physiology* **55**:1542–49.

Cotton, C.U., Boucher, R.C. and Gatzy, J.T. (1988) Bioelectric properties and ion transport across excised canine fetal and neonatal airways. *Journal of Applied Physiology: Respiratory Environmental and Exercise Physiology* **65**:2367–75.

Cotton, C.U., Weinstein, A. and Reuss, L. (1989) Osmotic water permeability of *Necturus* gallbladder epithelium. *Journal of General Physiology* **93**:649–79.

Cremaschi, D. and Hénin, S. (1975) Na^+ and Cl^- transepithelial routes in rabbit gallbladder. Tracer analysis of the transports. *Pflügers Archiv* **361**:33–41.

Diamond, J.M. (1968) Transport mechanisms in the gallbladder, in *Handbook of Physiology. Alimentary Canal. Bile; Digestion; Ruminal Physiology*, American Physiology Society, Washington DC, section 6, vol. V, pp. 2451–82.

Dörge, A. and Nagel, W. (1970) Effect of amiloride on sodium transport in frog skin. II Sodium transport pool and unidirectional fluxes. *Pflügers Archiv* **321**:91–101.

Duffey, M.E., Turnheim, K., Frizzell R.A. *et al.* (1978) Intracellular chloride activities in rabbit gallbladder: direct evidence for the role of the sodium-gradient in energizing 'uphill' chloride transport. *Journal of Membrane Biology* **42**:229–45.

Finkelstein, A. (1987) *Water Movements through Lipid Bilayers, Pores and Plasma Membranes: Theory and Reality*, Wiley, New York.

Frizzell, R.A., Dugas, M.C. and Schultz, S.G. (1975) Sodium chloride transport by rabbit gallbladder: direct evidence for a coupled NaCl influx process. *Journal of General Physiology* **65**:769–95.

Frömter, E. and Diamond, J.M. (1972) Route of passive ion permeation in epithelia. *Nature New Biology* **235**:9–13.

Fuchs, W., Larsen, E.H. and Lindemann, B. (1977) Current voltage curve of sodium channels and concentration dependence of sodium permeability in frog skin. *Journal of Physiology*, (London) **267**:137–66.

García-Díaz, J.F. and Armstrong W.M. (1980) The steady-state relationship between sodium and chloride transmembrane electrochemical potential differences in *Necturus* gallbladder. *Journal of Membrane Biology* **55**:213–22.

González-Mariscal, L., Contreras, R.G., Bolívar, J.J. *et al.* (1990) Role of calcium in tight junction formation between epithelial cells. *American Journal of Physiology* **259**:C978–C986.

Helman, S.L. and Miller, D.A. (1971). *In vitro* techniques for avoiding edge damage in studies of frog skin. *Science*, New York, **173**:146–8.

House, C.R. (1974) *Water Transport in Cells and Tissues*, Edward Arnold (Publishers) Ltd, London.

Hughes, B.A., Miller, S.S. and Machen, T.E. (1984) The effects of cAMP on fluid absorption and ion transport across frog retinal pigment epithelium. *Journal of General Physiology* **83**:875–99.

Jiang, C., Finkbeiner, W.E., Widdicombe, J.H. *et al.* (1993) Altered fluid transport across airway epithelium in cystic fibrosis. *Science* **262**:424–7.

Jovov, B., Wills, N.K. and Lewis, S.A. (1991) A spectroscopic method for assessing confluence of epithelial cell cultures. *American Journal of Physiology* **261**:C1196–C1203.

Jovov, B., Lewis, S.A., Crowe, W.E. *et al.* (1994) Role of intracellular Ca^{2+} in modulation of tight junction resistance in A6 cells. *American Journal of Physiology* **266**:F775–F784.

Koefoed-Johnsen, V. and Ussing, H.H. (1958) The nature of the frog skin potential. *Acta Physiologica Scandinavica* **42**:298–308.

Lewis, S.A. and Diamond, J.M. (1976) Na^+ transport by rabbit urinary bladder, a tight epithelium. *Journal of Membrane Biology* **28**:1–40.

Lewis, S.A., Eaton, D.C., Clausen, C. *et al.* (1977) Nystatin as a probe for investigating the electrical properties of a tight epithelium. *Journal of General Physiology* **70**:427–40.

O'Grady, S.M., Palfrey, H.C. and Field, M. (1987) Characteristics and functions of Na–K–Cl cotransport in epithelial tissues. *American Journal of Physiology* **253**:C177–C192.

Ostedgaard, L.S., Shasby, D.M. and Welsh, M.J. (1992) *Staphylococcus aureus* α-toxin permeabilizes the basolateral membrane of a Cl^--secreting epithelium. *American Journal Physiology* **263**:L104–L112.

Powell, D.W. (1987) Intestinal water and electrolyte transport, in *Physiology of the Gastrointestinal Tract*, (ed. L.R. Johnson), Raven Press, New York, pp. 1267–305.

Reuss, L. (1984) Independence of apical membrane Na^+ and Cl^- entry in *Necturus* gallbladder epithelium. *Journal of General Physiology* **84**:423–45.

Reuss, L. and Cotton, C.U. (1987) Isosmotic fluid transport across epithelia, in *Contemporary Nephrology*, Vol. 4 (eds S. Klahr and S. Massry), Plenum Publishing, New York, pp. 1–37.

Reuss, L. and Cotton, C.U. (1994) Volume regulation in epithelia: transcellular transport and cross-talk, in *Cellular and Molecular Physiology of Cell Volume Regulation*, (ed. K. Strange), CRC Press, Boca Raton, FL, pp. 31–47.

Reuss, L. and Finn, A.L. (1975) Electrical properties of the cellular transepithelial pathway in *Necturus* gallbladder. I. Circuit analysis and steady-state effects of mucosal solution ionic substitutions. *Journal of Membrane Biology* **25**:115–39.

Reuss, L. and Stoddard J.S. (1987) Role of H^+ and HCO_3^- in salt transport in gallbladder epithelium. *Annual Reviews of Physiology* **49**:35–49.

Reuss, L., Bello-Reuss, E. and Grady, T.P. (1979) Effects of ouabain on fluid transport and electrical properties of *Necturus* gallbladder: evidence in favor of a neutral basolateral sodium transport mechanism. *Journal of General Physiology* **73**:385–402.

Schultz, S.G. (1980) *Basic Principles of Membrane Transport*, Cambridge University Press.

Silva, P., Stoff, J., Field, M. *et al.* (1977) Mechanism of active chloride secretion by shark rectal gland: role of Na-K-ATPase in chloride transport. *American Journal of Physiology* **233**:F298–F306.

Slater, R.J. (ed.) (1990) *Radioisotopes in Biology. A Practical Approach*, Oxford University Press, New York.

Snouwaert, J.N., Brigman, K.K., Latour, A.M. *et al.* (1992) An animal model for cystic fibrosis made by gene targeting. *Science* **257**:1083–8.

Spring, K.R. and Kimura G. (1978) Chloride reabsorption by renal proximal tubules of *Necturus*. *Journal of Membrane Biology* **38**:233–54.

Ussing, H.H. (1949) The distinction by means of tracers between active transport and diffusion. The transfer of iodide across the isolated frog skin. *Acta Physiologica Scandinavica* **19**:43–56.

Ussing, H.H. and Zerahn, K. (1951) Active transport of sodium as the source of electric current in the short-circuited isolated frog skin. *Acta Physiologica Scandinavica* **23**:110–27.

Van Os, C.H., Wiedner, G. and Wright, E.M. (1979) Volume flows across gallbladder epithelium induced by small hydrostatic and osmotic gradients. *Journal of Membrane Biology* **49**:1–20.

Wiedner, G. (1976) Method to detect volume flows in the nanoliter range. *Review of Scientific Instruments* **47**:775–80.

5

Epithelial electrophysiology

Simon A. Lewis

At the beginning of this book we categorized epithelia as being either leaky (low electrical resistance) or tight (high electrical resistance). Chapter 4 provides a description of and uses for transepithelial electrical measurements. It is the purpose of this chapter to describe in detail the methods which allow one to characterize the electrical properties of an epithelium.

The simple measurement of transepithelial resistance is not sufficient to categorize a given epithelium as being either tight or leaky. As an example, the gastric mucosa has a low electrical resistance (100 Ωcm^2) and could be considered as a 'leaky' epithelium. However, this epithelium is capable of generating large spontaneous potentials and supporting large ionic concentration gradients, and has a low water permeability. These factors indicate that the gastric mucosa is a tight epithelium. The methods described in this chapter will permit a determination of the individual membrane resistances and the tight junction resistance which comprise the transepithelial resistance, and the calculation of the apical and basolateral membrane electromotive forces (e.m.f.). The determination of the e.m.f. will yield information about the ion selectivity of the individual membranes and how this selectivity might change as a function of altered transepithelial ion transport.

Epithelial Transport: A guide to methods and experimental analysis.
Edited by Nancy K. Wills, Luis Reuss and Simon A. Lewis.
Published in 1996 by Chapman & Hall, London. ISBN 978-94-010-7168-0

The purpose of this chapter is three-fold. First, a detailed description of the equipment needed to measure the basic electrical properties of an epithelium is provided. Next, procedures are presented for using this experimental set-up to measure the basic electrical properties of epithelia including the transepithelial potential difference, the transepithelial resistance, short-circuit current and capacitance. Last, we will outline the methods, assumptions and equations which allow one to determine the individual membrane resistances, the tight junctional resistance and the membrane equivalent electromotive forces (batteries), i.e. the ionic permeability of the individual membranes.

5.1 EQUIPMENT

The necessary equipment (see Appendix A for suppliers) needed to study any epithelium, including tissue-cultured epithelia, can be broken into three components:

1. A pair of hemi-chambers (typically termed Ussing chambers), in which the epithelium acts as a partition between the two hemi-chambers thus creating two separate compartments: the mucosal or luminal compartment and the serosal or blood side compartment.
2. Electrodes for sensing the epithelial voltage and for passing current.
3. Instrumentation which measures both voltage and current.

5.1.1 Chambers

Independent of the design details, the chambers must have some common features. These are as follows.

- A provision for controlling the temperature of the mucosal and serosal bathing solutions as well as bubbling both solutions with the gas mixture of choice.
- A design which minimizes damage to the cells that contact the inner circumference of the hemi-chambers. If not protected against, such 'edge damage' will result in a measured resistance the epithelium which is lower than the native tissue.
- Support for the epithelium on one side by a rigid but permeable layer. This reduces epithelial stretch and the possibility of cell or tight junctional damage.
- Placement of the voltage measuring electrodes on either side of the epithelium as close as is possible. This will reduce the magnitude of the solution series resistance which, if large, can compromise one's ability to voltage clamp the epithelium precisely (Chapter 4 and below).
- Placement of the current passing electrodes in the rear of the chambers, i.e. as far away from the epithelium as possible. This will assure that

there is a uniform current density across the epithelium. If the current density is not uniform, it will result in an overestimate of the epithelial resistance.

- The ability to change both of the bathing solutions easily and rapidly without interrupting electrical measurements and without altering the electrical properties of the epithelium.
- Multi-functional design. In addition to measuring transepithelial electrical properties, it should be constructed so that one can use microelectrodes or ion selective microelectrodes. The former electrodes are essential for determining the individual membrane resistances, while the latter electrodes are required to determine the individual membrane ionic permeabilities.
- Additional features include the ability to measure cell volume and use intracellular fluorescent dyes.

Three types of chamber design that are currently being used to study epithelial transport and are shown in Figure 5.1.

5.1.2 Electrodes

Electrodes are an essential component of any electrophysiological set-up since they provide the low resistance interface between the Ringer's solution and the electronic equipment. This section considers the electrodes one uses to sense the epithelial voltage and pass a transepithelial current. Although it might seem trivial, the type of electrodes to be used should be chosen carefully. Some guide lines are listed below.

There are three choices in voltage-measuring electrodes: silver/silver chloride (Ag/AgCl) electrodes, calomel electrodes or agar bridges. The latter must be connected to either Ag/AgCl electrodes or calomel electrodes (via a salt solution) which are then connected to a voltage measuring amplifier. These same electrode configurations can be used for the current passing limb of the electronics. The choice of electrode set depends upon the epithelium to be studied, as well as the composition of the solutions bathing the epithelium.

For instance, Ag/AgCl electrodes can only be used if the epithelium to be studied is not sensitive to trace levels of silver ions (as an example, toad urinary bladder epithelium is very sensitive to trace levels of silver) or if the chloride activities in the solutions bathing both sides of the epithelium are identical. If the chloride concentrations are not equal it will result in an asymmetrical voltage difference between the pair of voltage-measuring electrodes which will be summed into the transepithelial voltage, i.e. the measured voltage will be artifactually high or low depending upon the ratio of mucosal to serosal chloride concentrations.

If the epithelium is sensitive to trace levels of silver or if solutions

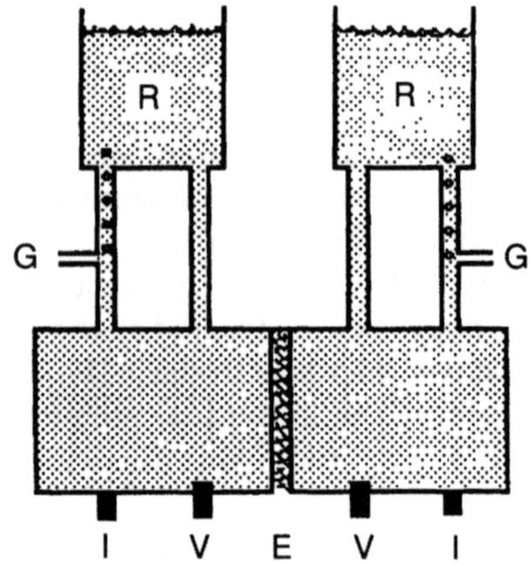

(a)

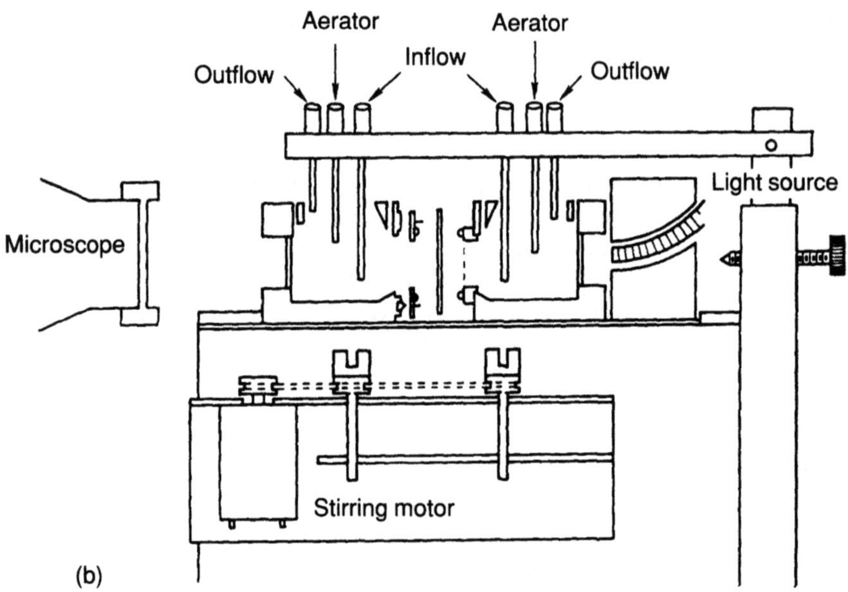

(b)

containing different levels of chloride are employed, then either agar bridges connected to Ag/AgCl electrodes or calomel electrodes must be used. The advantages of the agar bridge and Ag/AgCl electrode combination compared with a calomel electrode is that agar bridges are inexpensive and small. Thus it is easy to locate them close to the epithelial surface.

Calomel electrodes can be purchased from most scientific supply houses. Scintered Ag/AgCl pellets are commercially available from either Warner Instruments (Hamden CT, USA) or World Precision Instruments (Sarasota Fl, USA). Our preference is to use silver wire (cleaned with steel

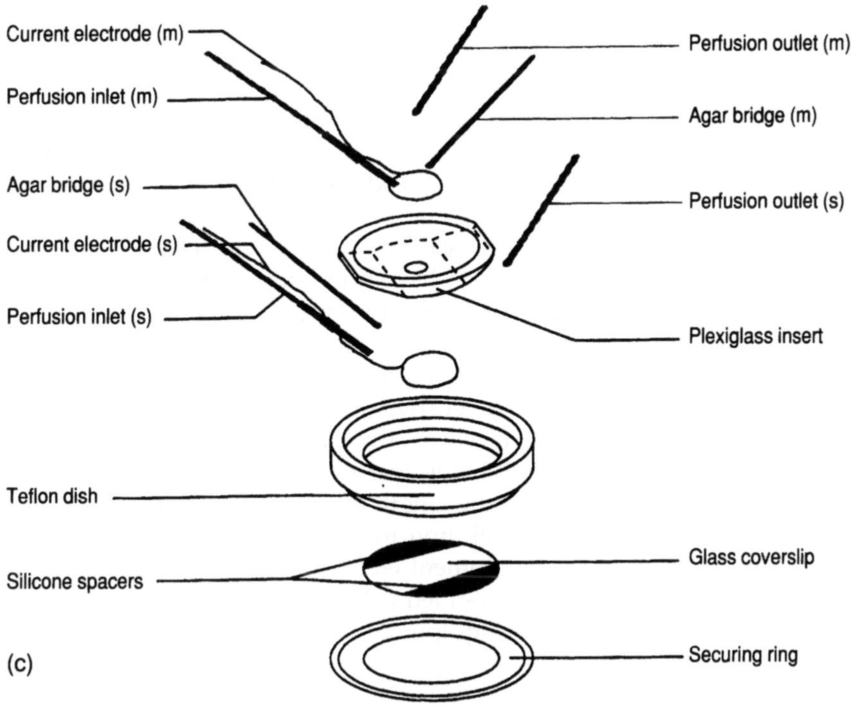

Figure 5.1 Three basic chamber designs for studying epithelial transport. (a) Original design by Ussing. This chamber circulates solution across the surface of the epithelium using a gas lift system (G) which also aerates the Ringers in the reservoirs (R) and can be temperature controlled. V = voltage measuring electrodes which are placed close to and on opposite sides of the epithelium. I = current passing electrodes. A disadvantage of this system is that it is closed and thus does not meet criteria (vi) and (vii) (see main text). In addition the original design caused significant edge damage (criterion (ii)). (b) Chamber was designed to meet all of the listed criteria except for (vi) and (viii). (c) Chamber designed to be used on an inverted microscope and meets all of the listed criteria including the ability to measure transepithelial electrical properties and monitor intracellular ionic activities using fluorescent dyes. A disadvantage is that solutions must be continuously flowed through the chamber and that this solution must be preheated. (Reproduced with permission from Crowe and Wills, 1991.)

wool and also available from Warner Instruments or World Precision Instruments), which can then be chlorided by either connecting the electrodes to the positive terminal of a power supply (with a reference silver wire connected to the negative terminal of the power supply) or by soaking the silver wires in full strength household bleach (sodium hypochlorite) for at least one hour. A successfully chlorided silver wire will have a brown-purple color. The presence of white spots indicates poor coating and the wires should be cleaned and re-coated. Agar bridges are easily constructed by heating a mixture of 3–5% agar with 1 M KCl (w/v). While still hot, the agar can then be drawn into the polyethylene tubing using a syringe or vacuum line. Since the polyethylene tubing is opaque, we have found it convenient to add a dye (methylene blue) to the agar/KCl solution; in this manner any discontinuity in the agar lead (break in the agar or an air bubble) can be easily visualized.

Current-passing electrodes must have a low interface resistance with the solution. If this interface resistance is too large, it will limit the current-passing capacity of the electronic equipment. Ideally a sheet or coil of Ag/AgCl wire should be used in the rear of each hemi-chamber. This sheet or coil will assure a uniform current density across the epithelium. If the tissue is sensitive to trace levels of silver, it will be necessary to use a wide-bore agar bridge placed at the rear of each hemi-chamber: every effort should be made to keep the length of the bridge short and fill the bridge with a low-resistance solution to maximize the current-passing capability of the electronic equipment (see below). A disposable pipette tip partially filled with agar, back-filled with a low resistance solution and then coupled to the electronic equipment with a Ag/AgCl wire, works very well.

5.1.3 Electronics

This section gives an overview of the electronic equipment needed to measure the properties of the epithelium being studied. The minimum equipment needed is as follows.

- A current/voltage clamp. This is essential to record the transepithelial voltage, the transepithelial resistance and the short circuit current (I_{sc}). Epithelial current/voltage clamps are commercially available (Appendix A).
- A pulse generator, either to pass current across the epithelium and record the change in the transepithelial voltage or to voltage clamp the epithelium to different levels and record the current that is passed to maintain this voltage. Either of these two measuring schemes allows one to calculate the transepithelial resistance. Some commercially available current/voltage clamps have built-in pulse generators.
- A two-channel paper chart recorder (three to four channels is better), to give a permanent record of the experiment. The recorder's pen speed should be fast enough to record the change in transepithelial voltage or current produced by the pulse generator. In recent years the cost of a computer system has become less than the cost of a paper chart recorder.

- An oscilloscope to measure the time-dependent changes in transepithelial voltage during a current pulse (Chapter 6). The oscilloscope can also be used to determine whether the current/voltage clamp is stable (i.e. it is not oscillating) as well as the response time of the clamp, e.g. whether the current or voltage pulse that is being passed is square or whether it has a finite rise time. This is important when one is estimating the epithelial surface area using capacitance (Chapter 6). An alternative to an oscilloscope is an analog-to-digital converter interfaced to a Macintosh® computer. Commercially available software allows this system to be used in an oscilloscope mode, as a chart recorder and as a pulse generator (MacLab®). Such a system will cost about the same amount as a conventional paper chart recorder, pulse generator and oscilloscope.
- A step up from the basic equipment listed above can be made using a computer, analog-to-digital (and digital-to-analog) converter and appropriate software. In this ideal case, the computer digitizes the data and can then display the data on the terminal screen, output it to a printer and store it on disk for later analysis etc. The advantage of using a computer interfaced to the current/voltage clamp is that the data (current and voltage) can be immediately analyzed by the computer for instant feedback on the viability of the epithelium.
- If microelectrodes are used, an anti-vibration table and a Faraday cage are necessary, as well as a microscope and micromanipulators.

When purchasing a current/voltage clamp, one must have an idea of the application for which it will be used and be aware of four features: compliance voltage of the current-passing amplifier; speed of the current and voltage circuits; series resistance compensation; and internal pulse generator. Compliance voltage of the current-passing amplifier (defined as the maximum voltage output of the amplifier) is important if agar leads are used as part of the current-passing circuit. Typically, most clamps use amplifiers which have a compliance voltage of 15 V. If the total resistance of the current-passing circuit is 10 000 Ω, the maximum current the clamp can pass is 1.5 mA. Although this current level is sufficient for high resistance (tight) epithelia, it is marginal for low resistance (leaky) epithelia or for where resistance has been decreased due to an experimental maneuver. Warner Instruments and World Precision Instruments sell a current/voltage clamp which uses a 120 V current-passing amplifier, and these instruments would be the choice for leaky epithelia. The disadvantage of clamps which have high compliance amplifiers is that they are slow (response time in tenths of seconds). In addition, the World Precision Instruments voltage clamp lacks a variable internal pulse generator. The Warner Instruments and the Physiologic Instruments current/voltage clamps are fast (with microsecond rise times) and have internal pulse generators and series resistance compensation circuitry and the compliance voltage can be specified. All of these clamps have outputs which can be easily connected to chart recorders, an oscilloscope and/or a computer.

The voltage-measuring and current-passing electrodes are first placed in the chambers and then connected to the current/voltage clamp. The clamp outputs are then connected to the chart recorder, oscilloscope and (if available) a computer. At this point one must check for asymmetries in the voltage measuring electrodes. This is done by placing both electrodes in the same bath and measuring the voltage. If this voltage is not zero (but only a few millivolts) it can be balanced out using a voltage offset found on most commercially available clamps. If the voltage is unstable or is large (> 10 mV) the electrodes must be replaced and the offset must be measured again. The recording mode to be used (voltage or current clamp) depends upon the experimental design. Advantages and disadvantages of these recording modes are as follows.

(a) Current clamp

In this mode the transepithelial current is typically clamped to zero (i.e. there is no net transepithelial current flow), the condition under which the tissue is exposed *in vivo*. This is the mode of choice for those interested in the net ionic transport across the epithelium. Another advantage is that if Ag/AgCl current-passing wires are being used, the possible contamination of the bathing solutions with trace levels of silver is reduced.

(b) Voltage clamp

In this mode the tissue's transepithelial voltage is set to a specified value (picked by the investigator) and the voltage clamp then passes a transepithelial current to maintain this voltage at the preset level. This mode is used when determining (by radio isotopic flux measurements or pharmacological ion transport blockers) the ions that are actively transported by the epithelium (Chapter 4).

5.2 MEASURING BASIC ELECTRICAL PROPERTIES

After mounting the epithelium in the chambers, connecting the current/voltage amplifier and eliminating asymmetries in the voltage measuring electrode, three basic properties of the epithelium can be measured: the transepithelial voltage, the transepithelial resistance and the short-circuit current (I_{sc}, a measure of the net active ion transport across the epithelium).

5.2.1 Transepithelial voltage

The first measurement is very straight forward and is simply read off from either the panel meter of the clamp or the paper chart recorder. It is important to make sure that there is no asymmetry potential between the voltage measuring electrodes. Since the transepithelial voltage is measured differentially, one must know which one of the two voltage measuring electrodes is considered zero (or ground). We routinely ground the serosal solution.

5.2.2 Transepithelial resistance

To measure the transepithelial resistance, either pass a current across the epithelium (ΔI) and measure the resultant voltage change (ΔV, under current clamp mode) or clamp the epithelium to a new voltage (ΔV) and measure the change in current (ΔI) required to hold the epithelium at that voltage (under voltage clamp mode). The resistance is then calculated using **Ohm's law**, which simply states that the resistance is equal to the change in the transepithelial voltage divided by the change in the transepithelial current:

$$R_{\text{meas.}} = (\Delta V_t / \Delta I_t)A \qquad (5.1)$$

where A is the area of the epithelium and the units of resistance are in ohms·cm^2. Since there is a finite distance between the voltage-measuring electrodes and the epithelium, the calculated resistance ($R_{\text{meas.}}$) is the sum of the transepithelial resistance (R_t) and the series resistance of the solution (R_s), i.e. the resistance of the bathing solution between the tissue and each of the voltage-measuring electrodes. This series resistance must be subtracted from $R_{\text{meas.}}$ to determine the actual transepithelial resistance ($R_t = R_{\text{meas.}} - R_s$). Some current/voltage clamps have a provision for automatically subtracting this series resistance (once it has been measured). Series resistance can be calculated by measuring the resistance of the chambers in the absence of the epithelium. Since solutions of different ionic composition have different resistivities, R_s must be determined for each solution used. Although for high resistance epithelia, series resistance is only a minor correction (e.g. 1–2% of $R_{\text{meas.}}$), for low resistance epithelia it can be 50% or greater of $R_{\text{meas.}}$. There is a second approach for measuring R_s in Chapter 6.

5.2.3 Short-circuit current

The measurement of the short-circuit current is deceptively simple. The short-circuit current (I_{sc}) is simply defined as the current that must be passed across the epithelium to reduce the transepithelial voltage to zero, i.e. it is the current that short-circuits the tissue. To measure I_{sc}, simply voltage clamp the epithelium to zero millivolts and read, from a panel meter, the applied current. When performing such a measurement, it is essential to compensate for the series resistance (R_s) by using the series compensate control on the clamp. If such a feature is not available then one can calculate the voltage which will reduce the transepithelial voltage to zero. This is done most simply by multiplying the spontaneous potential (V_t) by the ratio of R_s/R_t and then setting the clamp voltage at this value (its sign will be opposite to that of the V_t). An example of this procedure is illustrated in Circuit 1.

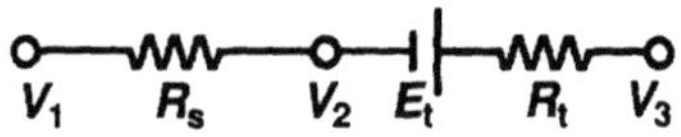

Circuit 1

In Circuit 1, the solution is modeled as a resistor (R_s = 50 Ω) and the epithelium has been modeled as a voltage source ($E_t = V_2 - V_3 = -10$ mV) in series with a resistor (R_t = 100 Ω); the voltage-measuring electrodes are at positions 1 and 3 and in this configuration (open circuit conditions) measure a voltage of (V_1–V_3) = –10 mV. (It is important to note that under these conditions V_2–V_3 = –10 mV.) Voltage clamping to 0 mV (between positions 1 and 3) reduces the transepithelial voltage (V_2–V_3) from –10 mV to –3.33 mV but increases the voltage across R_s (V_1–V_2) from 0 mV to +3.33 mV. To make the transepithelial voltage zero we must clamp the circuit (V_1–V_3) to +5 mV, i.e. $5\text{ mV} = -V_t(R_s/R_t)$. If either R_s or R_t changes, one must recalculate the voltage to which the circuit must be clamped to keep the transepithelial voltage at zero.

5.3 DETERMINING EPITHELIAL ELECTRICAL PROPERTIES

In the above section we introduced the concept that an epithelium can be modeled as an electrical circuit composed (in the most simple case) of a resistor and a voltage source. The justification for this electrical approach is that many epithelia produce a spontaneous potential even when bathed on both sides with identical solutions. Thus, this potential can be represented as a voltage source. Since an epithelium is capable of restricting (resisting) the movement of ions between two compartments, it has resistive properties. Therefore, it can be represented as a resistor. (Appendix B gives a step-by-step derivation of electrical equivalent circuits based on the morphology of the epithelium.) Circuit 2 is the most simple equivalent of an epithelium based on its morphology.

In this equivalent circuit, 'm' and 's' represent the mucosal and serosal compartments respectively, the cells of the epithelium are repre-

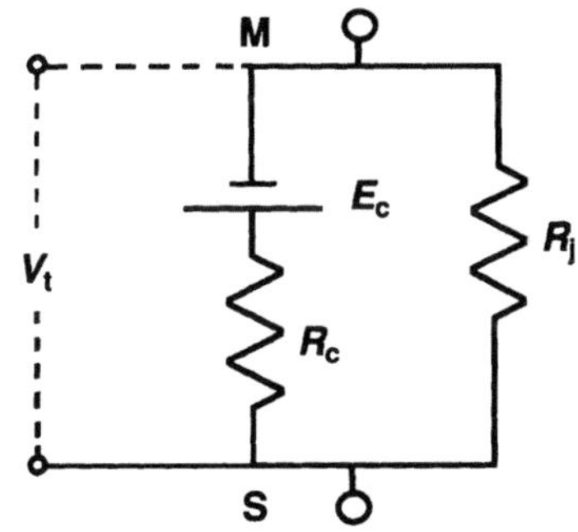

Circuit 2

sented by a resistor (R_c) in series with a voltage source (E_c), while the parallel tight junctions are represented by a simple resistor. This voltage source (also called the cellular electromotive force, e.m.f.) is a complex function of the conductive properties of the cell membranes and the composition of the ions in the bathing solution as well as the cell interior (see below). No voltage source is present in the tight junctions since the tissue is bathed by symmetric solutions. For simplicity, we have left out a series (solution) resistor. The transepithelial conductance (G_t, the conductance between mucosal solution 'm' and the serosal solution 's') is given by:

$$G_t = G_c + G_j \tag{5.2}$$

Since resistance is the inverse of conductance ($G_t = 1/R_t$), then the transepithelial resistance is:

$$R_t = \frac{R_c \times R_j}{R_c + R_j} \tag{5.3}$$

From Ohm's law, the current flow (i) around this circuit is:

$$i = \frac{E_c}{R_c + R_j} \tag{5.4}$$

Using Ohm's law we find that the voltage difference across the epithelium ($V_m - V_s = V_t$) is equal to the current flow through the tight junctions:

$$V_t = i \times R_j = \frac{E_c \times R_j}{R_c + R_j} \tag{5.5}$$

or through the cellular pathway:

$$V_t = E_c - i \times R_c = E_c\left(1 - \frac{R_c}{R_c + R_j}\right) = \frac{E_c \times R_j}{R_c + R_j} = \frac{E_c}{R_c} \times R_t \tag{5.6}$$

5.3.1 Transepithelial determination of epithelial parameters

One of the goals of this chapter is to develop methods for calculating the individual membrane and tight junctional resistances and the value of the cell voltage source. Inspection of the above equations shows three interesting features:

- The magnitude of the transepithelial potential is a function of both the cell and tight junctional resistance as well as the magnitude of the cell voltage source.
- The term (E_c/R_c) is equal to the short-circuit current (see above) and is indeed equal to the current generating capability of the cells.
- At a constant E_c, a decrease in R_c will result in an increase in V_t, thus as R_c approaches zero, V_t approaches E_c. Similarly, as R_c increases towards infinity, then R_t approaches R_j.

This relationship can be derived and formalized in the following manner (Wills, Lewis and Eaton, 1979):

$$V_t = \frac{E_c}{R_c} \times R_t \text{ rearranging gives } \frac{V_t}{E_c} = \frac{R_t}{R_c} \tag{5.7}$$

and

$$\frac{1}{R_t} = \frac{1}{R_c} + \frac{1}{R_j} \text{ rearranging gives } 1 = \frac{R_t}{R_c} + \frac{R_t}{R_j} \tag{5.8}$$

Substituting equation (5.7) into equation (5.8) gives:

$$\frac{V_t}{E_c} + \frac{R_t}{R_j} = 1 \tag{5.9}$$

If a perturbation which changes only the cell resistance is performed and the resulting paired values of V_t and R_t (during this perturbation) are plotted, there is a linear double intercept plot in which both R_j and E_c can be determined (Figure 5.2). A major question is how one knows that the perturbation has altered only the cell resistance and not the tight junctional resistance R_j or the cell voltage source E_c. The best indicator is that the plot is linear, thus a curvilinear plot will suggest that the assumption of a constant R_j and/or E_c has been violated.

A similar relationship can be derived if the experiments are performed under short-circuit conditions (Yonath and Civan, 1971). The derivation is shown below:

$$G_t = G_c + G_j \tag{5.10}$$

and

$$I_{sc} = E_c \times G_c \text{ rearranging gives } G_c = \frac{I_{sc}}{E_c} \tag{5.11}$$

Substituting equation (5.10) into (5.11) gives:

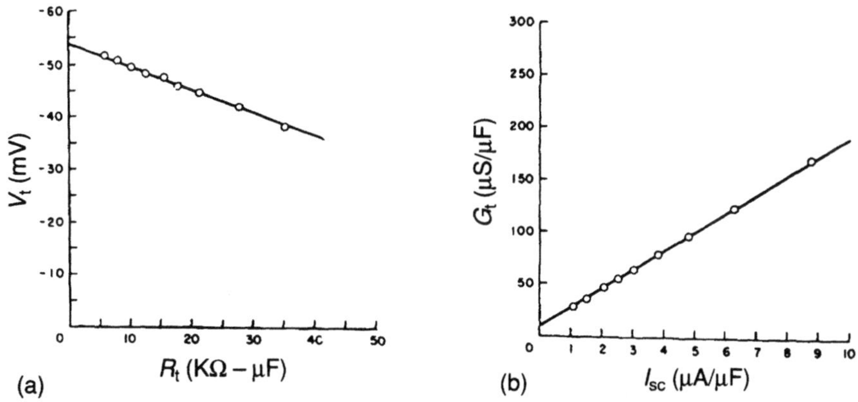

Figure 5.2 Plot of (a) V_t versus R_t and (b) G_t versus I_{sc}. In this example the cell membrane resistance (of rabbit urinary bladder epithelium) was decreased using the pore forming antibiotic gramicidin D. Of importance is that the mucosal solution was designed to mimic the ionic composition of the cell interior and as a consequence increasing the apical membrane resistance will not alter the cell emf. This is confirmed since both plots are linear and yield near identical values for E_c and R_j.

$$G_t = \frac{I_{sc}}{E_c} + G_j \tag{5.12}$$

Using this equation one plots the transepithelial conductance (G_t) versus the measured short circuit current (I_{sc}). Such a plot will have an intercept equal to the junctional conductance (the inverse junctional resistance) and a slope equal to the inverse cell voltage source (i.e. $1/E_c$). This equation has the same assumption as equation (5.9), i.e. the experimental perturbation must only change the cellular resistance (or conductance).

Experience has shown that a plot of V_t vs. R_t is more sensitive to changes in either R_j or E_c than is a plot of G_t vs. I_{sc}. This is because whereas I_{sc} is a measure of only the cellular pathway (it is not affected by the junctional resistance), V_t is a function of both the cellular pathway and the junctional resistance and is thus more sensitive to a change in either parameter. This is illustrated in Figure 5.3, in which the experimental conditions were such that both R_c and E_c are changing. Note that although the G_t vs. I_{sc} plot is linear, the V_t vs. R_t plot is non-linear.

To use this method one must be able to change only the resistive (conductive) properties of the cellular pathway and the change must be sufficiently large to result in a measurable change in the measured parameters, i.e. V_t, R_t, I_{sc} or G_t. As a consequence, this method is most profitably used on the so-called tight epithelia, since in this class a change in the cellular resistance will result in a significant change in R_t. In leaky

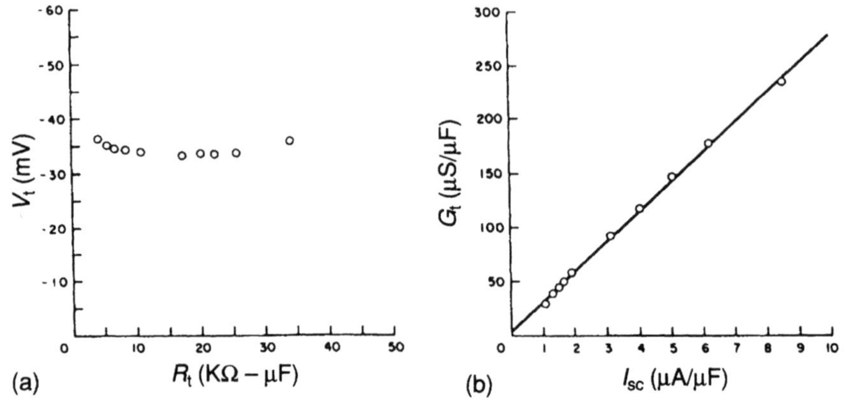

Figure 5.3 Plot of (a) V_t versus R_t and (b) G_t versus I_{sc}. In this example the cell membrane resistance was decreased using gramicidin D. In addition the mucosal solution was selected such that the cell emf (E_c) would change during gramicidin action on the cell resistance. This is illustrated by the fact that the plot of V_t versus R_t is non-linear, and shows that one of the assumptions of the equation has been violated (in this instance E_c is not constant). Note, however, that the plot of G_t versus I_{sc} is reasonably linear even though the assumption of a constant E_c has been violated.

epithelia, where R_j can be 10-fold lower than R_c, one must produce much larger perturbations (in this case a decrease) in R_c to obtain a reliable (measurable) decrease in R_t.

Three approaches have been used to alter R_c in a controlled manner: increasing the cell membrane conductance to a given ion using second messenger systems; decreasing the conductance to a given ion using pharmacological blockers; and artificially increasing the membrane conductance using pore-forming agents such as gramicidin D, nystatin or amphotericin B. When using pore-forming agents, one must use a mucosa solution which mimics the cell interior; thus the potassium content must be high while the calcium, sodium and chloride content must be low. Typically, one does an equimolar replacement of all of sodium with potassium and an equimolar replacement of chloride with a large monovalent anion such as gluconate. Such a solution has two advantages: first, since there is low chloride in this bathing solution, cell swelling due to KCl influx is reduced to a minimum; and secondly, since the mucosal and cell ion concentrations are matched, there will be no change in the apical membrane voltage source when the pore-forming agent is added, i.e. E_c will not be affected as R_c is decreased. In addition, since the ion concentrations are matched, the value of E_c will be approximately equal to the value of the voltage source of the basolateral membrane.

5.3.2 Dissecting the cellular pathway

In section 5.3.1 an epithelium was modeled as a cellular resistance in series with a voltage source and the cellular pathway was in parallel with junctional resistance. Although for the above analysis it was convenient to model the cellular pathway as a single resistor/voltage source, morphological considerations demand that we expand this model and consider that the cell pathway is actually composed of two membranes in series. Each of these membranes can be modeled with a series combination of a resistor and voltage source, as shown in Circuit 3.

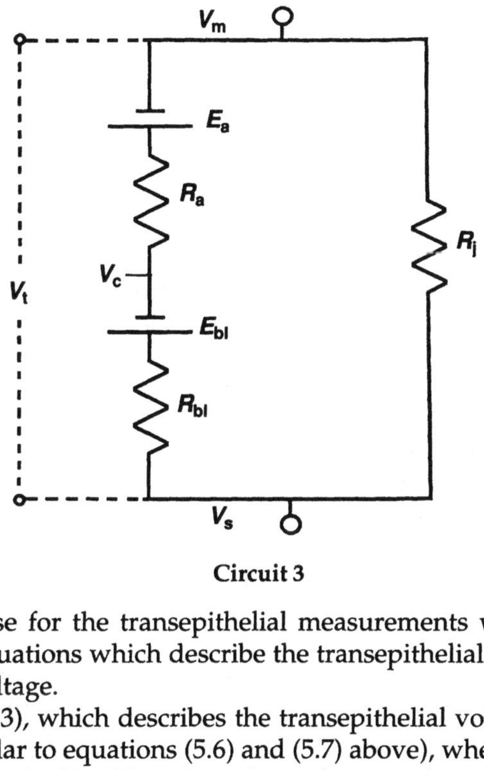

Circuit 3

As in the case for the transepithelial measurements we will start by deriving the equations which describe the transepithelial voltage and the intracellular voltage.

Equation (5.13), which describes the transepithelial voltage ($V_t = V_m - V_s$) is very similar to equations (5.6) and (5.7) above), where E_c is equal to the sum of the apical and basolateral membrane voltage sources (E_a and E_{bl} respectively) and R_c is equal to the sum of the apical and basolateral membrane resistors (R_a and R_{bl}, respectively).

$$V_t = \frac{E_a + E_{bl}}{R_a + R_{bl}} \frac{(R_a + R_{bl}) \times R_j}{R_a + R_{bl} + R_j} \tag{5.13}$$

For the basolateral membrane voltage ($V_{bl} = V_c - V_s$):

$$V_{bl} = \left(\frac{E_{bl}}{R_{bl}} - \frac{E_a}{R_a + R_j} \right) \frac{(R_a + R_j) \times R_{bl}}{R_a + R_{bl} + R_j} \tag{5.14}$$

This equation is interesting since it demonstrates that the magnitude of the basolateral membrane potential is a function of all the other circuit elements of the epithelium. That is to say, a change in either the apical membrane or junctional properties can cause a change in the basolateral membrane potential, without directly altering the properties of the basolateral membrane. To solve uniquely for the individual membrane voltage sources, it is obvious that one must measure the individual resistors of the above circuit.

In this section we will review the procedures used to calculate the values for each of the membrane resistors and membrane voltage sources. The approaches that are listed below require the use of intracellular microelectrodes. Because of this, we first briefly overview the technical aspects of microelectrode measurements, including criteria for the successful impalement of cells.

(a) Microelectrodes

The addition of microelectrodes adds another level of complexity to the experimental set-up. One needs a vibration-free table (a flotation table), a microscope (for positioning the microelectrode near the surface of the epithelium), a micromanipulator (to move the microelectrode into the cell in micrometer increments) and a high-impedance (high input resistance, preferably 10^{12} Ω) differential amplifier with a voltage offset control and a microelectrode resistance test circuit. Ancillary equipment includes a microelectrode puller, a compound microscope and microelectrode holders. The chamber design has already been discussed and should have a provision for allowing microelectrode impalement of the epithelial cells across the apical membrane (i.e. from the mucosal solution). Microelectrodes are fabricated using a puller and a length of glass capillary tubing. This electrode is then back-filled with 1 M KCl and placed in a microelectrode holder (World Precision Instruments or Warner Instruments) filled with 1 M KCl. The holder is placed in the micromanipulator and connected to the positive input of the differential amplifier. The negative input of the differential amplifier is connected to the voltage measuring electrode of the serosal chamber. The microelectrode resistance under these circumstances should be at least 20 MΩ. Using the micromanipulator, the microelectrode is advanced across the apical membrane of the cell into the cell interior. The following criteria are used to indicate a successful impalement:

- Upon impaling the cell, the voltage should rapidly settle to a new steady-state value.
- When in the cell, the membrane potential should remain stable (assuming that no perturbation has been performed).

- Upon removal of the electrode from the cell, the voltage recorded by the electrode should be the same as the pre-impalement value. Typically one sets the microelectrode voltage to the transepithelial voltage prior to impaling the cell.
- The value of the resistance ratio (see below) should rapidly settle to a steady state value and should not spontaneously change.

(b) Determining membrane resistances and voltage sources

Once the cell has been impaled, a voltage is recorded which is equal to the basolateral membrane potential, i.e. V_{bl}; thus the negative input of the microelectrode amplifier is connected to the serosal voltage measuring electrode. To calculate the individual membrane resistances and thus the values for the membrane voltage sources, there are in brief three approaches: changing an individual membrane resistor; altering a membrane voltage source; and cable analysis.

Changing membrane resistance

For this method we must be able to measure the transepithelial resistance (see above) and the ratio of the apical to the basolateral membrane resistances ($R_a/R_{bl} = \alpha$). To measure α, pass a transepithelial current and (using the microelectrode) measure the change in the basolateral membrane potential (ΔV_{bl}), then calculate the change in the apical membrane potential (ΔV_a). The change in the latter is calculated as the difference between the change in the transepithelial and the basolateral membrane potential:

$$\frac{\Delta V_a}{\Delta V_{bl}} = \frac{\Delta I_c R_a}{\Delta I_c R_{bl}} = \frac{R_a}{R_{bl}} = \frac{G_{bl}}{G_a} = \alpha \tag{5.15}$$

where I_c is a component of the transepithelial current that flows through the cellular pathway (the remaining current flows through the tight junctions).

So far we have two equations which describe the epithelial resistive properties. These are the transepithelial resistance (or conductance) and the resistance ratio (α). However, there are three unknowns in the circuit we are trying to solve: R_a, R_{bl} and R_j. Thus we do not have enough information to solve for the resistors.

The next step is to change only the apical membrane resistance and measure both α and R_t. This can be performed using one of the methods outlined for the transepithelial measurements. We now have four unknowns and four equations and can thus solve for the individual resistors using the following equations (Lewis, Eaton, Clausen and Diamond, 1977):

$$G_t = G_j + \frac{G_{bl}}{1 + \alpha} \tag{5.16}$$

and after changing the apical membrane resistance:

$$G_t' = G_j + \frac{G_{bl}}{1 + \alpha'} \tag{5.17}$$

where G_t' and α' are measured after altering the apical membrane conductance. Taking the difference between these two equations yields a value for G_{bl}. Then G_j is calculated from either equation (5.16) or (5.17), and G_a or G_a' are calculated from α or α', respectively.

A similar set of equations can be derived for a change in only the basolateral membrane resistance (or conductance):

$$G_t = G_j + \frac{G_a \cdot \alpha}{1 + \alpha} \quad G_t' = G_j + \frac{G_a \cdot \alpha'}{1 + \alpha'} \tag{5.18}$$

Measure G_t and α before and after a change in R_{bl} (Frömter and Gebler, 1977). Using the above equations, G_a, G_j and G_{bl} can be calculated both before and after the perturbation.

Changing membrane e.m.f.s

For this method we must measure not only R_t and α but also the change in the basolateral and transepithelial voltage caused by changing either the apical or basolateral membrane e.m.f. (or resistor). This method assumes that the perturbation will only change the e.m.f. (or resistor) of the membrane of interest. As an example, if we change the basolateral membrane e.m.f. (or resistor) this must not cause a change in the apical or junctional resistors or e.m.f.s.

When changing basolateral membrane e.m.f. only, measure the change in V_{bl} (ΔV_{bl} – this change is different from the one measured when passing current across the epithelium) and V_t (ΔV_t) caused by either a channel blocker or by stimulation of an electrogenic process (Lewis and Wills, 1982).

If we change E_{bl}, then:

$$\frac{\Delta V_{bl}}{\Delta V_t} = 1 + \frac{R_a}{R_j} = 1 + \gamma \tag{5.19}$$

where γ is the ratio of the apical membrane resistance to the tight junction resistance. Since we have measured α and R_t before and after the experimental perturbation, the three epithelial resistors can now be calculated. The equations are as follows.

For the values of the resistors before changing E_{bl}:

$$R_j = R_t\left(1 + \frac{\alpha}{\gamma(1 + \alpha)}\right) \tag{5.20}$$

$$R_a = R_j\gamma \tag{5.21}$$

$$R_{bl} = \frac{R_a}{\alpha} \tag{5.22}$$

and for the value of the resistors after changing E_{bl}:

$$R_j = R_t'\left(1 + \frac{\alpha'}{\gamma(1+\alpha')}\right) \tag{5.23}$$

$$R_a = R_j\gamma \tag{5.24}$$

$$R_{bl}' = \frac{R_a}{\alpha'} \tag{5.25}$$

A similar set of equations can be derived for changes in E_a, i.e. the apical membrane e.m.f.:

$$\frac{\Delta V_{bl}}{\Delta V_t} = \frac{R_{bl}}{R_j} = \frac{1}{\beta} \tag{5.26}$$

The values for the resistors before changing the apical membrane e.m.f. are:

$$R_j = R_t\frac{1+\alpha+\beta}{1+\alpha} \tag{5.27}$$

$$R_{bl} = \frac{R_j}{\beta} \tag{5.28}$$

$$R_a = \alpha R_{bl} \tag{5.29}$$

The resistor values after changing the apical membrane e.m.f. is then:

$$R_j = R_t\frac{1+\alpha'+\beta}{1+\alpha'} \tag{5.30}$$

$$R_{bl} = \frac{R_j}{\beta} \tag{5.31}$$

$$R_a' = \alpha' R_{bl} \tag{5.32}$$

Once the resistors for the epithelium have been determined, one can next calculate the value for the membrane voltage sources. Thus the basolateral membrane e.m.f. is calculated as:

$$E_{bl} = V_{bl} + V_t\frac{R_{bl}}{R_j} \tag{5.33}$$

and the apical membrane e.m.f. is calculated as:

$$E_a = V_t\left(1 + \frac{R_a}{R_j}\right) - V_{bl} \tag{5.34}$$

It is important to note that the calculated e.m.f.s are not necessarily those of a single ionic conductance but could very well be due to number of different ionic channels in parallel. One can dissect out the types of channel using pharmacologic blockers and determining the effect of these agents on the membrane conductance as well as the membrane e.m.f. (Appendix B).

5.4 ARTIFACTS IN ELECTRICAL MEASUREMENTS

It is worth noting that, for both of the above methods, it is best to measure a series of data points during the change in either the apical or the basolateral membrane resistances. This will allow one to plot G_t vs. $1/(1 + \alpha)$ or $\alpha/(\alpha + 1)$, where a linear relationship suggests that only the specified resistor has been changed. An example is shown in Figure 5.4. A non-linear relationship can have two interpretations. The first is that the applied perturbation altered not only the target membrane but also the tight junctions and/or the opposing membrane. A second interpretation is that the microelectrode damaged the cell, causing a reduction in the apical membrane resistance and thus an artifactually low value for α.

When using microelectrodes to record membrane potentials and the resistance ratio, use an oscilloscope to monitor the basolateral membrane potential as well as the voltage response of the microelectrode to a square transepithelial current pulse. Ideally, the voltage response should look like that shown in Figure 5.5. In this case the voltage has an exponential-like increase from the resting membrane potential to a new voltage which represents the sum of the resting membrane potential and the voltage response of the basolateral membrane to the current flow through this membrane. However, other waveforms are sometimes observed and are illustrated in Figures 5.6(c) and 5.7(c). These odd-looking waveforms are a result of the capacitance of the glass wall of the microelectrode (C_g; Figure 5.6(c)) and that in Figure 5.7(c) is due to the capacitance to ground of the input amplifier (C_e, i.e. the input impedance of the electrometer).

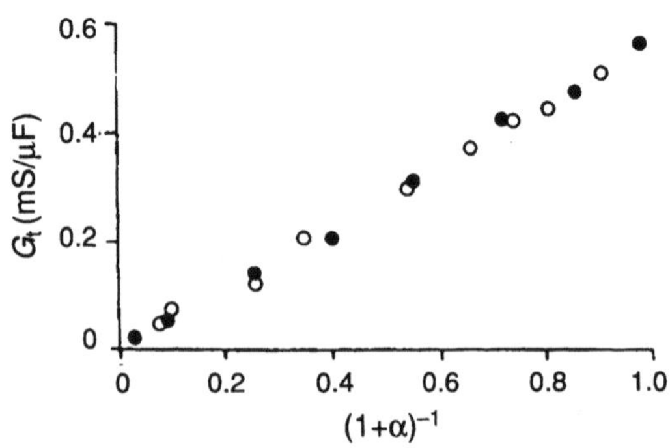

Figure 5.4 Plot of G_t versus $1/(1 + \alpha)$. The apical resistance was decreased using the polyene antibiotic nystatin. Apical bathing solution is NaCl (open circles) and potassium sulfate solution (closed circles). Note that the plot is linear, suggesting that there is no impalement damage produced by the microelectrode to the cell.

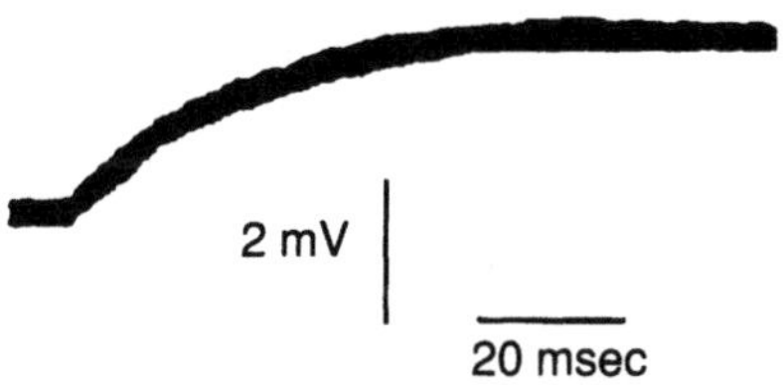

Figure 5.5 Recording of the time-dependent basolateral membrane potential response to a square current pulse. The initial small jump is due to solution series resistance, while the exponential increase is due to the charging of the membrane capacitance.

5.4.1 Glass capacitance

Glass capacitance has a marked effect on the microelectrode voltage response. Figure 5.6 shows a simplified equivalent circuit for the case of a microelectrode in a cell where the junctional resistance is assumed to be infinite compared with the cell membrane resistance. There is a significant capacitance of the glass of the microelectrode (C_g). Such a capacitance will result in a measured waveform which is the response of the basolateral membrane resistor and capacitor (RC) plus the difference between two other resistor/capacitor combinations: the apical membrane resistor and capacitor, and the microelectrode resistor (R_t – due to the resistance of the tip of the microelectrode) and capacitor (C_g – due to the glass capacitance). The voltage response is described by the equation shown in Figure 5.6. The most important aspects of this equation are as follows.

- The artifact caused by the microelectrode is biphasic. The apical membrane and microelectrode time constants (the produce of the paired values of resistor and capacitor), their ratio, the resistance ratio (α) and the basolateral resistance all contribute to the peak value of the artifact as well as the rate of rise and decay of the voltage.
- Neglecting the series resistance, this artifact does not alter the steady-state voltage response of the basolateral membrane, because the glass resistance is assumed to be large compared with the resistance of the tip of the electrode.
- Even though the time constant of the microelectrode is much less than that of the basolateral membrane, one must always be assured that during the current pulse the membrane voltage has reached a new steady state. If not, the value of α will be underestimated.

How can these problems be overcome? In tight or high resistance epithelia, there are a number of possible solutions. One of the most obvious is to reduce the value of either the glass capacitance, by shielding the electrode (Lewis and Wills, 1980), or the resistance at the tip of the

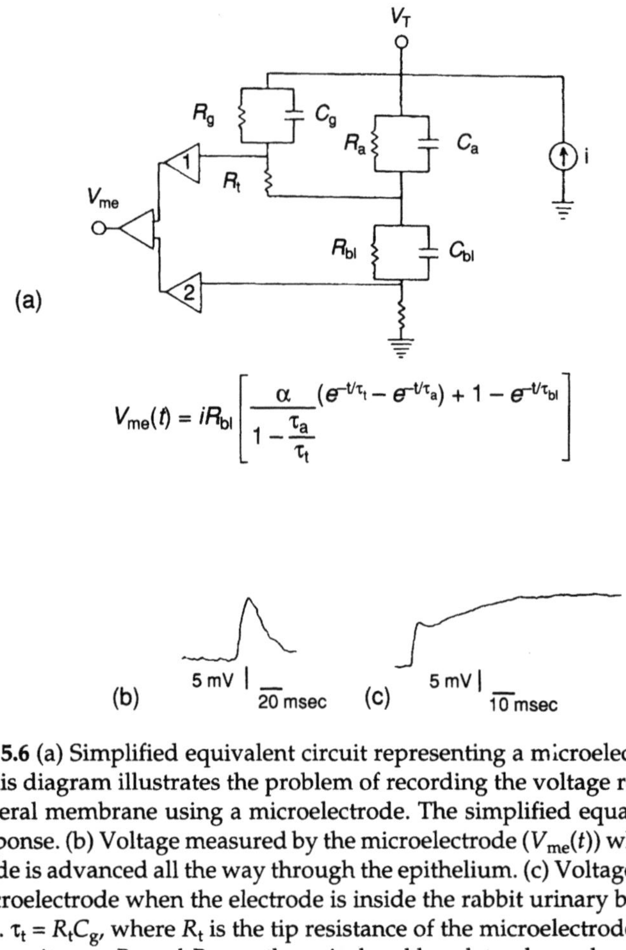

Figure 5.6 (a) Simplified equivalent circuit representing a microelectrode inside a cell. This diagram illustrates the problem of recording the voltage response of the basolateral membrane using a microelectrode. The simplified equation describes the response. (b) Voltage measured by the microelectrode ($V_{me}(t)$) when the microelectrode is advanced all the way through the epithelium. (c) Voltage measured by the microelectrode when the electrode is inside the rabbit urinary bladder epithelial cell. $\tau_t = R_tC_g$, where R_t is the tip resistance of the microelectrode and C_g is the glass capacitance. R_a and R_{bl} are the apical and basolateral membrane resistances, respectively; C_a and C_{bl} are the apical and basolateral membrane capacitors, respectively; τ_a is the apical membrane time constant and is equal to the product of R_a and C_a; τ_{bl} is the basolateral membrane time constant and is equal to the product of R_{bl} and C_{bl}.

microelectrode by beveling the tip of the electrode. Another (but less satisfactory) method is to reduce the resistance of the apical membrane. This method is not as effective as not only will it result in a decrease in the numerator of the first part of the equation but also it will decrease the denominator. A final method is to advance the microelectrode completely through the epithelium and measure the voltage response of the microelectrode to a current pulse (Figure 5.6). From this response, one can determine the time taken for the artifact to be over, and then use this as a minimum time for the duration of the current pulse.

5.4.2 Electrometer capacitance

The second and less significant problem is the impedance of the electrometer. This problem and the appropriate equivalent circuit are illustrated in Figure 5.7. Of particular interest is that one must now consider the difference in the response times of the individual amplifiers. Amplifier 2

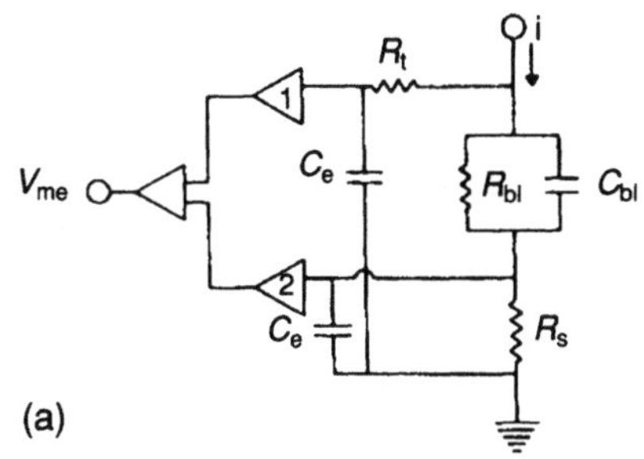

$$V_{me}(t) = iR_{bl}\left[1 - \frac{\tau_{bl}}{\tau_{bl} - \tau_e} e^{-t/\tau_{bl}} - \left(\frac{\tau_e}{\tau_e - \tau_{bl}} + \frac{R_s}{R_{bl}}\right) e^{-t/\tau_e}\right]$$

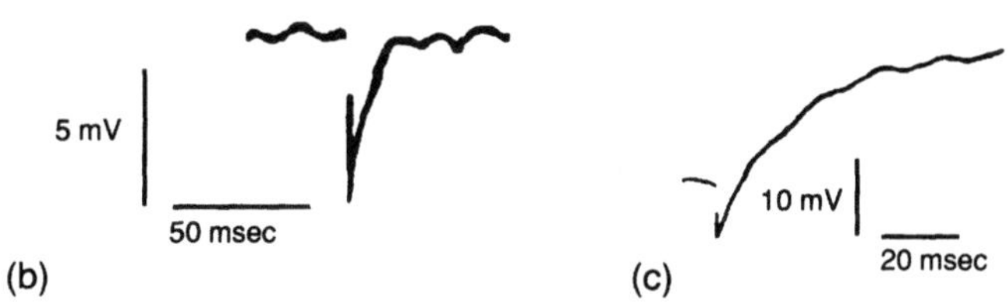

Figure 5.7 (a) Simplified equivalent circuit representing a microelectrode (with no glass capacitance) inside of an epithelial cell. This diagram illustrates the problem of the input impedance of the voltage follower amplifier (C_e) when it is connected to a high resistance microelectrode (R_t). The simplified equation describes the voltage response measured between the two amplifiers (V_{bl}). R_s is the resistance between amplifier 2 and the circuit ground. (b) Measured voltage response when the microelectrode is advanced all the way through the epithelium. (c) Measured voltage response when the electrode is in the cell. In this example R_t = 1000 MΩ and C_e = 1 pF. Note the rapid downward deflection: R_s is estimated from this deflection at 300 Ω.

(serosal solution amplifier) will have a rapid time response, the magnitude determined by the relatively small bath electrode resistance (about 300 Ω) to ground. On the other hand amplifier 1 (this is the amplifier to which the microelectrode is attached) will have a response which is described by the equation in Figure 5.7. The difference between the response of amplifiers 1 and 2 is then the measured voltage of the electrometer.

The equation which describes the response of the electrometers tells us that we can reduce this artifact by a number of methods. The first is to reduce the resistance of the tip of the microelectrode (again perhaps by beveling). Second, the magnitude can be reduced by decreasing the resistance to ground (R_s), since this resistance is caused by the solution and current-passing electrode resistance. Lastly, the artifact can be reduced by decreasing the capacitance to ground using a capacitative feedback network found on many electrometers.

To illustrate the magnitude of this artifact, a microelectrode was fabricated with a high tip resistance and a low capacitance. The electrode was then advanced all the way through the epithelium. The response of the microelectrode amplifier is shown in Figure 5.7b. As can be seen, there is a rapid downward deflection followed by a relaxation to the baseline. When divided by the current, the peak of this deflection is equal to the resistance to ground. The response of the voltage when the electrode is in a cell is shown in Figure 5.7c. Although this is an extreme example, one must remember that an increase in the tip resistance upon impaling a cell can cause considerable problems in analyzing the time response of the basolateral membrane potential to a square current pulse.

A common solution to these two artifacts is to reduce, as much as possible, the resistance of the tip of the microelectrode. An alternative method to beveling is to use a glass which has a thin wall. This will allow one to construct a microelectrode with a small tip diameter and a low tip resistance. However, the tip is more fragile and thus the life of the electrode will be reduced.

5.4.3 Electrode resistance

One must always be assured that the resistance of the microelectrode is small compared with the input resistance of the amplifier. If the microelectrode resistance is comparable to that of the electrometer, one will attenuate the voltage signal that the electrode is being used to sense. The amounts of attenuation can be easily calculated as the ratio of the input impedance of the amplifier to the sum of the electrode resistance and the amplifier resistance; thus when the electrode resistance equals the input impedance of the amplifier, the measured voltage is 50% that of the non-attenuated membrane voltage. As a rule of thumb it is desirable that the electrode resistance be at least 100 times lower than the input impedance of the electrometer. (This will cause a 1% error in the measured membrane voltage.)

REFERENCES

Crowe, W.E. and Wills, N.K. (1991) A simple method for monitoring changes in cell height using fluorescent microbeads and a Ussing-type chamber for the inverted microscope. *Pflügers Arch.* **419**:349–57.

Frömter, E. and Gebler, B. (1977) Electrical properties of amphibian urinary bladder epithelia. *Pflügers Arch.* **371**:99–108.

Lewis, S.A., Eaton, D.C., Clausen, C. and Diamond, J.M. (1977) Nystatin as a probe for investigating the electrical properties of a tight epithelium. *J. Gen. Physiol.* **70**:427–40.

Lewis, S.A. and Wills, N.K. (1980) Resistive artifacts in liquid-ion exchanger microelectrodes estimates of Na^+ activity in epithelial cells. *Biophys. J.* **31**:127–38.

Lewis, S.A. and Wills, N.K. (1982) Electrical properties of the rabbit urinary bladder assessed using gramicidin D. *J. Memb. Biol.* **67**:45–53.

Wills, N.K., Lewis, S.A. and Eaton, D.C. (1978) Active and passive properties of rabbit descending colon: a microelectrode and nystatin study. *J. Membr. Biol.* **45**:81–108.

Yonath, J. and Civan, M.M. (1971) Determination of the driving force of the Na^+ pump in toad bladder by means of vasopressin. *J. Memb. Biol.* **5**:366–85.

6

Impedance analysis of epithelia

Simon A. Lewis, Chris Clausen and Nancy K. Wills

In view of the importance of epithelial organization in vectorial ion transport, it is clear that a complete understanding of regulation of this process requires an understanding of the role of the structural features involved. For example, although ion channels are the basis of much of the electrical activity and the electrochemical driving forces for ion transport across epithelial cell membranes, we can never truly understand this process without understanding the role of the lateral intercellular spaces or other structures such as microvilli and crypts in regulating extracellular ion concentrations. Similarly, in attempting to understand how the density of conducting channels is regulated, it is important to have knowledge concerning the role of membrane insertion or removal in regulating membrane transport properties.

What methods are available for studies of epithelial structure and function? Morphometric analyses have been used with success but this method requires fixation of the tissue. Video analysis is another recent development but this technique can be cumbersome and, like morphometric analysis, it does not allow direct monitoring of both structural features

Epithelial Transport: A guide to methods and experimental analysis.
Edited by Nancy K. Wills, Luis Reuss and Simon A. Lewis.
Published in 1996 by Chapman & Hall, London. ISBN 978-94-010-7168-0

(such as membrane areas) and membrane conductances. Numerous electrical methods have been used to study epithelial transport properties, including transepithelial DC measurement methods and microelectrode techniques. However, these approaches have generally ignored tissue architecture and thus are also limited.

To date, impedance analysis is the only non-invasive electrical technique available for determining the role of epithelial structure in ion transport. A unique advantage of this method is that it allows measurement of membrane areas (expressed as capacitance where 1 μF ≈ 1 cm^2). As we will see below, it has been possible to use this method to derive meaningful morphological parameters, such as the resistance and width of lateral intercellular spaces and crypt dimensions.

6.1 WHAT IS IMPEDANCE ANALYSIS?

Impedance for a linear system is operationally defined as the 'resistance' that an electrical circuit (composed of resistors, capacitors or inductors) offers to a sinusoidal signal at a fixed frequency. Therefore, impedance is expressed in units of ohms and the results of this method describe the electrical properties of an epithelium in the frequency domain. It is traditionally determined by measuring the voltage resulting from a small current applied across the epithelium. In other words, impedance analysis methods focus on the frequency-dependent behavior of the epithelium.

Most past studies of the electrical properties of epithelia have used DC electrical methods and modeled the epithelium as a network of resistors. This approach is limited because it ignores the ability of biological membrane lipid bilayers to separate and store charge, i.e. biological membranes act as capacitors. Since capacitance is proportional to area, this capacitive property of membranes can provide important information about membrane area.

Unlike a resistor, the voltage response of a capacitor to a sinusoidal current signal is frequency dependent. For example, if one increases the frequency of a current sinusoid of fixed amplitude across a capacitor, its impedance decreases, i.e. the voltage amplitude decreases as frequency increases. Another essential attribute of a capacitor is that a time or phase delay occurs between the peak in the sinusoid input and the sinusoid output response. For any frequency of an applied sinusoid current, the voltage sinusoid will lag behind the current sinusoid by 90 degrees (i.e. one quarter of the period of the sinusoid). This lag is called the phase shift and for an ideal capacitor the phase shift is always –90 °.

In impedance analysis methods, the epithelium is modeled as a network of resistors and capacitors. The apical and basolateral membranes are each considered as a parallel combination of a resistive element which has frequency-independent response to sinusoidal input signals and a capacitive component which responds in a frequency-dependent manner. Thus, the impedance of the epithelium will depend not only on the values of the specific circuit elements but also on how they are physically interconnected, i.e. the structure of the electrical circuit itself.

The remainder of this chapter focuses on impedance analysis of so-called 'flat-sheet' epithelia (i.e. non-tubular epithelia) and, in particular, the analysis of epithelia which are composed of linear resistors and capacitors. We will not consider cases in which the membrane area is constantly changing or in which ionic channels (comprising the conductive properties of the membrane) are being activated or inactivated by small membrane voltage perturbations. For a discussion of the latter point, refer to Warncke and Lindemann (1985).

6.2 ADVANTAGES AND LIMITATIONS OF IMPEDANCE ANALYSIS

Before describing the technical aspects of measuring epithelial impedance, we will first outline the types of questions that impedance analysis can answer, as well as possible pitfalls that one must be careful to avoid when using this method.

6.2.1 Advantages

Impedance analysis yields values for the actual areas of the apical and basolateral membranes and is capable of determining extracellular parameters such as the geometry of lateral intercellular spaces or epithelial crypts, or the electrical properties of extraepithelial structures. These methods are also capable of rapidly yielding estimates of membrane conductances without having to perturb a membrane resistance or perform the time-intensive measurement of epithelial cable properties (as is required for microelectrode DC analysis of membrane conductances).

Since estimates of membrane areas and resistances do not require external intervention, the effects of pharmacological alteration of membrane transport processes on membrane resistances and capacitances (surface areas) can be directly determined. This type of measurement allows one to address such questions as whether an increase in membrane transport properties is regulated by alterations in net membrane area (i.e. net endocytosis or exocytosis of membrane containing transport proteins) or by an activation or inactivation of transporter units (e.g. channels). Because impedance analysis allows measurement of extracellular structures, such as the lateral intercellular spaces, the experimenter can determine whether these structures change their geometry as a function of transport rate across the epithelium or are affected by pharmacological manipulations.

Impedance analysis is also a useful method for independently verifying membrane resistance values calculated using DC methods, e.g. microelectrode measurements of membrane resistance ratios. One can also assess the effects of extracellular structures such as the lateral intercellular space (LIS) resistance on these measurements.

Lastly, an important advantage of impedance analysis is that it permits computation of specific membrane conductances, or membrane conductances normalized to area. This feature allows the experimenter to compare different epithelia to determine whether differences in resistance measured *in vitro* are due to differences in specific membrane conductances, membrane areas, or tight junctional resistance. Using this approach, one can compare different types of epithelia or assess the source of variability between experiments for a single type of epithelium.

6.2.2 Limitations and pitfalls

As with any other technique, the experimenter must be aware of the possible limitations and pitfalls encountered in using impedance analysis.

First, impedance analysis requires the use of an equivalent circuit model of resistors and capacitors, which is based on the morphology of the epithelium being studied. Therefore, the results of the method are valid only if the model is physiologically meaningful. In addition, resistor and capacitor values are determined by fitting the equivalent circuit model to the data and are only as accurate as the data. As discussed by Clausen (1989), care must be taken to ensure against contamination of the data by any extraneous phase shifts inherent to the electronic measuring circuit.

Another problem is that electrical circuits often have non-unique solutions. Therefore, use of impedance analysis requires the experimenter to obtain an estimate of at least one additional epithelial (circuit) parameter using an alternate method (cf. Clausen, 1989). This parameter is typically the ratio of apical to basolateral membrane resistances obtained using microelectrode methods, or an estimate of the junctional resistance, apical membrane resistance or basolateral membrane resistance, obtained from other techniques.

A possible pitfall in impedance analysis arises from the complexity of the equivalent circuit model. For data that are poorly fit by a particular model, it is easy to improve the fit by simply increasing the complexity of the model and thereby increasing the number of adjustable parameters. However, the addition of an extra parameter to the model is only meaningful if it is based on an actual tissue property such as its structure.

An important limitation of the method is that transepithelial impedance measurements, like DC transepithelial resistance measurements, are less sensitive to cell membrane properties in so-called 'leaky' epithelia (cf.

Gordon *et al.*, 1990). For the case where the junctional resistance is much lower than the cellular resistance, one must use intracellular microelectrode impedance analysis techniques. The technical and theoretical aspects of this measurement have been reviewed by Kottra and Frömter (1982) and Lim *et al.* (1984).

For epithelia composed of a heterogeneous cell population, transepithelial impedance measurements alone are not able to determine which cell type is reacting to a specific stimulus or the relative proportion of different cell types in the epithelium. In any case, it is important to have previous information concerning tissue morphology and the effects of pharmacological tools in order to validate the results of impedance analysis.

In summary, as in any other method, impedance analysis has certain limitations. Of these limitations, the most crucial is the selection of an appropriate equivalent circuit model by which to fit the data to extract meaningful values of membrane areas and conductances. The main advantage of impedance analysis is the ability to measure rapidly and in a non-obtrusive manner not only the individual membrane conductances but also the individual membrane areas.

6.3 CHOICE OF A CURRENT WAVEFORM

6.3.1 Voltage response of equivalent circuit elements

(a) The square current pulse

In the preceding discussion we have defined impedance of circuit elements (resistors and capacitors) in terms of their response to an alternating current (a sinusoid). However, one is not limited to a sinusoid waveform; indeed any waveform can be used to determine the impedance of a circuit. Figure 6.1 outlines the response of a given circuit element or network of elements to a square current pulse along with the characteristic equation which describes the voltage response to the current input signal. It should be noted that the voltage response to a square current waveform is time-dependent. Consequently, this response can be analyzed in the time domain. Using Fourier analysis, the input current waveform and the voltage response can each be described as a sum of sinusoids of increasing frequency and variable magnitude.

In theory, square pulse analysis and sinusoidal analysis are equivalent. In practice, the square current pulse method is more limited since high frequency information is lost. This loss occurs because the amplitudes of the high frequency sinusoids calculated for the square current pulse are contaminated by the intrinsic noise of the voltage sensing amplifiers. Because of this inaccuracy, square pulse analysis is best analyzed in the time domain.

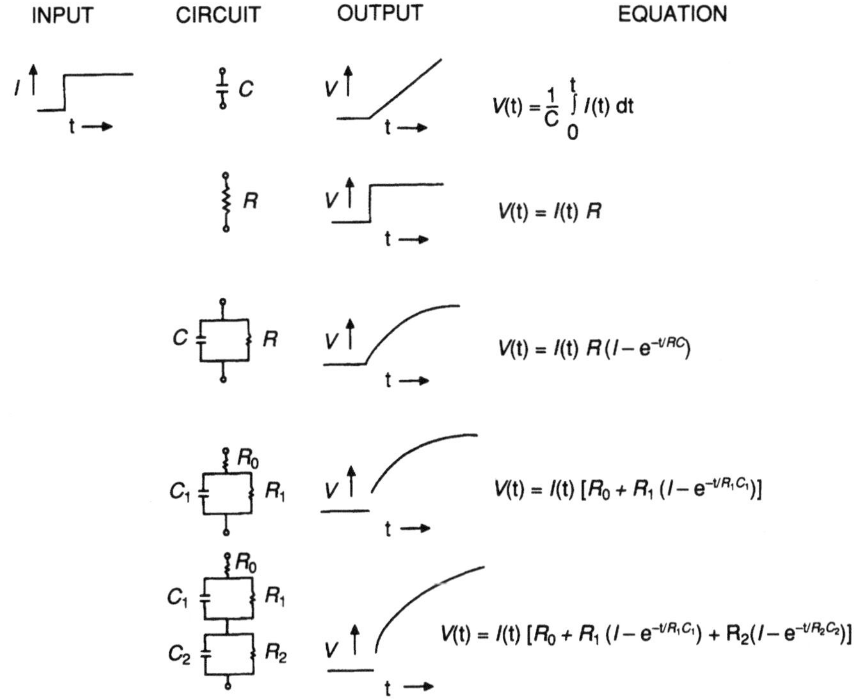

Figure 6.1 Response of simple equivalent circuits to a square current input, and the generalized equation which describes the voltage output. The output response of the last circuit is shown as two easily discernible exponential traces. In this example, the value of the two time constants (R_1C_1 and R_2C_2) were selected to be greatly different. If the time constants are approximately equal, this circuit will show a single exponential response similar to the single RC with series resistor. The voltage (V), current (I) and time (t) scales are in arbitrary units.

As shown in Figure 6.1a, the shape of the voltage response through a capacitor is a ramp. The response through a resistor is an identical square step (Figure 6.1b) and the response through a parallel resistor capacitor network is a hyperbolic curve (Figure 6.1c). Figure 6.1d shows the response through a network of a resistor in series with a parallel resistor capacitor. The response is a voltage step (current flow through the series resistor) followed by a hyperbolic curve.

Figure 6.1e shows the response through a series resistor and two parallel resistor capacitor combinations arranged in series. This response is characterized by an initial voltage jump followed by the sum of two hyperbolic curves, each curve being a function of an individual resistor capacitor network. The latter equivalent circuit represents the minimal configuration of resistors and capacitors which is required to describe the passive electrical properties of any epithelium. Indeed, we will see that even this circuit is an oversimplification when we consider the morphology of epithelia in general.

(b) Sinusoidal current waveforms

Consider a linear circuit composed of resistors and capacitors (or inductors) being driven by a sinusoidal source of constant current. It can be easily shown (Desoer and Kuh, 1969) that at steady state, the individual currents through the circuit elements, as well as the circuit element voltages, will also be sinusoidal and will have the same frequency as the current source.

Now consider the current through, and the voltage across, a single current element:

$$i(t) = I_m\cos(\omega t + \psi) \qquad (6.1a)$$

$$v(t) = V_m\cos(\omega t + \theta) \qquad (6.1b)$$

where I_m and V_m are the respective amplitudes, ψ and θ are the respective phase angles and ω is the angular frequency, ($2\pi f$, where f is frequency in hertz). For the analysis that follows, it is easier to express the currents and voltages from equations (6.1a) and (6.1b) in exponential notation. Using Euler's relationship:

$$i(t) = \mathrm{Re}[Ie^{j\omega t}] \qquad (6.2a)$$

$$v(t) = \mathrm{Re}[Ve^{j\omega t}] \qquad (6.2b)$$

where Re[.] denotes the 'real part', $j = \sqrt{-1}$, and I and V are the 'imaginary part' consisting of the following complex quantities:

$$I = I_m e^{j\omega\psi} \qquad (6.3a)$$

$$V = V_m e^{j\omega\theta} \qquad (6.3b)$$

I and V are termed the phasors and, in essence, express the magnitude and phase angle information of the sinusoids at angular frequency ω. The equivalence between equations (6.1a) and (6.2a) or (6.1b) and (6.2b) may not be apparent. However, taking i(t) as an example, we see that:

$$i(t) = \mathrm{Re}[Ie^{j\omega t}] = \mathrm{Re}[I_m e^{j\psi}e^{j\omega t}] = \mathrm{Re}[I_m e^{j(\omega t+\psi)}]$$

and using Euler's relationship (i.e. $e^{jx} = \cos x + j\sin x$):

$$i(t) = \mathrm{Re}[I_m\cos(\omega t + \psi) + jI_m\sin(\omega t + \psi)] = I_m\cos(\omega t + \psi).$$

Hence, equations (6.1a) and (6.2a) are truly equivalent equations, as are (6.1b) and (6.2b).

The impedance of a circuit element is a complex quantity, and is defined as the ratio of the voltage phasor to the current phasor at a given frequency, namely:

$$Z(j\omega) = V/I \qquad (6.4)$$

Substituting equations (6.3a,b) into the definition of impedance, equation (6.4), yields:

$$Z(t) = Ve^{j\psi}/Ie^{j\theta} = (V_m/I_m)e^{j(\psi-\theta)} = Zr + jZx \tag{6.5}$$

where Z_r and Z_x (the real and imaginary impedance components respectively) are defined as:

$$Z_r = (V_m/I_m)\cos(\psi - \theta) \tag{6.6a}$$

$$Z_x = (V_m/I_m)\sin(\psi - \theta) \tag{6.6b}$$

This readily illustrates that impedance can be expressed in terms of the magnitude ($|Z| = V_m/I_m$) and its phase angle ($\angle Z = \psi - \theta$), as well as by its real (Z_r) and imaginary (Z_x) parts. Phase angle and magnitude values can be computed from the real and imaginary parts using the following two relationships:

$$|Z| = \sqrt{(Z_r^2 + Z_x^2)} \tag{6.7a}$$

$$\angle Z = \tan^{-1}(Z_x/Z_r) \tag{6.7b}$$

6.3.2 Impedance of resistors and capacitors

In the case of a resistor, Ohm's law defines the current-voltage relationship, i.e.:

$$V(t) = R \cdot i(t) \tag{6.8}$$

Substituting equations (6.2a,b) into equation (6.8) yields:

$$\mathrm{Re}[Ve^{j\omega t}] = R \cdot \mathrm{Re}[Ie^{j\omega t}] = \mathrm{Re}[RIe^{j\omega t}] \tag{6.9}$$

This implies that:

$$V = RI \tag{6.10}$$

so the impedance of a resistor (Z_r) is:

$$Z_r = R \tag{6.11}$$

and equals the resistor value itself.

In the case of a capacitor, the current–voltage relation is:

$$i(t) = C\,dV(t)/dt \tag{6.12}$$

and substituting for i(t) and v(t) yields:

$$\mathrm{Re}[Ie^{j\omega t}] = C\frac{d}{dt}\mathrm{Re}[Ve^{j\omega t}] = \mathrm{Re}[Cj\omega Ve^{j\omega t}] \tag{6.13}$$

Therefore, the impedance of a capacitor (Z_c) is:

$$Z_c = 1/(j\omega C) = -j/(\omega C) \tag{6.14}$$

Unlike Z_r, which exhibits constant magnitude ($|Z| = R$) and zero phase angle ($\angle Z_r = 0$), the impedance of a capacitor decreases with frequency ($|Z_c| = 1/(\omega C)$), and the voltage and current sinusoids are one quarter of a cycle out of phase ($\angle Z_c = \tan^{-1} -\infty = -90°$).

6.3.3 Circuits under sinusoidal steady-state conditions

Under steady-state conditions for a sinusoidal current signal, Kirchoff's Current law and Kirchoff's Voltage law can be written directly in terms of current phasors (Is) and voltage phasors (Vs), respectively. Therefore, the impedance of a circuit composed of several elements connected in series is equal to the sum of the impedances of each of the elements. For example, the impedance of a series resistor–capacitor circuit is:

$$Z = Z_r + Z_c = R + (-j/(\omega C)) = R - j/(\omega C) \tag{6.15}$$

When determining the impedance of circuits composed of elements connected in parallel, it is convenient to consider the so-called admittance of the elements. Admittance ($Y(j\omega)$) is defined simply as the reciprocal of impedance, i.e.:

$$Y(j\omega) = 1/Z(j\omega) \tag{6.16}$$

From this definition, one can easily see that the magnitude of admittance equals the reciprocal of the impedance magnitude ($|Y|(j\omega) = 1/|Z|(j\omega)$), and the phase angle of admittance is equal to the negative of the impedance phase angle ($\angle Y(j\omega) = -\angle Z(j\omega)$).

The admittance of a circuit composed of elements connected in parallel is simply equal to the sum of the admittances of each of the elements. A parallel resistor–capacitor (RC) circuit therefore has the admittance:

$$Y = (1/R) + j\omega C = G + j\omega C \tag{6.17}$$

where G is the conductance of the resistor. The admittance of a parallel RC circuit is of great importance since it forms the basic equivalent circuit of a biological membrane; the conductance represents the membrane ionic conductance and the capacitance represents the membrane capacitance which is proportional to membrane area.

The impedance of a biological membrane (a parallel RC circuit) is given by:

$$Z = 1/Y = (R(1 - j\omega RC))/(1 + \omega^2 R^2 C^2) \tag{6.18}$$

and exhibits magnitude and phase angle equal to:

$$\angle Z = -\tan\omega RC \tag{6.19}$$

$$|Z| = R/\sqrt{1+ \omega^2 R^2 C^2} \tag{6.20}$$

From the above, it is apparent that the phase angle starts at 0° under DC conditions ($\omega = 0$) and increases to $-90°$ as $\omega \to \infty$. The magnitude, however, is equal to R for the DC condition, and progressively decreases with increasing frequency. This decrease is expected since as frequency increases, more current flows through the capacitor (recall that the current flowing through the capacitor is proportional to $\mathrm{d}V/\mathrm{d}t$).

Equations (6.18), (6.19) and (6.20) suggest two possible graphic approaches that can be used to display impedance data. For example, the amplitude of the real part (abscissa) can be plotted against the amplitude of the imaginary part (ordinate) for each of the angular frequencies (ω; see equation (6.18)). This arrangement is called a **Nyquist** or **Cole–Cole plot**. Alternatively, one can plot the magnitude versus frequency (as a log–log plot for spatial resolution; equation (6.20)) and the phase angle versus frequency (as a linear–log plot; equation (6.19)). These plots are collectively called a **Bode plot**. Figure 6.2 shows both of these plotting formats for a series of equivalent circuits along with the characteristic equations which describe each of these circuits.

The choice of the graphical representation is entirely up to the investigator. We prefer the Bode plots. First, the phase angle is a much more sensitive indicator of circuit values (Valdiosera *et al.*, 1974) than is a change in the real and imaginary parts. Second, the Nyquist plot represents the impedance on a linear scale, thus giving high weight to low frequency data because of their high values, where as Bode plots of logarithmic magnitude are more evenly weighted. Last, the logarithmic frequency scale of the Bode plots does not give uneven weighting to the low frequency data, while the Nyquist representation tends to cluster middle and high frequency data towards the intercept of the real and imaginary axis.

In the preceding section we have shown that there are two choices of waveform that can be used to determine the impedance of an epithelium. We also commented that the square wave method yields good information in the low frequency range but not in the high frequency range. The alternative method is to apply discrete sinusoids at a series of frequencies spanning a frequency range of three to four orders of magnitude. The latter approach is highly accurate but it can be very time consuming: a typical impedance run can take 15 minutes (Clausen *et al.*, 1979). Thus when performing impedance analysis using discrete sinusoids, the epithelial preparation must be in a steady state, i.e. the membrane conductance and capacitances must be time invariant.

In recent years, methods have been developed in a number of laboratories which combine the speed of transient analysis with the accuracy of sinusoidal analysis. The basic approach is to synthesize a waveform which, after Fourier transform, is composed of a progression of sinusoids that span a frequency range of three to four orders of magnitude and are nearly equal in amplitude. This synthesized waveform can be a sequential

CIRCUIT	IMPEDANCE	NYQUIST	BODE
	$Z = \frac{-j}{\omega C_1}$	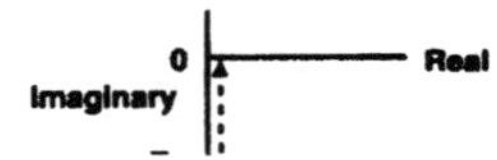	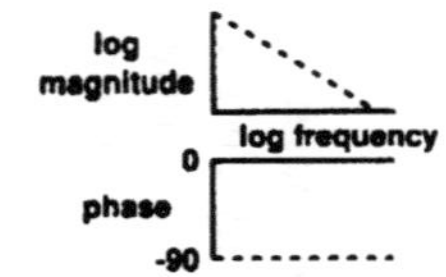
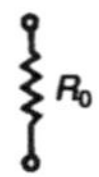	$Z = R_0$	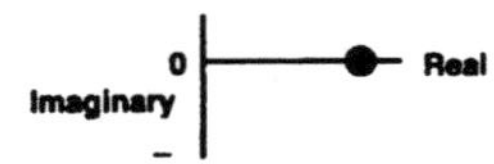	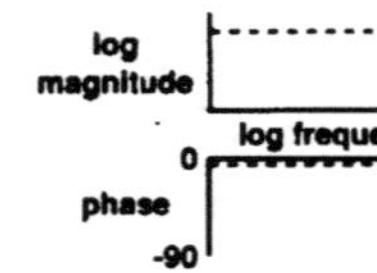
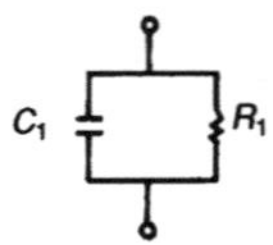	$Z = \frac{R_1}{[1 + (\omega R_1 C_1)^2]} - j\frac{\omega R_1^2 C_1}{[1 + (\omega R_1 C_1)^2]}$	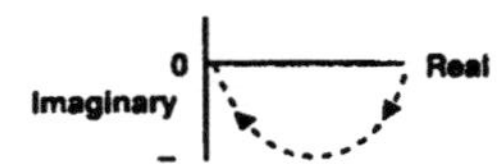	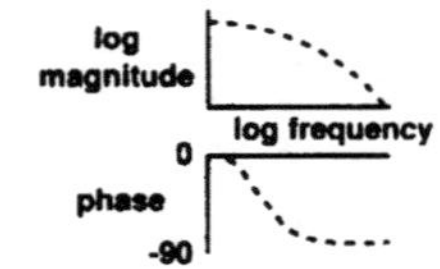
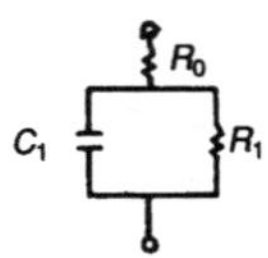	$Z = R_0 + \frac{R_1}{[1 + (\omega R_1 C_1)^2]} - j\frac{\omega R_1^2 C_1}{[1 + (\omega R_1 C_1)^2]}$	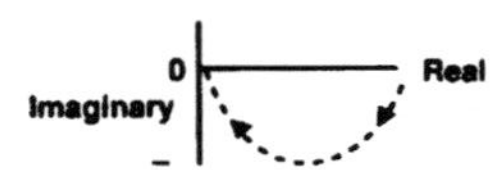	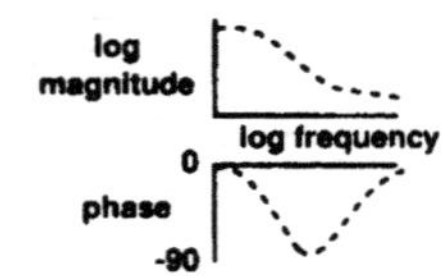
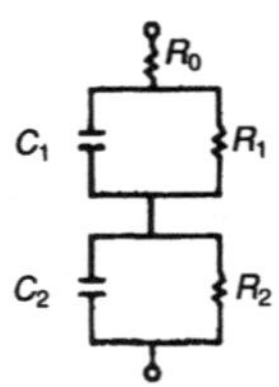	$Z = R_0 + \frac{R_1}{[1 + (\omega R_1 C_1)^2]} + \frac{R_2}{[1 + (\omega R_2 C_2)^2]} - j\left(\frac{\omega R_1^2 C_1}{1 + (\omega R_1 C_1)^2} + \frac{\omega R_2^2 C_2}{1 + (\omega R_2 C_2)^2}\right)$	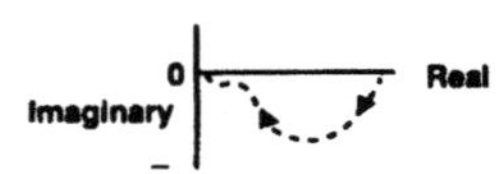	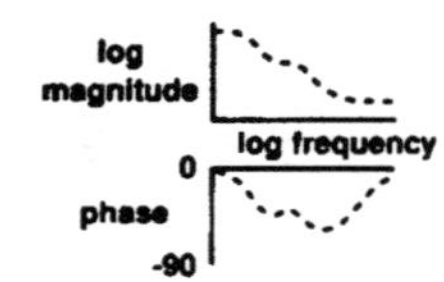

sum of sinusoids (called a sine wave burst method; Warncke and Lindemann, 1979), white noise (Mathias, Rae and Eisenberg, 1979), or a pseudo-random binary noise (Poussart and Ganguly, 1977; Clausen and Fernandez, 1981).

Regardless of the method used, the manipulation of the data consists of the following steps. The first is to filter, record (digitize) and store the input on a computer – we will call this signal $x(t)$. The $x(t)$ data is Fourier transformed and the resulting data (we will call this $X(f)$) is stored in the computer's memory. Next the input waveform $x(t)$ is imposed across the epithelium and the system's response (which we will call $y(t)$) is recorded. The $y(t)$ data is then Fourier transformed, resulting in $Y(f)$. The impedance (or its inverse, admittance) is then calculated at each frequency using a transfer function ($H(f)$). In the equation below, impedance is calculated when Y is the voltage and X is the current and admittance is calculated when Y is the current and X is the voltage:

$$H(f) = X^{*}Y / X^{*}X$$

where X^{*} is the complex conjugate of the Fourier transformed $x(t)$. Note that for complex numbers, the complex conjugate of $a + jb$ is $a - jb$. In the transfer function above, this results in an equation in which j appears only in the numerator, i.e. the transfer function has the form $H(f) = a' + jb'$. This transfer function is then composed of real and imaginary parts which are used to calculate the magnitude and phase at each frequency.

An advantage of the use of synthesized waveforms is that data can be collected very rapidly. For example, in less than 30 seconds, one can typically collect 2048 paired phase and magnitude data points. The number of data points can then be reduced for further analysis. In general, we have found it convenient to reduce the data to 100 points (i.e. pairs of real and imaginary numbers) equally spaced on a logarithmic scale over the frequency range of 2 Hz to 9.1 KHz. This sample size can be readily

Figure 6.2 The impedance of simple equivalent circuits are shown along with the corresponding equations and representative Nyquist and Bode plots. The two time constants of the last circuit were selected to be greatly different in value such that one can observe two semicircles in the Nyquist representation, or two peaks in the phase–frequency and two plateaus in the magnitude–frequency of the Bode representation. Time constants with similar values reduce this impedance equation to that represented by the model of a single RC in series with a resistor. The Nyquist representation has units of ohms (or $\Omega \cdot cm^2$) for both the real and imaginary axis; however, in these examples the scale are in arbitrary units. Direction of the arrows is for increasing frequency. The Bode representation of the magnitude plot is shown as an arbitrary scale with units of ohms or $\Omega \cdot cm^2$ and frequency in hertz. The phase representation is in degrees and for the circuits shown can vary between 0° and –90°.

handled by the computer for fitting an appropriate equivalent circuit model to the data.

6.3.4 Choice of waveform dictates information content

As may already be clear, sinusoidal analysis methods are more accurate and, using recent advances, are as fast as transient analysis. However, from the point of view of cost and set-up time, transient analysis is easily implemented and reasonably inexpensive when compared with sinusoidal analysis. We must stress that the information content of transient analysis is much less than sinusoidal analysis. As we will see below, transient analysis is restricted to using the so-called lumped equivalent circuit, and distributed effects due to more complex morphological features are only observed at high frequencies, i.e. frequencies at which transient data is inaccurate.

As a rule of thumb, if one is interested in answering the question of whether the membrane area changes during transport alterations, transient analysis is a reasonable approach to use as a first approximation. However, if one wants to quantitate the area and area changes of both the apical and basolateral membranes, and assess distributed resistance effects, transient analysis is inadequate (see figure 3 in Clausen *et al.*, 1979) and one must use sinusoidal analysis.

6.4 DATA COLLECTION AND INSTRUMENTATION

This section very briefly describes the instrumentation involved in measuring the response of the epithelium to the input waveform. In the discussions so far we have applied a current waveform and measured the voltage output. However, one could also apply a voltage waveform and measure the current output (Warncke and Lindemann, 1981; Gogelein and Van Driessche, 1981).

In terms of the output measuring equipment, the measurement amplifier must have a high input impedance and negligible input capacitance. In general the amount of 'stray' capacitance, which can be introduced by either the measuring amplifiers or the measuring electrodes, must be kept to a minimum (preferably zero). As mentioned previously, the distance between the voltage sensing electrodes and the epithelium must be as small as possible, since we are interested in measuring the impedance of the epithelium and not the impedance of the Ringer's solution bathing both sides of the epithelium. Lastly, the current-passing electrodes must be placed such that there is a uniform current density at the surface of the epithelium. This distance will depend upon the resistivity of the bathing solution but should not be less than a few millimeters away from the epithelial surface.

The current-passing equipment is usually a stimulus isolation unit or voltage source (e.g. a pseudo-random noise generator) connected via a high resistance (1 or 10 MΩ) carbon resistor to a Ag/AgCl wire in one chamber and a grounded Ag/AgCl wire in the opposing chamber. For square pulse analysis, the resulting voltage output is measured by a differential voltage amplifier of a four-electrode voltage clamp. For the synthesized waveform signal, a high input impedance differential amplifier (e.g. model 113A, Princeton Applied Research NJ) is used. For both techniques, the response is digitized using a 12-bit A/D converter and stored in computer memory.

For the synthesized waveform signal experiments, it is necessary first to filter the voltage output before digitizing, using an antialiasing filter. Typically during data collection, the recording amplifier is AC coupled, high-pass filtered at 0.3 Hz, low-pass filtered at 300 kHz and the gain set so that the peak-to-peak response is ca. 4–6 V. This waveform is then low-pass filtered to avoid aliasing artifacts (see Sherman-Gold, 1993, pp. 255-257, for an in-depth description of aliasing). The band pass is set to less than half the maximum frequency of the input signal. As an example, for the pseudo-random binary signal generator described by Clausen and Fernandez (1981), two bandwidths are used with frequency ranges for 2.2 Hz to 2.16 kHz and 21 Hz to 21.6 kHz. The low-pass filter settings for these bandwiths were 1 kHz and 10 kHz, respectively. Cauer-elliptic filters are useful since they offer the maximal roll-off (120 dB/octave). Data above ca. 9.1 kHz are dropped due to the finite roll-off of the filter.

To increase the signal-to-noise ratio, signal averaging is also typically employed, e.g. five to ten sequential blocks of voltage output traces are averaged per experimental run. An advantage of the synthesized waveform is that it is periodic. Consequently, the input signal (in this case current) needs only to be recorded at the beginning of the experiment. Typically this is performed by measuring and storing in computer the filtered and digitized voltage drop across a calibrated 1 kΩ carbon resistor. Additional precautions include the measurement of the impedance of a saline filled chamber (prior to mounting the epithelium) to be assured that there is minimal phase shift caused by the voltage-measuring electrodes and amplifier. Any instrument or electrode dependent phase shift (which is typically less than 0.5°) is subtracted from the experimental phase data prior to analysis.

6.5 DATA ANALYSIS

6.5.1 Square step (transient) analysis

The complete protocol for transient analysis has been described in detail by Lewis and deMoura (1984). In brief, the 'on' voltage response to a square current pulse is digitized (at 100 μsec/point with a resolution of 0.05 mV) and stored in computer memory. These data are converted to an 'off' voltage response by subtracting the time-dependent 'on' voltage response from the pre-stimulus voltage. The absolute value of this voltage change ($|\Delta V|$) is stored together with the corresponding time (t, where $t = 0$ is the time at which the current is applied). Value $|\Delta V|$ is then fitted to the sum of exponentials using a non-linear curve fitting routine (e.g. NFIT, Island Products, Galveston TX, 77555). The number of exponentials is limited by the equivalent circuit used. In general, this type of analysis will yield two voltage terms, which are converted to resistances by dividing by the amplitude of the current pulse, and two time constants, each of which is the product of a resistor and a capacitor. The relationship between the best-fit values and the actual epithelial resistors and capacitors depends upon the equivalent circuit model used to represent the epithelium. As discussed below, the choice of model is one of the most crucial aspects of extracting meaningful parameters for either transient or impedance analysis.

6.5.2 Sinusoids

The analysis procedure for impedance measurements using synthesized waveforms has been described in detail by Clausen and Fernandez (1981). In brief, the impedance is calculated as the Fourier transform of the epithelial voltage response to the current, divided by the Fourier transform of the current. As is obvious from the impedance equations shown in Figure 6.2, curve fitting a model to the data must use a non-linear curve fitting routine.

The procedure for determining the equivalent circuit parameters, is outlined below. The first step is to choose an adequate equivalent circuit model. Next the data sets (phase and magnitude) are normalized to their respective maximum values (this procedure reduces uneven weighting of one parameter set over the other). The selected model is then fitted to the impedance data by means of a derivative-free, non-linear least-squares algorithm (Brown and Dennis, 1972) which minimizes the errors between the theoretical curves (phase and magnitude) and the experimental data by adjusting the parameter values for the theoretical curves. Convergence (i.e. an acceptable fit of model to data) is arbitrarily defined as the point at which successive changes in parameter values are less than 1% of the parameter's value. To test for false or local minima in the curve fitting algorithm, differing initial guesses for the circuit parameters should be used. Convergence to the same best-fit parameter set reduces the possibility of local or false minima. To test for the significance of extracted parameters, one calculates the standard deviation of each fit parameter with respect to the parameter value (Hamilton, 1964). When the standard deviation is an appreciable fraction of the parameter value (e.g. 10%) the parameter is then poorly determined and the data is either refitted by a different model or it is discarded.

The last, and perhaps most important, aspect of curve fitting is to determine which of a selection of equivalent circuits models best describes the data. This is important, since as one increases the complexity of a model, additional parameters must be fitted. It is well known that the addition of a random parameter will improve (or not change) the fit of a model to data. Fortunately, one can determine whether the fit has been improved due to the addition of a 'significant' parameter as opposed to an improved fit due to the addition of a random parameter. This determination is carried out by computing the Hamilton's R-factor (which expresses the percentage misfit of theoretical curve to data, also called the 'sum of squares of the residuals') for each individual model (Clausen *et al.*, 1979), and then calculating the ratio of the R-factors of the two fits (i.e. normal, compared with normal plus additional parameter). Using a modified F-distribution of the R-ratio, the number of parameters for each model and the number of data points, one determines whether the added parameter yields a statistical improvement of fit when compared with that expected from the addition of a random parameter.

6.6 CHOICE OF A MODEL

In almost all of the above sections, we have stated that one of the most important aspects of epithelial impedance analysis is the ability to choose an appropriate model to which one fits the data. From a historical perspective, there has been an evolution of equivalent circuit models to describe epithelial impedance. Nine basic types of model which have been used are show in Figure 6.3 (models I–IX).

The first four of these models were derived in an attempt to describe the data adequately without offering a morphological basis for the individual placement of the circuit parameters. Model I was used by Teorell and Wersall (1945) to describe the impedance of frog gastric mucosa; these authors warned against identifying a particular element to a biological structure. Model II, in which a frequency-sensitive impedance was placed in parallel to a resistor, was constructed by Cole (1932) and later by Brown and Kastella (1965); Brown and Kastella (1965) speculated that this impedance was a non-ideal capacitor since the locus of the semicircle (for Nyquist plots) was below the real axis and not on the axis as would occur if that impedance was an ideal capacitor. Model III is included because of its equivalence to models V and VII as described by Schwan (1957) and more recently by Suzuki *et al.* (1982). Hogben (1978) and Rehm and Tarvin (1978) used Model IV to describe the voltage response of the gastric mucosa to a square current pulse. This circuit was generated from a more

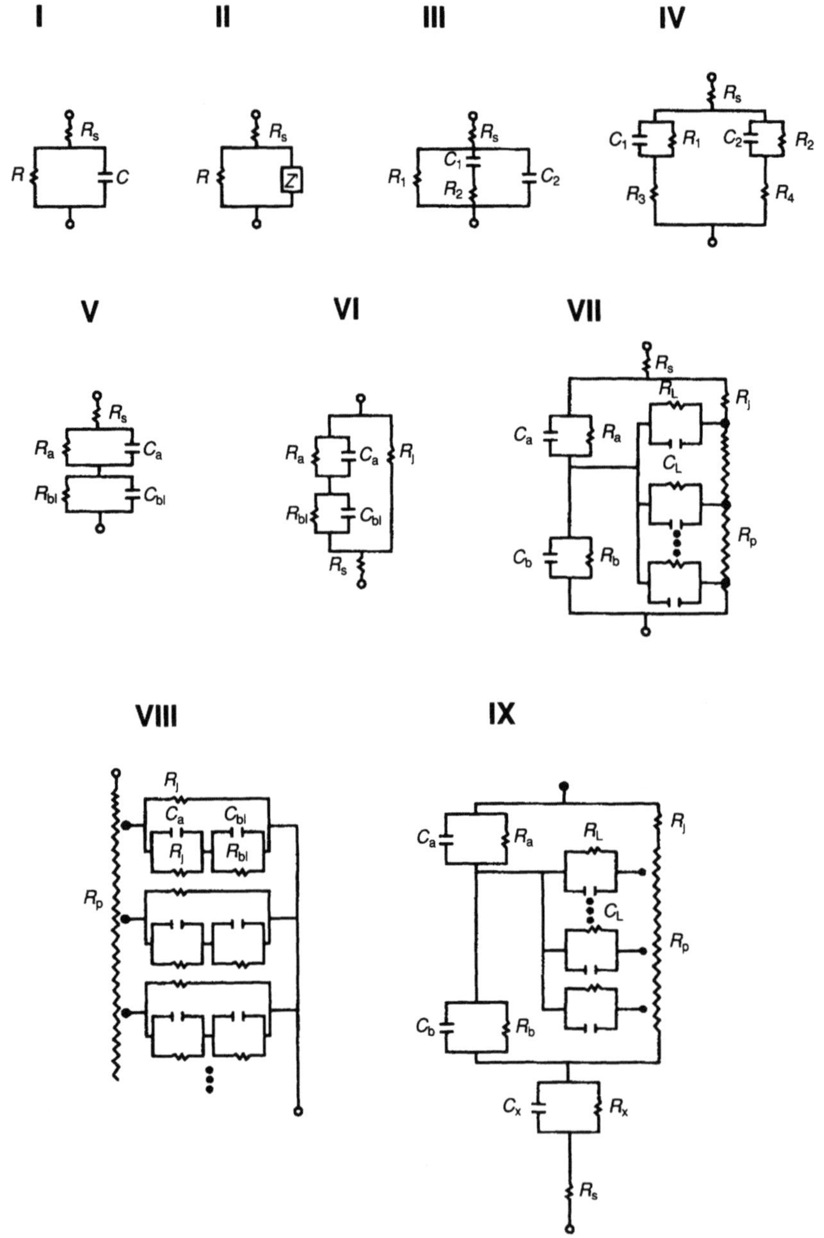

Figure 6.3 A series of equivalent circuits used to describe the impedance properties of epithelial tissues. In all of these models the resistor, R_s, is the resistance of the solution between the voltage measuring electrodes. See main text for details.

complex model (two circuits V in parallel) assuming that the series RCs of each limb had near equal values. The resistors R_3 and R_4 are lumped equivalents of the luminal resistance of the gastric gland and the antiluminal resistance of the gastric gland respectively.

The first 'morphologically correct' circuit that bore at least a resemblance to epithelial architecture is represented by model V and was applied to frog skin by Smith (1971) and Noyes and Rehm (1970). In this, the apical and basolateral membranes were modeled as a series combination of two parallel resistor–capacitor (RC) circuits (R_a and C_a, and R_{bl} and C_{bl}, respectively). This series combination was placed in series with a resistor which represented the solution resistance.

This minimal circuit was adequate to account for the cellular impedance of an epithelium, but it ignored the resistive pathway which is in parallel with the cellular pathway, i.e. the tight junctions. The latter circuit is termed the **lumped model** (model VI) and is the most simple of circuits which takes into account the basic morphology of an epithelium. In addition to the cellular structures, this model also considers the conductive properties of the tight junctions. It is generally argued that the existence of a capacitative component in the tight junctions is unlikely because of the small surface area of this structure when compared with the surface area of a cell (Lewis, Clausen and Diamond, 1976).

The next evolution of equivalent circuits came when it was realized that the lateral intercellular space (the space which extends from the tight junctions to the basal membrane of the epithelial cell and is bounded by the lateral membrane of the epithelium) possessed an impedance at high frequencies (i.e. frequencies greater than 500 Hz) which is comparable to the impedance of the lateral membrane. This circuit is called the distributed resistance model and is represented in model VII, in which the impedance of the lateral membrane (i.e. lateral resistance R_l and capacitance C_l) is distributed along the length of the lateral intercellular space, while the impedance of the basal membrane (i.e. basal resistance R_b and capacitance C_b) is considered as a single unit. This circuit was found to describe adequately the impedance of epithelia with reasonably simple morphology, i.e. a smooth apical membrane surface with no large invaginations. Thus for epithelia such as rabbit urinary bladder (Clausen, Lewis and Diamond, 1979) and frog skin, this distributed model and in a few instances the simpler lumped model were adequate to describe epithelial impedances.

These models failed when they were used to describe the impedance of more geometrically complex tissues such as gastric mucosa, which contains gastric pits and apical tubulovesicles (Clausen, Machen and Diamond, 1983), or rabbit descending colon which contains crypts (Wills and Clausen, 1987; model VIII). The latest model for epithelial impedance has included the addition of an extraepithelial impedance (R_x and C_x in model IX) (Clausen, Reinach and Marcus, 1986; Lewis, Clausen and Wills, 1991; Wills, Purcell and Clausen, 1992) for frog corneal epithelium, toad urinary bladder and cultured renal monolayers. Morphologically, this extraepithelial RC most probably reflects the corneal endothelium, the underlying smooth muscle layers and peritoneal epithelium of the toad urinary bladder and (for cultured monolayers) basolateral membrane intrusions into the filter support, respectively.

Again, here is an example where one must have a intimate understanding of the morphology of the tissue under study, and this morphology must include not only the epithelium but also associated extraepithelial structures. When choosing an electrical equivalent circuit to fit to the data, one must first know the complete structure of the tissue under study. It is then necessary to develop a circuit to correspond to this structure and generate the appropriate impedance equation so that one can fit the model to the data. Finally, one must determine whether the calculated parameters are physically reasonable.

6.7 NON-UNIQUENESS OF MODELS

It is important to emphasize that a good fit to a model does not necessarily mean that this particular model is the correct representation of the epithelium. We will consider two cases: a non-morphological model that can 'adequately' describe epithelial impedance data, and two morphologically based models that produce equally good fits to the same data set and yet yield dramatically different parameter values. An example of a non-morphologically based model adequately describing an impedance data set was noted by Suzuki *et al.* (1982) (Figure 6.3, model III). The equation which describes the impedance of this circuit has the same form as the equation which describes models V and VI (Figure 6.3). On a morphological basis, one would not use a model such as circuit III to extract meaningful parameter values from epithelial impedance data.

Let us now consider models V and VI in more detail. The relationships between resistors and capacitors in these circuits are described by the same equations (Lewis and deMoura, 1984) as follows:

$$C_1C_2/(C_1 + C_2) = C_aC_{bl}/(C_a + C_{bl})$$

$$(R_1 + R_2)/R_1C_1R_2C_2 = (R_a + R_{bl})/R_aC_aR_{bl}C_{bl}$$

$$1/R_1C_1 + 1/R_2C_2 = (R_a + R_j)/R_aR_jC_a + (R_{bl} + R_j)/R_{bl}R_jC_{bl}$$

$$1/R_1C_1R_2C_2 = (R_a + R_{bl} + R_j)/R_aR_{bl}R_jC_aC_{bl}$$

Note that a five-parameter model (for simplicity we ignore the solution resistance R_s) can be determined by four parameters. To curve fit this five-parameter model to the impedance data, one must have an independent estimate of one of the parameters. For example, Clausen *et al.* (1979) measured the ratio of the apical to the basolateral membrane resistances (see Chapter 5 for method of measurement) for rabbit urinary bladder, while Wills and Clausen (1987) independently measured the resistance of the tight junctions using either antibiotics or the sodium channel blocker amiloride. Other examples of circuit equivalence are given in Clausen, Machen and Diamond (1983) who showed that a morphologically based nine-parameter model is equivalent to a less realistic (but possibly morphologically accurate) seven-parameter model.

Finally, we note that for certain parameter values, circuits V and VI can be reduced to the form of circuit I of Figure 6.3. One example is the situation in which the time constants (the inverse product of the resistor and capacitor) of the two series RCs are equal to each other. The second example is when the junctional resistance in circuit VI has a value much less (one tenth or lower) than either of the membrane resistances. To separate the individual parameter values for the first situation, one must measure (using an independent method) the ratio of the membrane resistors, the ratio of the membrane capacitors, or an absolute value of one of the membrane resistors or capacitors (Clausen, Lewis and Diamond, 1979). To resolve the individual values for the second situation, one must measure the impedance of the cellular pathway using a microelectrode. This is a technically difficult procedure; for details, refer to the work by Frömter and colleagues (Lim *et al.*, 1984; Suzuki *et al.*, 1978; Schifferdecker and Frömter, 1978).

In summary, the use of a morphologically based model is essential for impedance analysis of epithelia. Due to the non-unique characteristics of circuits, one must be able to provide an independent estimate of one of the circuit parameters of the chosen equivalent circuit. Lastly, common sense must be used in assessing whether best-fit parameters are reasonable. For instance, if an investigator fits the data by circuit VI using an independently measured value for the junctional resistance and finds that the best-fit values for the basolateral membrane resistance and capacitance are 2 ohm-cm^2 and 50 microfarads/cm^2 respectively, it is highly likely that these are incorrect. Such a result could suggest that the two time constants are nearly equal so that the curve fitting routine simply returns unreasonable values. Again, it is important to remember that the parameter values returned by the curve fitting routine are only as accurate as the original data set.

6.8 INTERPRETATION OF PARAMETERS

This section explores the utility of impedance analysis on flat-sheet epithelia by using examples from the literature. Table 6.1 is a compilation of epithelia on which impedance analysis has been performed, the models used to extract membrane parameter values and the type of waveform used to determine the impedance. Listed below are some interesting observations concerning these epithelia.

6.8.1 Membrane surface areas

Although it was generally thought that the ratio of surface areas for an epithelium would be about 1 to 5 (apical to basolateral), there are three exceptions. Rabbit descending colon has a larger apical area than basolateral area, which suggests that the area amplification offered by the microvilli and crypts increases the apical surface area some 20-fold over the nominal equivalent surface area of a disk. Although the basolateral membrane area is also increased, this area amplification is not as dramatic as that offered by the microvilli. A similar phenomenon is found for frog gastric mucosa, i.e. a larger apical than basolateral membrane surface area and a greatly enhanced apical area due to microvilli and dense packing of gastric pits. The third example is frog corneal epithelium. This epithelium is composed of five to seven cell layers connected by low resistance gap junctions. The basolateral surface area is then the total area offered by all of these cell layers; as a consequence a ratio of 1 to 36 (apical to basolateral) is expected.

6.8.2 Membrane resistances

In agreement with the original observations of Frömter and Diamond (1972), the large variability in transepithelial resistances measured among epithelia is, in general, due to the value of the tight junctions. However, impedance has shown that an epithelium which has a low transepithelial resistance (e.g. frog stomach) does not necessarily have a low junctional resistance. The low transepithelial resistance of frog stomach is due to its large surface area. Thus normalizing the apical resistance to the apical surface area, the apical membrane of frog stomach has a specific resistance of some 30 000 $\Omega \cdot \mu F$. This value is one of the highest apical membrane resistances (for control conditions) so far reported. The large range of apical membrane resistances of the epithelia shown in Table 6.1 is dramatically reduced after normalizing to the apical membrane surface area. Similarly, normalizing the basolateral membrane resistance to the basolateral membrane surface area dramatically reduces the range in basolateral membrane resistances.

In experiments involving transepithelial impedance measurements and DC microelectrode measurements, Clausen *et al.* (1986) were able to determine the relationship between cell current and apical membrane conductance for frog cornea epithelium. From this relationship, these investigators were able to calculate the cell chloride and potassium

Table 6.1 Summary of impedance properties of flat-sheet epithelia

Animal	*Tissue*	R_a Ωcm^2	C_a $\mu F/cm^2$	R_{bl} Ωcm^2	C_{bl} $\mu F/cm^2$	R_J Ωcm^2	R_p Ωcm^2	R_x Ωcm^2	C_x $\mu F/cm^2$	*Method*	*Model*	*Comments*	*References*
Rabbit	Urinary bladder	12500	1.8	1020	8.6	≥100000	130			S	D		Clausen *et al.* (1979)
	Descending colon	283	20.7	79	10.9	677	28			R	D		Wills and Clausen (1987)
		273	22.4	78	11.1	677	22				C	R_p in crypt resistance	
Toad	Urinary bladder	2601	1.12	3000	8.4	—	—			S	L		Warncke and Lindemann (1985)
		—	≅2.0	39000	≅13.0	>18000	—			P	L	Na^+-gluconate	Lewis *et al.* (1985)
		—	1.97	—	—	—	—			P	L	K^+-depolarized	Palmer and Lorenzen (1983)
		21410	0.9	560	4.1	12000	104	94	3.8	R	D-2C		Lewis *et al.* (1991)
Turtle	Urinary bladder	2940	3.8	217	8.8	∞	290					Mucosal amiloride	Clausen and Dixon (1986)
Frog	Stomach	150	200	41	99	—	R_a, 24 R_{bl}, 12			S	DD	Resting	Clausen *et al.* (1983)
	Corneal Epithelium	554	2.6	910	94	2330	331	19	5.1		D-2C		Clausen, Reinach and Marcus (1986)
Necturus	Gallbladder	—	7	—	18	—	—			S	L		Schifferdecker and Frömter (1978)
		1220	8	201	26.3	91	—			P	L		Suzuki *et al.* (1982)
		2310	7	87	32.4	145	91			R	D		Lim *et al.* (1984)
Xenopus laevis	Kidney A6 cells	2703	1.2	333	4.8	3030	43	42	6.9	R	D	Millipore HA filters	Wills *et al.* (1993)
	2F3 cells	1266	1.4	285	4.2	4349	65	35	4.5	R	D		Wills *et al.* (1993)

R_a and C_a = apical membrane resistance and capacitance R_{bl} and C_{bl} basolateral membrane resistance and capacitance; R_J = junctional resistance; R_p (unless otherwise noted) = distributed resistance along the lateral intercellular space (LIS); R_x and C_x = extra epithelial resistor and capacitor. 'Methods' refers to the waveform used: P = square current pulse; S = discrete sinusoids; R = random noise. Model: L = lumped model; D = distributed model; DD = double distributed model (one for apical membrane, the other for the basolateal membrane): C = crypt model; D-2C = distributed two-cell model. (See Figure 6.3 for geometry of these equivalent circuit models.) Except where noted in comments, normal bathing solutions were used (see appropriate reference for experimental details).

activities as a function of Cl^- secretion rate. In these impedance studies, the use of DC microelectrode methods served a dual function: they provided an independent estimate of the membrane resistance ratio for impedance circuit analysis and information concerning the electrical driving force across each membrane. Together these methods provide a powerful means for assessing changes in membrane conductances and driving forces affecting net ion movements across epithelial cell membranes.

6.8.3 Geometric considerations

All of the epithelia listed in Table 6.1 were best-fitted by equivalent circuit models containing a distributed resistance parameter (R_p). In one instance (rabbit descending colon) two models (Figure 6.3; circuits VII and VIII) yielded equally good fits to the data. Although this precludes the selection of the 'best' equivalent circuit model to describe the colon, one still has confidence in the best-fit values of the membrane parameters since they were not significantly different between the two models. A second point is that the impedance data of frog stomach required the use of a double distributed model – one for the basolateral membrane and the other for the apical membrane. Whether the latter distributed resistance is due to the gastric pits, luminal infolding of the apical membrane of the acid secreting cells or the anti-luminal infolding of the apical membrane (i.e. the distributed resistance of an apical membrane-bound cytoplasmic space) is not known (Clausen *et al.*, 1983).

As already mentioned, one of the advantages of impedance analysis is that it is capable of yielding morphological information about the epithelium under study. In addition to the membrane areas noted above, one can obtain information about the geometry of the lateral intercellular space. Clausen *et al.* (1979) reasoned that R_p is the product of the bathing solution resistivity (actually the resistivity of the solution in the lateral intercellular space) and a geometric term composed of the length of the lateral intercellular space divided by the space's cross-sectional area. Assuming a simple cuboidal shape for the rabbit urinary bladder cells of 20 microns per side, these authors calculated that the LIS had a width of about 7 nm, a value in good agreement with that determined from electron micrographs of 10 nm (for details, see Clausen *et al.*, 1979).

Although these distributed resistor values seem small when compared with the resistance of the basolateral membrane, they can be of significant value. For example, the gallbladder and turtle urinary bladder LIS resistances are equal in value to the basolateral membrane resistances. Although this is not a significant problem for turtle urinary bladder, where the junctional resistance has a large value, gallbladder has a low junctional resistance and, consequently, transepithelial current pulses generate large voltage drops along the lateral intercellular space. Such voltages are sensed by microelectrodes. Thus, the ratio of apical to basolateral membrane resistances (which is measured as the ratio of the apical to basolateral membrane voltage change caused by a transepithelial current pulse) will be artifactually low since the microelectrode is sensing the voltage changes caused by current flow across both the basolateral membrane and the lateral intercellular space. For tight epithelia, the measuring error of this resistance ratio is small. For the colon, the error is approximately 10–15%. However, for gallbladder, this error can be large – approximately 500–2300%. This problem has been considered in detail by Boulpaep and Sackin (1980) and quantitated by Frömter and colleagues using both transepithelial and microelectrode impedance analysis (Lim *et al.*, 1984). It is important to stress that these errors will occur only if the change in the LIS width occurs along the full length of the LIS. Thus a local constriction of the LIS near the tight junctions will cause an increase in the paracellular resistance without significantly altering the measured value of the apical to basolateral membrane resistance (Stoddard and Reuss, 1988). Of importance here is that not only can morphological information be extracted from impedance analysis, but also that DC microelectrode data can be in error if the incorrect model is used to determine membrane resistance values.

6.8.4 Area changes

There are three basic mechanisms by which epithelia can regulate the rate of ion transport: activation of pre-existing quiescent transporters already resident in the membrane; modulation of the kinetics of a single transporter unit; and the insertion of transporters from a cytoplasmic pool into either or both the apical and basolateral membranes. If the latter mechanism is involved, it might be predicted that the membrane area into which the transporter is inserted might increase. Thus by measuring the impedance of an epithelium during alteration of transport, it might be possible to determine whether insertion is a mechanism for transport regulation.

Table 6.2 summarizes the epithelial preparations which showed a change in membrane area during a change in transport rate. One of the most dramatic examples of an area increase is frog stomach which upon stimulation of proton secretion by histamine increased its apical membrane area by more than 200% with no change in basolateral membrane area (Clausen *et al.*, 1983). Inhibition of acid secretion (by cimetidine, a H_2 receptor blocker) resulted in a 70% decrease in apical area. Similarly, membrane area changes were noted when proton secretion in turtle urinary bladder was inhibited by acetazolamide. In the latter example, the decrease in apical membrane area is thought to reflect changes in the mitochondrial-rich cells. Because these represent only a

Table 6.2 Membrane area changes associated with transport regulation

Tissue	*Modulator*	*Transported species*	*% ΔC_a*	*% ΔC_b*	*Method*	*Comment*	*Reference*
Frog stomach	Histamine	H^+ ↑	>200%	0	S		Clausen *et al.* (1983)
	Cimetidine	H^+ ↓	–69%	0	S		Clausen *et al.* (1983)
Toad urinary bladder	Amiloride	Na^+ ↓	0	0	R		Lewis *et al.* (1991)
	ADH	Na^+ ↑, H_2O	30%	33%	B		Warncke and Lindemann (1981)
	ADH+methohexital	Na^+ ↑	0	—	P	H_2O response inhibited by methohexital	Stetson *et al.* (1982)
Rabbit urinary bladder	Amiloride	Na^+ ↓	0	0	S		Clausen *et al.* (1979)
	Mechanical stretch	Na^+ ↑	>21%	—	P		Lewis and de Moura (1984)
	Cell swelling	Na^+ ↑	80%	—	P	$^1/_2$ osmotic-stretch	Lewis and de Moura (1984)
Turtle urinary bladder	Acetazolamide	H^+ ↓	–8%	0	R		Clausen and Dixon (1986)
Renal A6 Cells	PGE_2	Cl^- ↑, Na^+ ↑	58%	>200%	R		Mo and Wills (1995)
	Forskolin	Cl^- ↑, Na^+ ↑	83%	>200%	R		Mo and Wills (1995)

See Table 6.1 for tissue identification and method glossary. ↑ = increase and ↓ = decrease in rate of transported species. Percentage change in capacitance (% C_x) = [(exp. – control)/control] × 100. 'Modulator' = mechanism used which resulted in a change in membrane capacitance.

small percentage of the total number of cells in the epithelium, the percentage decrease of their surface area will be much larger than the measured 8% decrease of apical surface area.

Of course, a change in surface area with a change in transport does not necessarily imply cause and effect. An example of this is illustrated by toad urinary bladder. In this particular epithelium, the serosal addition of antidiuretic hormone results in an increase in apical membrane sodium permeability, water permeability and membrane surface area. Does this increase in area result in an increase in sodium and/or water permeability? Stetson *et al.* (1982), using square current pulses (transient analysis), were able to demonstrate that selective inhibition of the ADH-induced increase in water permeability (without altering the increase in sodium permeability) was associated with an inhibition of the increase in apical surface area.

In recent studies using transient analysis, membrane insertion and withdrawal has been proposed as a mechanism by which epithelia (in this case rabbit urinary bladder) can increase their surface area during mechanical distention (for details, see Lewis and deMoura, 1984). In these studies, impedance analysis was used to determine which cytoskeletal structures were responsible for vesicle insertion and withdrawal. This approach allows one to determine not only whether transport is regulated by vesicle insertion and withdrawal, but also the cytoskeletal structures which might be responsible for the vesicle translocation.

REFERENCES

Boulpaep, E.L. and Sackin, H. (1980) Electrical analysis of intraepithelial barriers, in *Current Topics in Membrane Transport*, Vol. 13 (ed. E.L. Boulpaep), Academic Press, New York.

Brown, A.C. and Kastella, K.G. (1965) The AC impedance of frog skin and its relation to active transport. *Biophys. J.* **5**:591–607.

Brown, K.M. and Dennis, J.E. (1972) Derivative free analogues of the Levenberg–Marquardt and Gauss algorithms for nonlinear least squares approximation. *Numer. Math.* **18**:289–97.

Clausen, C. (1989) Impedance analysis in tight epithelia. *Methods in Enzymology* **171**:628–42.

Clausen, C. and Dixon, T.E. (1986) Membrane electrical parameters in turtle bladder measured using impedance techniques. *J. Memb. Biol.* **92**:9–19.

Clausen, C. and Fernandez, J.M. (1981) A low-cost method for rapid transfer function measurements with direct application to biological impedance analysis. *Pflügers Arch.* **390**:290–5.

Clausen, C., Lewis, S.A. and Diamond, J.M. (1979) Impedance analysis of a tight epithelium using a distributed resistance model. *Biophys. J.* **26**:291–318.

Clausen, C., Machen, T.E. and Diamond, J.M. (1983) Use of AC impedance analysis to study membrane changes related to acid secretion in amphibian gastric mucosa. *Biophys. J.* **41**:167–78.

Clausen, C., Reinach, P.S. and Marcus, D.C. (1986) Membrane transport parameters in frog corneal epithelium measured using impedance analysis. *J. Memb. Biol.* **91**:213–25.

Cole, K.S. (1932) Electrical phase angle of cell membranes. *J. Gen. Physiol.* **15**:641–9.

Desoer, C.A. and Kuh, E.S. (1969) *Basic Circuit Theory*. McGraw-Hill Book Company, New York.

Frömter, E. and Diamond, J. (1972) Route of passive ion permeation in epithelia. *Nat. New Biol.* **235**:9–13.

Gogelein, H. and Van Driessche, W. (1981) Capacitive and inductive low frequency impedances of *Necturus* gallbladder epithelium. *Pflügers Arch.* **389**:105–13.

Gordon, L.G.M., Kottra, G. and Frömter, E. (1990) Methods to detect, quantify, and minimize edge effects in Ussing chambers. *Methods in Enzymology* **171**:628–42.

Hamilton, W.C. (1964) *Statistics in Physical Science*. Ronald Press, New York.

Hogben, C.A. (1978) The transient nature of equivalent circuits: in the sea, on the land, on the meaning of epithelial conductance. *Acta. Physiol. Scand.* Special Supplement:**111**–30.

Kottra, G. and Frömter, E. (1982) A simple method for constructing shielded, low-capacitance glass microelectrodes. *Pflügers Arch.* **395**:156–8.

Lewis, S.A. and deMoura, J.L.C. (1984) Apical membrane area of rabbit urinary bladder increases by fusion of intracellular vesicles: an electrophysiological study. *J. Memb. Biol.* **82**:123–36.

Lewis, S.A., Butt, A.G., Bowler, M.J. *et al.* (1985) Effects of anions on cellular volume and transepithelial Na^+ transport across toad urinary bladder. *J. Memb. Biol.* **83**:119–37.

Lewis, S.A., Clausen, C. and Diamond, J.M. (1976) Appendix, in Na^+ transport by rabbit urinary bladder, a tight epithelium. *J. Memb. Biol.* **28**:35–40.

Lewis, S.A., Clausen, C. and Wills, N.K. (1991) Modulation of membrane properties of the toad urinary bladder by amiloride and anions. *Biochim. Biophys. Acta* **1070**:99–110.

Lim, J.J., Kottra, G., Kampmann, L. and Frömter, E. (1984) Impedance analysis of *Necturus* gallbladder epithelium using extra- and intracellular microelectrodes. *Cur. Top. Memb. Trans.* **20**:27–46.

Mathias, R.T., Rae, J.L. and Eisenberg, R.S. (1979) Electrical properties of structural components of the crystalline lens. *Biophys. J.* **25**:181–201.

Mo, L. and Wills, N.K. (1995) Modulation of membrane impedances in cultured renal (A6) epithelia. *FASEB J.*, in press.

Noyes, D.H. and Rehm, W.S. (1970) Voltage response of frog gastric mucosa to direct current. *Am. J. Physiol.* **219**:184–92.

Palmer, L.G. and Lorenzen, M. (1983) Antidiuretic hormone-dependent membrane capacitance and water permeability in the toad urinary bladder. *Am. J. Physiol.* **244**:F195–F204.

Poussart, D. and Ganguly, U.S. (1977) Rapid measurement of system kinetics an instrument for real-time transfer function analysis. *Proc. IEEE* **65**:741–7.

Rehm, W.S. and Tarvin, J.T. (1978) Interpretation of the voltage response of epithelial tissue to step currents. *Acta. Physiol. Scand.* Special Supplement:143–54.

Schifferdecker, E. and Frömter, E. (1978) The AC impedance of *Necturus* gallbladder epithelium. *Pflügers Arch.* **377**:125–33.

Schwan, H.P. (1957) Electrical properties of tissue and cell suspensions, in *Advances in Biological and Medical Physics*, Vol. 5 (ed. J.J. Lawrence and C.A. Tobias), Academic Press, New York.

Sherman-Gold, R. (1993) *The Axon Guide for Electrophysiology and Biophysics Laboratory Techniques*, Axon Instruments, Inc., Foster City, CA 94404, USA.

Smith, P.G. (1971) The low-frequency electrical impedance of the isolated frog skin. *Acta Physiol. Scand.* **81**:355–366.

Stetson, D.L., Lewis, S.A., Alles, W. and Wade, J.B. (1982) Evaluation by capacitance measurements of antidiuretic hormone induced membrane area changes in toad bladder. *Biochim. Biophys. Acta* **689**:267–74

Stoddard, J.S. and Reuss, L. (1988) Voltage- and time-dependence of apical membrane conductance during current clamp in *Necturus* gallbladder epithelium. *J. Membr. Biol.* **103**:191–204.

Suzuki, K., Kottra, G., Kampmann, L. and Frömter, E. (1982) Squarewave pulse analysis of cellular and paracellular conductance pathways in *Necturus* gallbladder epithelium. *Pflügers Arch.* **394**:302–12.

Suzuki, K., Rohlicek, V. and Frömter, E. (1978) A quasi-totally shielded low-capacitance glass microelectrode with suitable amplifiers for high-frequency intracellular potential and impedance measurements. *Pflügers Arch.* **378**:141–8.

Teorell, T. and Wersall, R. (1945) Electrical impedance properties of surviving gastric mucosa of the frog. *Acta Physiol. Scand.* **10**:243–57.

Valdiosera, R., Clausen, C. and Eisenberg, R.S. (1974) Circuit models of the passive electrical properties of frog skeletal muscle fibers. *J. Gen. Physiol.* **63**:432–59.

Warncke, J. and Lindemann, B. (1979) A sinewave burst method to obtain impedance spectra of transporting epithelia during voltage clamp. *Pflügers Arch.* **382**:R12.

Warncke, J. and Lindemann, B. (1981) Effect of ADH on the capacitance of apical epithelial membranes, in *Physiology of Non-excitable Cells* (ed. J. Salanki), Vol. 3 of Advances in Physiological Science, pp. 129–33.

Warncke, J. and Lindemann, B. (1985) Voltage dependence of Na channel blockage by amiloride: Relaxation effects in admittance spectra. *J. Memb. Biol.* **86**:255–65.

Wills, N.K. and Clausen, C. (1987) Transport-dependent alterations of membrane properties of mammalian colon measured using impedance analysis. *J. Memb. Biol.* **95**:21–35.

Wills, N.K., Purcell, K. and Clausen, C. (1992) Impedance properties of cultured renal epithelia. *J. Membr. Biol.* **125**:273–85.

Wills, N.K., Purcell, K., Clausen, C. and Millinoff, L. (1993) Effects of aldosterone on the impedance properties of cultured renal epithelia. *J. Membr. Biol.* **133**:17–27.

7

Measurements and interpretation of cytoplasmic ion activities

Guillermo A. Altenberg and Luis Reuss

Understanding the mechanisms of ion transport across epithelial barriers requires knowledge of intracellular ion activities. Ion transport across cell membranes is mediated by carriers, channels or pumps, and the ion fluxes mediated by these transport proteins depend on ion activities and/or electrical potential. Studies of ion transport mechanisms in epithelial cells are complex because of the differences in transport properties of the apical and basolateral membranes, i.e. the transporters expressed at the two membrane domains are different.

Measurements of apical and basolateral membrane voltages and of intracellular ion activities are necessary to determine the driving forces for ion transport across each membrane and to identify the mechanisms involved in ion transport. This chapter emphasizes the methodologies available to study cytoplasmic ion activities, but does not discuss

Epithelial Transport: A guide to methods and experimental analysis.
Edited by Nancy K. Wills, Luis Reuss and Simon A. Lewis.
Published in 1996 by Chapman & Hall, London. ISBN 978-94-010-7168-0

measurements of ion activities in intracellular compartments other than the cytoplasm (for ion-activity measurements in organelles, see Rizzuto *et al.*, 1992, and Hofer and Machen, 1993). Also, it does not address the use of the measurements of intracellular ion activities to estimate changes in cell volume, which is presented in Chapter 8. This chapter focuses on non-destructive, high time-resolution methods and provides examples of the use of measurements of intracellular ion activities to identify specific transport mechanisms.

At this point it is important to consider the meaning of the term **ion activity** and its distinction from free concentration. The activity of the ion j is its **effective concentration**, responsible for the chemical potential of the ion in solution and the rates of its chemical reactions. The ion activity is different from the free concentration measured by chemical methods. The ion activity is lower (except at very high dilution) because the mobility of ions in solution is restricted by ion–ion and ion–solvent interactions. The correction factor required to convert free concentrations to activities is the **activity coefficient**. At constant temperature, the activity coefficients of ions in biological solutions depend on the ionic strength of the solution. The relationship is quantitated by the Debye–Hückel equation (e.g. Robinson and Stokes, 1959).

From the practical point of view it is important to keep in mind that most electrophysiologists express data as intracellular ion activities, but most data gathered with ion-sensitive indicators are expressed as free concentrations.

7.1 ELECTROCHEMICAL DRIVING FORCES AND THE STUDY OF ION TRANSPORT MECHANISMS

The electrochemical driving force for a specific ion across a membrane depends on the ion activities at both sides of the membrane and the membrane voltage. The voltage at which the driving force for net ion movement across a membrane is zero is the **equilibrium potential** (E_j), given by the Nernst equation (Chapter 2):

$$E_j = -2.3\, RT/zF \log (j)_i/(j)_o \tag{7.1}$$

where *R*, *T*, *z* and *F* have their usual meanings, and $(j)_i$ and $(j)_o$ are the intracellular and extracellular activities of the ion *j*, respectively. For monovalent ions, the product 2.3 RT/zF is positive (cations) or negative (anions), and its absolute value is 58 or 61 mV at 20 or 37°C, respectively. For example, if the ratio $(c)_i/(c)_o$, where *c* is an univalent cation, is 0.1, the calculated equilibrium potential at 37°C will be ca. 61 mV. If the measured membrane voltage (V_m) is equal to the calculated equilibrium potential (E_j), the driving force for net ion movement across the membrane is zero (driving force = $E_j - V_m$, with positive values favoring influx into the cells).

If the membrane voltage is different from the equilibrium potential, then there is a driving force for net ion movement. For the example above, if the membrane voltage is –80 mV, then the electrochemical driving force for net flux of *c* from the extracellular medium into the cell is equivalent to 19 mV ($E_j - V_m$ = –61 mV –(–80 mV)). On the other hand, if the measured membrane voltage is –40 mV, then the driving force favorable for net movement of *c* from the cell into the extracellular medium will be –21 mV ($E_j - V_m$ = –61 mV –(–40 mV)).

The discussion above applies to any cell. However, for epithelial cells, the electrochemical driving forces across the apical and basolateral membranes can be, and frequently are, different. This occurs because of different ion transport properties of apical and basolateral membranes, and because of differences in composition of the external solutions (i.e. gastric fluid vs. interstitial 'serum-like' fluid) can cause trans-epithelial differences in voltage and/or ion activities. This is evident in oxyntic cells of the gastric epithelium. In these cells, the ion transport mechanisms can establish a transepithelial H^+ concentration ($[H^+]$) ratio ($[H^+]_{lumen}/[H^+]_{blood}$) in excess of 10^6.

Of the three parameters needed to calculate electrochemical driving force for ion transport, the determination of extracellular ion concentrations is relatively simple, but measurements of membrane voltage and intracellular ion activities are more difficult. In this chapter, we will focus on methods of measurement of intracellular ion activities. Membrane voltages, which can be measured with intracellular micro-electrodes or estimated with voltage-sensitive indicators, are discussed in detail in Chapter 5 (see also Waggoner, 1979; Altenberg *et al.*, 1990; Loew, 1993).

Table 7.1 lists the methods available for rapid and non-destructive determination of intracellular ion activities, and these methods are discussed in the following sections. Other methods require destruction of the sample or are too slow for single-membrane studies of epithelial cells. Destructive methods are based on the calculation of intracellular ion concentrations from measurements of total ion content, total and extracellular water, and subtraction of the extracellular ion content of the sample. Extracellular water is determined using membrane-impermeable indicators such as inulin or sucrose. These methods, as well as the most sophisticated destructive measurements by X-ray microanalysis and nuclear magnetic resonance spectroscopy, will not be discussed further (for details, see MacKnight and Leader, 1989; Beck *et al.*, 1987; Hansen *et al.*, 1993; Rick, 1994). These techniques are especially useful for the determination of ions in subcellular compartments such as organelles, or to compare ion composition in neighboring but different cell types, but are not the most useful for accurate identification and quantitation of membrane transport mechanisms in epithelial cells. In the following sections we

Table 7.1 Rapid and continuous methods of measurement of cytoplasmic ion activities

Ion-sensitive microelectrodes	*Ion-sensitive indicators*
Glass	Ion-sensitive fluorescent dyes
Liquid membrane	Ion-sensitive photoproteins
– Ion-exchanger	Ion-sensitive indicators followed by absorbance changes
– Neutral carrier	

discuss the importance of kinetic determinations of intracellular ion activities and the methods that allow for these determinations.

7.2 IMPORTANCE OF RAPID MEASUREMENTS OF INTRACELLULAR ION ACTIVITIES TO IDENTIFY ION TRANSPORT MECHANISMS

A transmembrane ion flux requires both a net driving force and a finite permeability. For example, alpha-intercalated cells from the kidney do not express Na^+ transport proteins in the apical membrane. Therefore, although there is a significant electrochemical driving force for Na^+ influx, there is no Na^+ transport. In addition, cell membranes can perform active ion transport, i.e. transport that opposes the existing electrochemical driving force. An example is the Na^+,K^+-ATPase present in the basolateral membranes of most epithelial cells. Although the Na^+ electrochemical driving force favors Na^+ influx across the basolateral membrane, the Na^+,K^+-ATPase actively extrudes Na^+ against its electrochemical driving force (Chapter 1). These examples indicate that, although important, the determination of the electrochemical driving force is not definitive in studies of the mechanisms of ion transport across apical or basolateral membranes of epithelial cells. To identify and characterize the ion transport mechanisms at a particular membrane, kinetic measurements of intracellular ion activities and cell membrane voltages are mandatory. This notion is stressed by the example in Figure 7.1, which depicts two different hypothetical epithelial cells that perform transepithelial (apical-to-basolateral) NaCl transport. In cell (a), Na^+ influx is mediated by a NaCl cotransporter. In cell (b), Na^+ entry across the apical membrane is via two parallel and independent carriers, Na^+/H^+ and Cl^-/HCO_3^- exchangers. Coupling between the cation and the anion exchanger is via intracellular pH. The Na^+/H^+ exchanger operates as a Na^+ influx mechanism driven by the Na^+ gradient, and elevates intracellular pH. The consequent elevation in $[HCO_3^-]_i$ helps drive Cl^- influx through the anion exchanger. In both cells, NaCl efflux across the basolateral membrane is via the Na^+,K^+-ATPase and a Cl^- channel, and the primary energy source

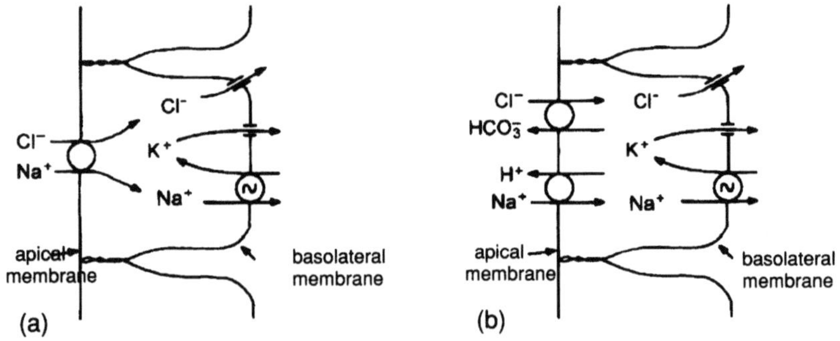

Figure 7.1 Two transport models for NaCl absorptive epithelial cells: (a) cotransport model; (b) double ion-exchange model. (Modified from Reuss and Altenberg, 1995, with permission.)

for transport is the Na^+,K^+-ATPase, which maintains $[Na^+]_i$ low and $[K^+]_i$ high. Recycling of K^+ across the basolateral membrane occurs via K^+ channels.

To ascertain the mechanism of Na^+ influx, one could measure $[Na^+]_i$ upon removal of Cl^- from the apical bathing solution. In cell (a), removal of Cl^- from the apical bathing solution causes a fall in $[Na^+]_i$ by abolishing Na^+ influx via the NaCl cotransporter. In cell (b), Cl^- removal reverses the gradient for Cl^- transport via the Cl^-/HCO_3^- exchanger, and increases pH_i. The decrease in $[H^+]_i$ abolishes Na^+ entry via the cation exchanger, $[Na^+]_i$ falls and NaCl reabsorption ceases. If measurements of transepithelial Na^+ and Cl^- transport and $[Na^+]_i$ and $[Cl^-]_i$ are performed only at the steady state, then it is not possible to distinguish between models (a) and (b). On the other hand, if rapid measurements of intracellular Na^+ and Cl^- are performed, then the decreases in intracellular Na^+ and Cl^- concentrations can be dissociated. The result of such an experiment in the epithelium of the *Necturus* gallbladder is illustrated in Figure 7.2. Following Na^+ removal from the apical bathing solution, the initial rate of fall in intracellular Na^+ activity is much faster than following Cl^- removal from the same solution, a result inconsistent with transport model (a) (which predicts similar rates of fall of intracellular Na^+ activity in both instances; Reuss, 1984). Additional experiments showed that the coupling of Na^+ and Cl^- fluxes across the apical membrane is indirect and mediated by changes in pH_i (Reuss, 1991). This example emphasizes the importance of rapid and continuous measurements of intracellular ion activities.

Continuous measurements of intracellular ion activities are invaluable not only to identify ion transport mechanisms, but also to study the regulation of ion transport (Reuss, 1984; Chaillet *et al.*, 1985; Reuss, 1987) and to quantify ion transport rates across specific membrane transporters (Reuss, 1984; Negulescu *et al.*, 1990). These measurements are not possible with

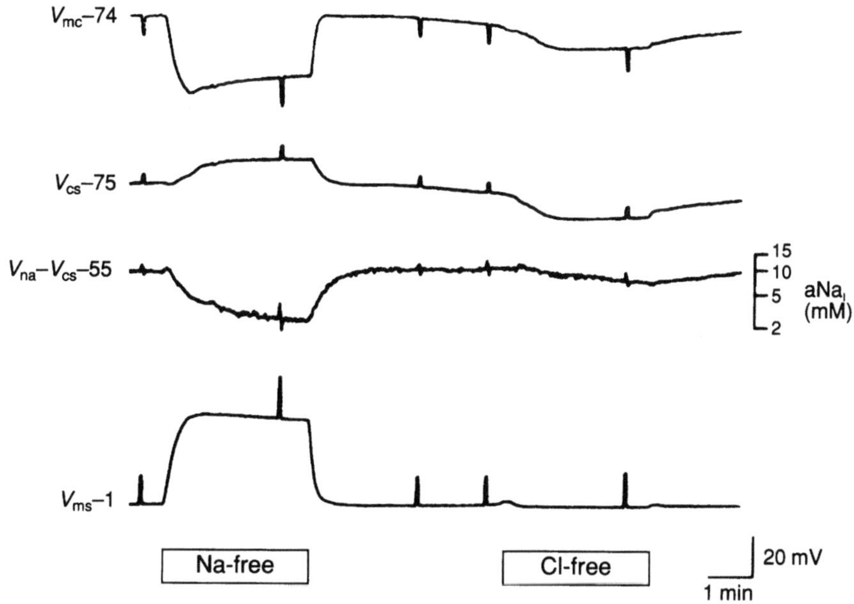

Figure 7.2 Effects of removal of Na^+ (left) and Cl^- (right) from the apical bathing solution on membrane voltages and intracellular Na^+ activity in a *Necturus* gallbladder epithelial cell. (Reproduced from Reuss, 1984, with permission.)

destructive techniques or with methodologies with poor time resolution.

In conclusion, dynamic measurements of intracellular ion activities are necessary to establish the mechanisms of ion transport across the apical and basolateral membranes of epithelial cells, as well as their regulation. In the following sections, we explain the bases for dynamic measurements of intracellular ion activities using electrophysiological (ion-sensitive microelectrode) and optical (ion-sensitive indicator) techniques.

7.3 MEASUREMENTS OF INTRACELLULAR ION ACTIVITIES WITH ION-SENSITIVE MICROELECTRODES

Due to the miniaturization required for intracellular measurements, determinations of intracellular ion activities are generally performed with liquid-membrane microelectrodes (Ammann, 1986; Altenberg *et al.*, 1990). Although Na^+-, K^+- and pH-sensitive microelectrodes constructed with ion-sensitive glass have been used in the past, they have been replaced by those based on liquid-membrane sensors (ion-exchangers or carriers). Details on the specific ion-sensitive resins are beyond the scope of this chapter (see Tsien and Rink, 1980; Ammann, 1986; Ammann *et al.*, 1987; Kondo *et al.*, 1989; Altenberg *et al.*, 1990).

7.3.1 Ion-sensitive microelectrodes: general properties and experimental use

An important practical aspect of the use of ion-sensitive microelectrodes is that the resistivity of the ion-sensitive liquid membranes is high, and as a result a high-input impedance electrometer is required for accurate measurements. Typically, the resistance of useful ion-sensitive microelectrodes is 10^{10} to 10^{11} ohms. For this reason, an electrometer with an input impedance > 10^{14} Ω and a leakage current of less than 10^{-14} A must be employed (additional details on the instrumentation required for studies with ion-sensitive microelectrodes are given in Chapter 5). Another interesting point brought about by the high electrical resistance of the ion-sensitive resin is the possibility of electrical shunting across the wall of the microelectrode tip (Lewis and Wills, 1980). This problem can be solved by insulating the glass, using thick glass or using a glass with high resistivity. The most frequently used ion-sensitive microelectrodes and some relevant references are listed in Table 7.2.

The voltage output of the ion-sensitive microelectrode is equal to the sum of the electrical potential difference between the tip of the microelectrode and the reference (ground) electrode (in the extracellular solution) and a potential proportional to the ion activity of the solution adjacent to the microelectrode tip. Hence, to measure intracellular ion activities it is necessary to subtract the membrane voltage from the output of the ion-sensitive microelectrode. There are two ways to do this: measure the cell membrane voltage by independent impalements with conventional microelectrodes (Chapter 5) and subtract the measured voltage from the voltage output of the ion-sensitive microelectrode; or use double-barrel ion-sensitive microelectrodes. The latter is preferable because both intracellular ion activity and cell membrane voltage are measured simultaneously in the same cell. However, it is very difficult to impale small cells with double-barrel microelectrodes without producing damage. In epithelial preparations, it is still possible to obtain the cell membrane voltage either from simultaneous impalement of another cell with a conventional microelectrode or from the average voltage measured in several cells of the same preparation.

The use of two single-barrel microelectrodes (conventional and ion-selective) is acceptable when the epithelial cells under study are electrically coupled via gap junctions. In this case, the cell membrane voltage under control conditions or upon experimentally induced changes is virtually the same in neighboring cells. When intracellular ion activities are determined with single-barrel microelectrodes the impalements should be validated (e.g. it has to be shown that both microelectrodes are positioned in identical compartments, and that the conductance and

Table 7.2 Ion-sensitive sensors most commonly used to measure cytoplasmic free ion concentrations

Ion measured	*Indicator*	*Detection method*	*References*
Na^+	ETH 227	Electrometric	Ammann, 1986; Altenberg *et al.*, 1990
	SBFI	Fluorescence	Harootunain *et al.*, 1989; Negulescu and Machen, 1990
K^+	Potassium tetrakis p-(chlorophenylborate)	Electrometric	Ammann, 1986; Altenberg *et al.*, 1990
	PBFI	Fluorescence	Kasner and Ganz, 1992
H^+	Tridodecylamine	Electrometric	Ammann, 1986; Altenberg *et al.*, 1990
	BCECF	Fluorescence	Thomas *et al.*, 1979; Buckler and Vaughan-Jones, 1990; Negulescu and Machen, 1990
	SNARF-1	Fluorescence	Owen, 1992
	Me_2CF	Absorbance	Chaillet and Boron, 1985
Ca^{2+}	ETH 1001	Electrometric	Tsien and Rink, 1980; Ammann, 1986
	ETH 129	Electrometric	Ammann *et al.*, 1987
	Fura-2	Fluorescence	Grynkiewicz *et al.*, 1985; Negulescu and Machen, 1990; Williams and Fay, 1990; Gillis and Gailly, 1994
	Fluo-3	Fluorescence	Kao *et al*, 1989; Minta *et al.*, 1989
	Indo-1	Fluorescence	Grynkiewicz *et al.*, 1985;Owen, 1993
	Aequorin	Luminescence	Blinks, 1989; Borle, 1994
Mg^{2+}	ETH 1117	Electrometric	Lanter *et al.*, 1980
	ETH 5214	Electrometric	Hu *et al.*, 1989
	Mag-fura-2	Fluorescence	Hurley *et al.*, 1992; Illner *et al.*, 1992
Cl^-	Corning Cl^- exchanger	Electrometric	Ammann, 1986; Altenberg *et al.*, 1990
	5,10,15,20-tetraphenyl-21H, 23H-porphin manganese (III) chloride	Electrometric	Kondo *et al.*, 1989
	SPQ	Fluorescence	Chao *et al.*, 1990; Verkman *et al.*, 1992; Vasseur *et al.*, 1993
	MQAE	Fluorescence	Verkman *et al.*, 1989; Verkman, 1990

SBFI = sodium-binding benzofuran isophthalate; PBFI = Potassium-binding benzofuran isophthalate; BCECF = 2′,7′-bis-(2-carboxyethyl)-5-(6)-carboxyfluorescein; SNARF -1 = carboxy-seminaphthorhodafluor-1; Me_2CF = 4′,5′-dimethyl-5-(6)carboxyfluorescein; SPQ = 6-methoxy-N-[3-sulfo-propylquinolinium; MQAE = N-(6-methoxyquinolyl)acetoethyl ester.

selectivity of the punctured membranes are unchanged). There are two useful ways of doing this:

- Show that the voltage deflections produced by transepithelial current pulses are the same in the conventional and ion-sensitive microelectrode.
- Ascertain that rapid changes of the cell membrane voltage in response to bathing-media ion substitutions (e.g. K^+ for Na^+) are the same.

The first approach can only be employed in epithelia mounted as monolayers with separate apical and basolateral solutions, and the second approach needs an adequate system to exchange the bathing solution in order to measure rapid changes in membrane voltage before changes in intracellular ion activities take place. An example of validation of separate impalements by changing bathing-solution composition is depicted in Figure 7.3.

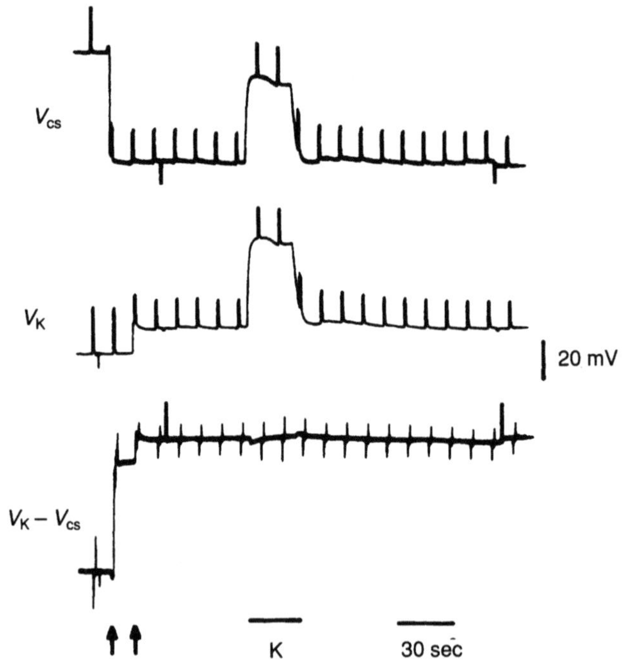

Figure 7.3 Validation of separate impalements with conventional and K^+-sensitive microelectrodes. Times of impalements indicated by arrows. During the period indicated by the bar, $[K^+]$ was increased from 2.5 to 92.5 mM. V_{cs} = basolateral membrane voltage; V_K = voltage recorded by the K^+-sensitive microelectrode; $V_K - V_{cs}$ = differential trace. The steady-state depolarization elicited by high K^+ was virtually the same in both cells.

7.3.2 Calculations of intracellular ion activities from ion-sensitive microelectrode data

Intracellular ion activities can be calculated using the Nicolsky–Eisenman equation:

$$V_i^* = E_o + S \log (a_i + k_{ij}\, a_j) \qquad 7.2$$

where V_i^* is the difference between the voltage output of the ion-sensitive microelectrode and the cell membrane voltage, E_o is a potential that depends on the temperature and includes the junction and reference electrode potentials, S is the slope of the electrode (ideally 2.3 RT/zF), a_i and a_j are the activities of the ions i and j, and k_{ij} is the selectivity coefficient for j over i.

The Nicolsky–Eisenman equation is employed to calculate intracellular ion activities when S and k_{ij} are constant under the experimental conditions. If k_{ij} is negligible, then equation (7.2) reduces to the Nernst equation. For example, the Nernst equation can be used to calculate pH with H^+-sensitive microelectrodes, due to the negligible interference by other ions. On the other hand, under many biologically relevant conditions, S and/or k_{ij} are not constant for some electrodes (i.e. Na^+- and quaternary ammonium-sensitive microelectrodes), and equation (7.2) does not describe their behavior adequately. In these instances, empirical calibrations with solutions mimicking the intracellular milieu are recommended. Details on the determination of S and k_{ij} and procedures for empirical calibrations can be found elsewhere (Ammann, 1986; Altenberg *et al*, 1990).

7.4 MEASUREMENTS OF INTRACELLULAR ION ACTIVITIES WITH ION-SENSITIVE FLUORESCENT DYES

The use of ion-sensitive fluorescent dyes to measure intracellular ion activities has exploded during the last decade. In principle, intracellular Na^+, K^+, Cl^-, Ca^{2+}, Mg^{2+} and H^+, as well as other less frequently measured ions, can be determined with ion-sensitive fluorescent dyes (Table 7.2). Ion-sensitive fluorescent dyes can also be used to measure changes in cell-water volume (Chapter 8).

7.4.1 General properties of ion-sensitive dyes

Ion-sensitive fluorescent dyes share many of the advantages of ion-sensitive microelectrodes, including continuous recording, high time resolution and adequate sensitivity. Because of their rapid response to changes in intracellular ion activities, ion-sensitive fluorescent dyes allow for identification of specific ion transport mechanisms at the single membrane level (see above). In fact, the measurement of intracellular ion activities with ion-sensitive fluorescent dyes is the method of choice in most cases because, while ion-sensitive microelectrodes are accurate and sensitive for most intracellular ion activity measurements, their use requires extensive training, and fluorescent indicators are considerably easier to use. In addition, microelectrodes are invasive, and in practice their use is limited to relatively large cells. An advance that facilitated the use of ion-sensitive fluorescent dyes was the development of cell-permeable hydrophobic acetate esters of aromatic alcohols or acetoxymethylesters of carboxylic acids. Once in the cells, de-esterification is accomplished by esterases, which results in release of the ion-sensitive fluorescent dye. In most cases, the free forms of the dyes permeate cell membranes very slowly. Cell-permeable compounds of the indicators Fura–2, Quin–2, Fluo–3, BCECF, SNARF–1, SBFI and others are now commercially available. In cases where cells cannot be easily loaded with the cell-permeable forms of the dyes, cell-impermeable compounds can still be loaded by a variety of methods: electroporation, gravity loading, scrape loading, microinjection and ATP permeabilization (McNeil, 1989). One of these loading methods is also required when using dextran-bound ion-sensitive fluorescent dyes to prevent compartmentalization (Bright *et al.*, 1989; see below).

7.4.2 Ratiometric and non-ratiometric ion-sensitive dyes

Ratiometric dyes are currently the most frequently used ion-sensitive fluorescent dyes. Binding to the specific ion produces a spectral shift in excitation and/or emission spectra. Examples of these dyes are Fura–2 and BCECF, which are Ca^{2+}- and pH-sensitive, respectively. Figure 7.4a shows the fluorescence excitation spectrum of Fura–2, measured in a cuvette. The emission intensity changes with free [Ca^{2+}] in such a way that the ratio of emitted light at excitation wavelengths 340 and 380 nm increases with elevation in free [Ca^{2+}]. The main advantage of ratiometric dyes such as Fura–2 is that ratio measurements are independent of dye concentration, dye leak, photobleaching (irreversible destruction of the excited fluorophore), changes in cell volume, thickness of the sample and, in principle, all factors that do not affect the spectra or ion-binding properties of the dye. The factors above will modify the fluorescence emission at all wavelengths, and will have no effect on the ratiometric determinations. For comparison, Figure 7.4b shows the response of the non-ratiometric fluorescent Ca^{2+} indicator Fluo–3 to changes in free [Ca^{2+}]. For this dye, there is no excitation or emission spectral shift upon binding Ca^{2+}.

The advantage of ratiometric over non-ratiometric measurements is further illustrated *in vivo* in Figure 7.5, which shows the absence of effect of the continuous decrease in BCECF fluorescence (in this case produced by photobleaching) on the calculated pH_i. The emitted light intensities at both 440 and 495 nm excitation decrease continually as a function of time,

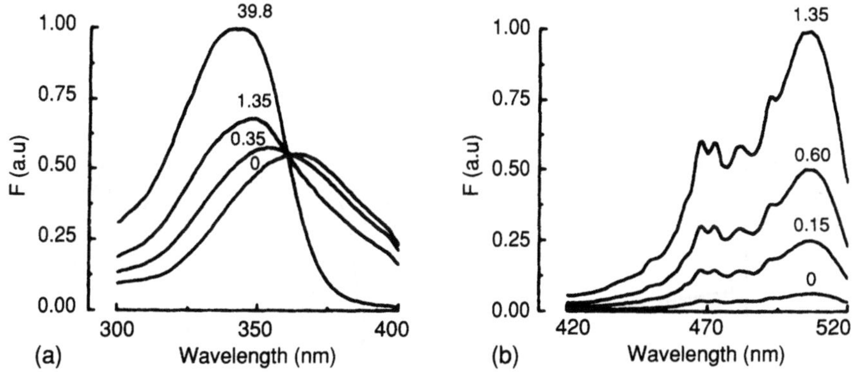

Figure 7.4 (a) Fura-2 and (b) Fluo-3 excitation spectra at different free [Ca^{2+}] *in vitro*. Measurements were carried out in cuvettes using a spectrofluorometer. Fluorescence emission was measured at 510 and 530 nm, respectively. Numbers indicate values of free [Ca^{2+}] in micromoles. Composition of the solutions (containing 1 mM Mg^{2+}) was taken from Grynkiewicz *et al.* (1985).

but the ratio, and hence the calculated pH_i, do not change. Although photobleaching does not introduce errors in ratiometric determinations, breakdown of fluorescence dyes by light can release toxic compounds (Negulescu and Machen, 1990). Thus, it is always advisable to minimize excitation light and to use the lowest possible dye concentrations, depending on experimental requirements (cell type, single cells vs. cell populations, cuvette vs. microscopy experiments, detection method).

In the case of non-ratiometric ion-sensitive fluorescent dyes (single excitation, single emission), decreases in fluorescence due to photobleaching and/or leakage will influence calculations of intracellular ion activities. Examples of non-ratiometric dyes are the Cl^--sensitive dye SPQ and the Ca^{2+}-sensitive dye Fluo–3. For these dyes, the ion concentration is derived from the Stern–Volmer equation:

$$F_0/F = 1 + K\,(1/[Q]) \qquad (7.3)$$

where F_0 and F are the fluorescence intensities in the absence and presence of the quencher, K is the Stern–Volmer quenching constant, and $[Q]$ is the concentration of the quencher, e.g. free [Ca^{2+}] for Fluo–3. **Fluorescence quenching** refers to decrease in fluorophore emission.

Changes in intracellular ion activities change fluorescence intensity of non-ratiometric dyes without spectral shifts. Therefore, a continuous decrease in fluorescence due to photobleaching and/or leakage could be interpreted as a change in intracellular free concentration (increase in intracellular free [Cl^-] with SPQ and decrease in intracellular free [Ca^{2+}] with Fluo–3, respectively). The implication is that for non-ratiometric

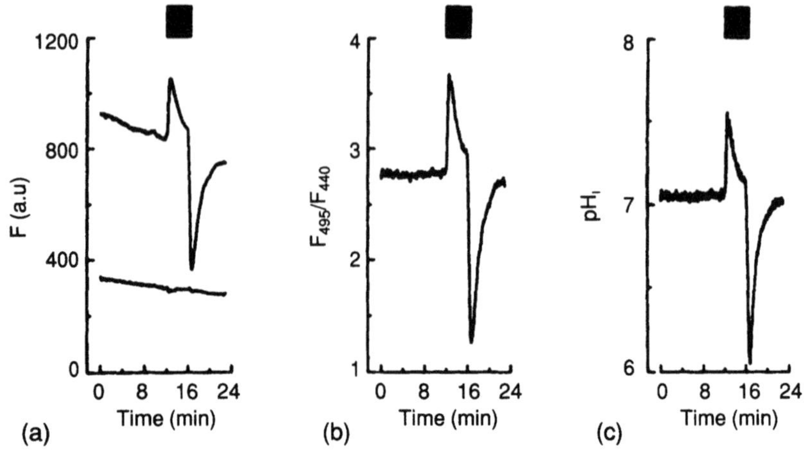

Figure 7.5 Example of ratiometric determination of pH_i with BCECF in a human mammary epithelial cell: (a) fluorescence emission at 495 (top trace) and 440 nm (bottom trace) excitation wavelengths; (b) ratio of the traces in (a); (c) estimation of pH_i from the ratiometric fluorescence measurement in (b) and *in situ* calibration (not shown; see Figure 7.7 and text). The bars denote exposure to 25 mM NH_4Cl, a maneuver known to acid-load most cells. Data were obtained from a monolayer of 150–200 cells mounted in the stage of an inverted microscope. Emitted fluorescence was measured at 535 nm.

dyes it is essential to minimize and correct for decreases in fluorescence brought about by mechanisms independent of changes in intracellular ion activity.

7.4.3 Pitfalls in the use of ion-sensitive dyes

Potential pitfalls in the use of ion-sensitive fluorescent dyes are intracellular compartmentalization, photobleaching and dye leakage. The determination of the intracellular distribution of the dye is of fundamental importance because subcellular compartments are not homogeneous in regard to ion activities. For instance, pH is different in cytoplasm, lysosomes and mitochondria, and the fluorescence emitted by a pH-sensitive dye will then vary according to its distribution in the different compartments.

Dye compartmentalization can be evaluated by a variety of techniques, but it is still one of the major drawbacks in the use of ion-sensitive fluorescent dyes. Evaluation methods include:

- Direct visualization and photography or conventional imaging, with or without co-localization studies using dyes of known intracellular distribution (Harootunian *et al.*, 1989).

- Semiquantitative estimation of compartmentalized dye by measuring dye release upon selective permeabilization of the plasma membrane with digitonin or alpha-toxin (Harootunian *et al.*, 1989; Crowe *et al.*, 1995).
- Confocal fluorescence microscopy (Kurtz and Emmons, 1993).

The latter method requires expensive instrumentation but allows both direct measurement of intracellular inhomogeneities, and assessment of cytosolic fluorescence even if there is substantial compartmentalization of the ion-sensitive fluorescent dye. Although the mechanisms are not well understood, this compartmentalization can be minimized by using the non-ionic detergent Pluronic F-127, loading the cell at room temperature (instead of 37°C) and in the absence of serum, cooling the cells to 4°C before exposure to acetoxymethylesters, or using blockers of organic anion transport.

The degrees of photobleaching and leakage vary considerably among ion-sensitive fluorophores. If ratiometric dyes are employed, the decrease in fluorescence by photobleaching or leakage has no effect on the calculation of intracellular ion activities because the fraction of change in the concentration of ion-free and ion-bound dye should be the same.

If photobleaching is a problem, then the frequency of the measurements should be reduced to a minimum, avoiding exposure of the dye to excitation light between measurements. This is routinely accomplished when fluorescence imaging systems are used. However, image averaging is frequently necessary. With non-imaging quantitative fluorescence systems based on photomultiplier detection, dye exposure to low-intensity excitation light can be maintained to 1/10 of a second or less in many conditions, using computer-driven shutters. Unfortunately, in most commercial instruments there is continuous light exposure independently of the frequency of the measurements. Cells studied in a cuvette experience less photobleaching than cells studied under a microscope. In cuvette experiments fluorescence is measured from a large number of cells (e.g. 10^6 cells) and thus the intensity of the excitation light can be reduced. In fluorescence microscopy, the high numerical aperture objectives needed because of the low intensity of the emitted light (i.e. single-cell studies), focus the excitation light in a small area and therefore increase photobleaching. Nevertheless, fluorescence microscopy studies have several advantages:

- Measurements in different cell types in the same epithelium.
- Direct evaluation of dye-compartmentalization artifacts.
- Simultaneous electrophysiological measurements.
- Rapid and easy changes in solution composition.

For some dyes, leakage is a problem that can be solved or minimized.

For example, leak of anionic ion-sensitive fluorescent dyes mediated by organic anion transporters can be blocked by addition of competing organic anions such as probenecid or sulfinpyrazone (Di Virgilio *et al.*, 1990). In any case, corrections are needed to estimate intracellular ion activities accurately when non-ratiometric dyes are used. Depending on the speed of the fluorescence decay produced by photobleaching or leakage of the dye, linear or exponential corrections can be employed (Chao *et al.*, 1990; Altenberg *et al.*, 1994).

Figure 7.6 illustrates these points by depicting the continuous decrease in fluorescence of the Cl^--sensitive dye SPQ. Interaction of this dye with Cl^- reduces fluorescence emission by collisional quenching (elevations in

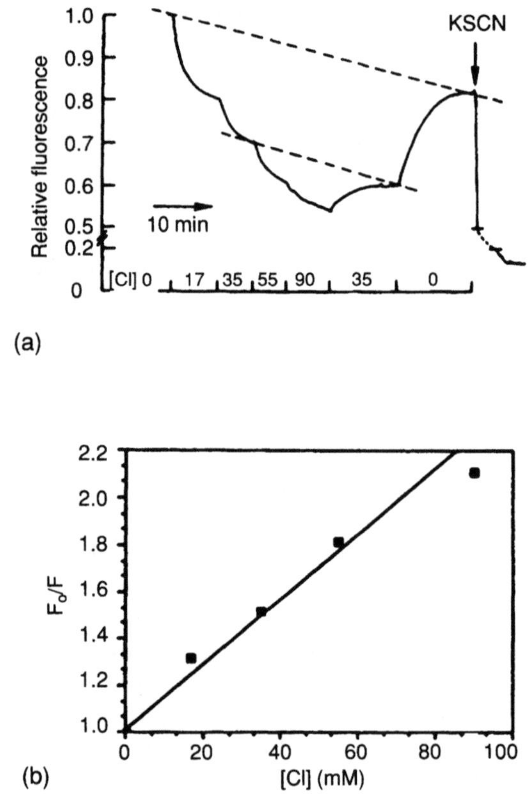

Figure 7.6 Example of SPQ calibration *in situ.* (a) Monolayers of primary cultures of tracheal epithelial cells were superfused with solutions containing high $[K^+]$, varying $[Cl^-]$ (indicated on x-axis in millimoles), and the ionophores nigericin (5 μM) and tributyltin (10 μM). Dashed lines represent fall in SPQ fluorescence due to leakage. (b) Stern–Volmer plot of data from (a). F_o/F is SPQ fluorescence in the absence of Cl^- divided by that in the presence of Cl^-. The fitted Stern–Volmer constant was 13 M^{-1}. (Reproduced from Chao *et al.*, 1990, with permission.)

fluorescence indicate decreases in intracellular free [Cl^-]). In the example shown (from Chao *et al.*, 1990), the continuous decrease in fluorescence under constant experimental conditions is linear, allowing for a simple correction.

7.4.4 Calculations of intracellular ion activities from ion-sensitive fluorescent-dye data

Calibrations of ion-sensitive fluorescent dyes *in situ* are preferred to *in vitro*. For *in situ* calibration, ionophores or plasma-membrane permeabilizing agents (e.g. digitonin at low concentration) are used to alter the intracellular free-ion concentration and the change in the fluorescence signal is measured. Specific procedures depend on the ion under study (Malgaroli *et al.*, 1987; Blinks, 1989; Chao *et al.*, 1990; Negulescu and Machen, 1990; Williams and Fay, 1990; Hurley *et al.*, 1992; Kasner and Ganz, 1992). Here, we will briefly discuss the two most frequently used approaches: direct calibration at the end of the experiment; and assessment of the fluorescence signal at one ion concentration and calculation of experimental values from a complete calibration curve obtained in other cells. The first approach is most frequently employed when using the pH-sensitive dye BCECF, and the second one is commonly used for the Cl^- and Ca^{2+}-sensitive dyes SPQ and Fura–2, respectively.

Figure 7.7 illustrates a typical calibration in BCECF-loaded cells using the high-K^+/nigericin technique (Thomas *et al.*, 1979; Chaillet and Boron, 1985). The calibration solutions have different pH values and contain a high [K^+], ideally identical to the intracellular [K^+], and the ionophore nigericin, which under these experimental conditions acts as a K^+/H^+ exchanger. At equilibrium, if the incorporation of nigericin in the membranes is sufficient, the transmembrane K^+ and H^+ gradients will be the same; $[K^+]_o/[K^+]_i = [H^+]_o/[H^+]_i$, and hence if $[K^+]_o = [K^+]_i$, then $pH_o = pH_i$ (subscripts o and i denote extracellular and intracellular compartments, respectively). Figure 7.7 shows that changes in extracellular pH are accompanied by rapid changes in the ratio of emitted fluorescence at 440 and 495 nm excitation. From this kind of experiment, a calibration curve (fluorescence ratio vs. pH) is constructed, and employed to convert experimental determinations of fluorescence ratio values to pH_i. The assumption is that the high-K^+/nigericin calibration solutions equilibrate the pH across the cell membranes. This calibration seems to be appropriate to assess changes in pH_i, but is somewhat questionable for determination of absolute pH_i values. In the latter case, it is advisable to confirm the pH_i determinations using a different method (G. Boyarsky, personal communication).

Figure 7.6 depicts a calibration of SPQ using the tributyltin/nigericin technique (Chao *et al.*, 1990). The calibration solutions have the same pH

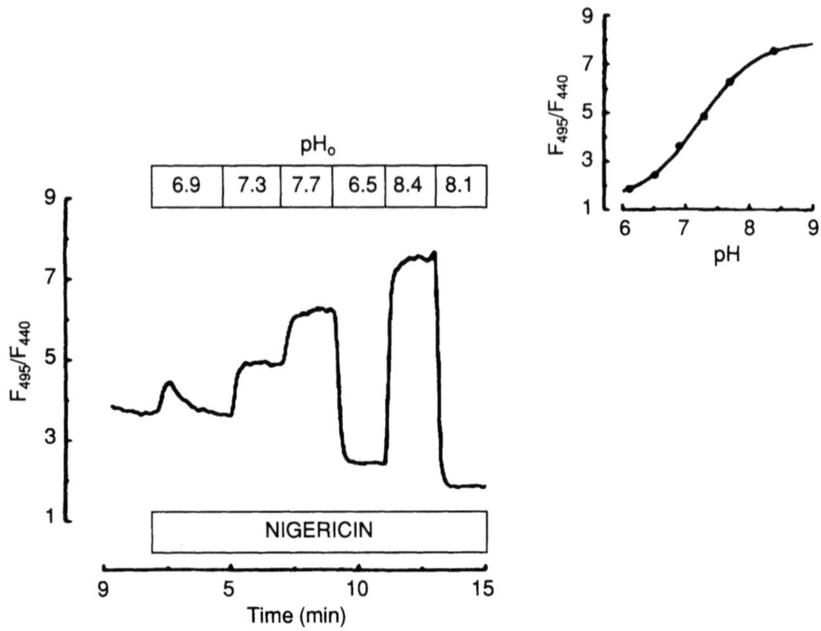

Figure 7.7 Example of calibration of BCECF *in situ* in human breast cancer cells. Nigericin was used at at a concentration of 10 μM. pH_o = pH of the superfusate. Data are expressed as ratio of measured fluorescence intensity at excitation wavelengths 495 and 440 nm (F_{495}/F_{440}). Initial elevation in F_{495}/F_{440} is due to removal of HCO_3^-/CO_2 (replaced with Hepes). See Figure 7.5 for additional details.

and high [K^+], and contain variable concentrations of Cl^- and two ionophores, nigericin and tributyltin, which perform K^+/H^+ and Cl^-/OH^- exchange, respectively. In this two-ionophore calibration, intracellular and extracellular [Cl^-] are equilibrated by tributyltin if the pH changes induced by the same ionophore are in turn short-circuited by nigericin (see BCECF calibration above). Since SPQ leaks from the cells, appropriate corrections are needed, which makes it impractical to perform calibrations in the same cell(s) used for the experiments. Calibration data are generally obtained from several cells and average values are used to convert experimental fluorescence intensities to $[Cl]_i$, using at least one reference point obtained in each specific experiment. Most investigators determine the maximal fluorescence obtained in the absence of Cl^-, and the background fluorescence obtained by quenching SPQ with SCN^-. SCN^- is more efficient than Cl^- in quenching SPQ (the SCN^- Stern–Volmer constant *in vitro* is ≈218 M^{-1} vs. ≈111 M^{-1} for Cl^-).

Calibration procedures for Ca^{2+}-sensitive dyes are similar to the one for SPQ, i.e. one or two reference points are obtained for each specific experiment, but complete calibrations are not performed in the cells after the experiment (Negulescu and Machen, 1990).

7.4.5 Measurements of intracellular free Ca^{2+} with Ca^{2+}-sensitive photoproteins

Aequorin is the most frequently employed Ca^{2+}-sensitive photoprotein. For details on the origin and properties of the protein, as well as technical aspects of the measurements, see Blinks (1989) and Borle (1994). The main advantages of the use of aequorin are its high sensitivity and absence of Ca^{2+}-buffering effects. Aequorin has a Ca^{2+} apparent K_m of 10 μM and has a high sensitivity in the physiological $[Ca^{2+}]_i$ range. Since it can be used at low concentrations, an advantage of aequorin is its negligible Ca^{2+}-buffering effect. In contrast, Ca^{2+} buffering can be an important problem with Fura–2 and other Ca^{2+}-sensitive fluorescent dyes. On the other hand, the use of aequorin has several disadvantages. The most important is that because the cell membrane is impermeable to the protein, it is necessary to load the cell by transient permeabilization or microinjection. In addition, the reaction of aequorin with Ca^{2+} is irreversible.

In conclusion, continuous measurements of intracellular ion activities with adequate time resolution are needed to identify the mechanisms of transport of ions across the apical and basolateral membranes of epithelial cells and their regulation. The available methods include ion-sensitive microelectrodes and ion-sensitive indicators. Intracellular microelectrode techniques have good time-resolution, are accurate and provide cell membrane voltage information that can be very useful. However, microelectrode techniques are difficult in small mammalian cells. Ion-sensitive indicators also have good time resolution, and although in most cases are not as accurate as ion-sensitive microelectrodes, they are in most instances the method of choice because they are easy to use. Continuing developments in optical techniques as well as synthesis of new indicators will certainly bring improvements in optical measurement of intracellular ion activities.

REFERENCES

Altenberg, G., Copello, J., Cotton, C. *et al.* (1990) Electrophysiological methods for studying ion and water transport in *Necturus* gall bladder epithelium. *Methods in Enzymology*, **192**:650–82.

Altenberg, G.A., Deitmer, J.W., Glass, D.C. *et al.* (1994) P-glycoprotein-associated Cl^- currents are activated by cell swelling but do not contribute to cell volume regulation. *Cancer Research*, **54**:618–22.

Ammann, D. (1986) *Ion-selective Microelectrodes*, Springer-Verlag, Berlin.

Ammann, D., Bührer, T., Schefer, U. *et al.* (1987) Intracellular neutral carrier-based

Ca^{2+} microelectrode with subnanomolar detection limit. *Pflügers Archiv*, **409**:223–28.

Beck, F.-X., Dorge, A., Rick, R. *et al.* (1987) Effect of potassium adaptation on the distribution of potassium, sodium and chloride across the apical membrane of renal tubular cells. *Pflügers Archiv*, **409**:477–85.

Blinks, J.R. (1989) Use of calcium-regulated photoproteins as intracellular Ca^{2+} indicators. *Methods in Enzymology*, **172**:164–202.

Borle, A.B. (1994) Ca^{2+}-bioluminescent indicators. *Methods in Toxicology*, **18**:315–327.

Bright, J.R., Whitaker, J.E., Haugland, R.P. *et al.* (1989) Heterogeneity of the changes in cytoplasmic pH upon serum stimulation in quiescent fibroblasts. *Journal of Cellular Physiology*, **141**:410–19.

Buckler, K.J. and Vaughan-Jones, R.D. (1990) Application of a new pH-sensitive fluoroprobe (carboxy-SNARF-1) for intracellular pH measurement in small, isolated cells. *Pflügers Archiv*, **417**:234–9.

Chaillet, J.R. and Boron, W.F. (1985) Intracellular calibration of a pH-sensitive dye in isolated, perfused salamander proximal tubules. *Journal of General Physiology*, **86**:765–94.

Chao, A.C., Widdicombe, J.H. and Verkman, A.S. (1990) Chloride conductive and cotransport mechanisms in cultures of canine tracheal epithelial cells measured by an entrapped fluorescent indicator. *Journal of Membrane Biology*, **113**: 193–202.

Crowe, W.E., Altamirano, J., Huerto, L. and Alvarez-Leefmans, *J.* (1995) Calcium-dependent volume changes in single N1E-115 neuroblastoma cells measured with a fluorescent probe. *Neuroscience*, in press.

Di Virgilio, F., Steinberg, T.H. and Silverstein, S.C. (1990) Inhibition of Fura-2 sequestration and secretion with organic anion transport blockers. *Cell Calcium*, **11**:57–62.

Gillis, J.M. and Gailly, Ph. (1994) Measurement of $[Ca^{2+}]_i$ with diffusible Fura-2 AM: Can some potential pitfalls be evaluated? *Biophysical Journal*, **67**:476–7.

Grynkiewicz, G., Poenie, M. and Tsien, R.Y. (1985) A new generation of Ca^{2+} indicators with greatly improved fluorescence properties. *Journal of Biological Chemistry*, **260**:3440–50.

Hansen, L.L., Rasmussen, J., Friche, E. *et al.* (1993) Method for determination of intracellular sodium in perfused cancer cells by ^{23}Na nuclear magnetic resonance spectroscopy. *Analytical Chemistry*, **214**:506–10.

Harootunian, A.T., Kao, J.P.Y., Eckert, B.K. *et al.* (1989) Fluorescence ratio imaging of cytosolic free Na^+ in individual fibroblasts and lymphocytes. *Journal of Biological Chemistry*, **264**:19458–67.

Hofer, A.M. and Machen, T.E. (1993) Technique for *in situ* measurement of calcium in intracellular inositol 1,4,5-trisphosphate-sensitive stores using the fluorescent indicator mag-fura-2. *Proceedings of the National Academy of Sciences*, **90**:2598–602.

Hu, Z., Bührer, T., Müller, M. *et al.* (1989) Intracellular magnesium ion selective microelectrode based on a neutral carrier. *Analytical Chemistry*, **61**:574–6.

Hurley, T.W., Ryan, M.P. and Brinck, R.W. (1992) Changes of cytosolic Ca^{2+} interfere with measurements of cytosolic Mg^{2+} using mag-fura-2. *American Journal of Physiology*, **263**:C300–C307.

Illner, H., McGuigan, J.A.S. and Lüthi, D. (1992) Evaluation of mag-fura-2, the new fluorescent indicator for free magnesium measurements. *Pflügers Archiv*, **422**:179–84.

Kao, J.P.Y., Harootunian, A.T. and Tsien, R.W. (1989) Photochemically generated cytosolic calcium pulses and their detection by Fluo–3. *Journal of Biological Chemistry* **264**:8179–84.

Kasner, S.E. and Ganz, M.B. (1992) Regulation of intracellular potassium in mesangial cells: a fluorescence analysis using the dye, PBFI. *American Journal of Physiology*, **262**:F462-F467.

Kondo, Y., Bührer, T., Seiler, K. *et al.* (1989) A new double-barrelled, ionophore-based microelectrode for chloride ions. *Pflügers Archiv*, **414**:663–8.

Kurtz, I. and Emmons, C. (1993) Cell biological applications of confocal microscopy. *Methods in Cell Biology*, **38**:183–93.

Lanter, F., Erne, D., Ammann, D. *et al.* (1980) Neutral carrier based ion-selective electrode for intracellular magnesium activity studies. *Analytical Chemistry*, **52**:2400–2.

Lewis, S.A., and Wills, N.K. (1980) Resistive artifacts in liquid-ion exchanger microelectrode estimates of Na activity in epithelial cells. *Biophysical Journal*, **31**:127–38.

Loew, L.M. (1993) Confocal microscopy of potentiometric fluorescent dyes. *Methods in Cell Biology*, **38**:195–209.

MacKnight, A.D.C. and Leader, J.P. (1989) Volume regulation in epithelia: experimental approaches. *Methods in Enzymology*, **17**:744–90.

Malgaroli, A., Milani, D., Meldolesi, J. *et al.* (1987) Fura–2 measurement of cytosolic free Ca^{2+} in monolayers and suspensions of various types of animal cells. *Journal of Cell Biology*, **105**:2145–55.

McNeil, P.L. (1989) Incorporation of macromolecules into living cells. *Methods in Cell Biology*, **29**:153–73.

Minta, A., Kao, J.P. and Tsien, R.Y. (1989) Fluorescent indicators for cytosolic calcium based on rhodamine and fluorescein chromophores. *Journal of Biological Chemistry*, **264**:8171–8.

Negulescu, P.A. and Machen, T.E. (1990) Intracellular ion activities and membrane transport in parietal cells measured with fluorescent dyes. *Methods in Enzymology*, **192**:38–81.

Owen, C.S. (1992) Comparison of spectrum-shifting intracellular pH probes 5′(and 6′)-carboxy-10-dimethylamino-3-hydroxyspiro[7H-benzo[c]xanthene-7, 1′(3′H)-isobenzofuran]-3′-one and 2′,7′-biscarboxyethyl-5(and 6)-carboxyfluorescein. *Analytical Biochemistry*, **204**:65–71.

Owen, C.S. (1993) Simultaneous measurement of two cations with fluorescent dye Indo-1. *Analytical Biochemistry*, **215**:90–5.

Reuss, L. (1984) Independence of apical membrane Na^+ and Cl^- entry in *Necturus* gallbladder epithelium. *Journal of General Physiology*, **84**:423–45.

Reuss, L. (1987) Cyclic AMP inhibits Cl^-/HCO_3 exchange at the apical membrane of *Necturus* gallbladder epithelium. Journal of General Physiology, **90**:173–96.

Reuss, L. (1991) Salt and water transport by gallbladder epithelium, in *Handbook of Physiology. The Gastrointestinal System*, vol. IV. (ed. American Physiological Society). Oxford University Press, New York, pp. 303–22.

Reuss, L. and Altenberg, G.A. (1995) cAMP-activated Cl^- channels: regulatory role in gallbladder and other absorptive epithelia. *News in Physiological Sciences*, **10**:86–91.

Rick, R. (1994) pH_i determines rate of sodium transport in frog skin: results of a new method to determine pH_i. *American Journal of Physiology*, **266**:F367-F374.

Rizzuto, R., Simpson, A.W.M., Brini, M. *et al.* (1992) Rapid changes of mitochondrial Ca^{2+} revealed by specifically targeted recombinant aequorin. *Nature*, **358**:325–7.

Robinson, R.A., and Stokes, R.H. (1959) *Electrolyte Solutions*, Butterworths, London.

Thomas, J.R., Buchsbaum, R.N., Zimniak, A. *et al.* (1979) Intracellular pH measurements in Ehrlich ascites tumor cells utilizing spectroscopic probes generated in situ. *Biochemistry*, **81**:2210–18.

Tsien, R.Y. and Rink, T.J. (1980) Neutral carrier ion-selective microelectrodes for measurement of intracellular free calcium. *Biochimica et Biophysica Acta*, **599**:623–38.

Vasseur, M., Fragne, R. and Alvarado, F. (1993) Buffer-dependent pH sensitivity of the fluorescent chloride-indicator dye SPQ. *American Journal of Physiology*, **264**:C27–C31.

Verkman, A.S. (1990) Development and biological applications of chloride-sensitive fluorescent indicators. *American Journal of Physiology* **259**:C375–388.

Verkman, A.S., Chao, A.C. and Hartmann, T. (1992) Hormonal regulation of Cl transport in polar airway epithelia measured by a fluorescent indicator. *American Journal of Physiology*, **262**:C23–C31.

Verkman, A.S., Sellers, M.C., Chao, A.C. *et al.* (1989) Synthesis and characterization of improved chloride-sensitive fluorescent indicators for biological applications. *Analytical Biochemistry*, **178**:355–361.

Waggoner, A.S. (1979) Dye indicators of membrane potential. *Annual Review of Biophysics and Bioengineering*, **8**:47–68.

Williams, D.A. and Fay, F.S. (1990) Intracellular calibration of the fluorescent calcium indicator Fura–2. *Cell Calcium*, **11**:75–83.

8

Measurement of epithelial cell volume

Luis Reuss, Calvin U. Cotton and Guillermo A. Altenberg

8.1 WATER PERMEATION AND CELL VOLUME

Cell volume measurements are necessary to understand a variety of problems pertaining to epithelial transport, namely: cell volume maintenance and regulation; mechanisms and regulation of water transport at the single-cell and single-membrane levels; mechanisms, modulation and regulation of ion transport processes (because net ion fluxes frequently result in net water fluxes); and mechanisms of intracellular signalling and cross-talk between apical and basolateral membranes of epithelial cells.

8.1.1 Water permeates cell membranes by diffusion and osmosis

Water can move across animal cell membranes by two basic processes: diffusion and osmosis. (Filtration does not play a significant role because of the low prevailing hydrostatic pressure differences.)

Diffusion can be understood at the molecular level as a process of random walk of the diffusing molecules at rates proportional to the chemical potential. At a constant temperature, the water chemical potential

Epithelial Transport: A guide to methods and experimental analysis.
Edited by Nancy K. Wills, Luis Reuss and Simon A. Lewis.
Published in 1996 by Chapman & Hall, London. ISBN 978-94-010-7168-0

depends on the water concentration, which is inversely proportional to the solute concentration. If two aqueous solutions (1 and 2) are separated by a membrane that is permeable to water (diffusive permeability = P_d), there will be water diffusion across the membrane, i.e. water molecules will cross from side 1 to side 2 and vice versa. The unidirectional fluxes will be proportional to the concentrations of water in the two compartments, i.e. if the water concentrations are the same on both sides, then the unidirectional diffusive water fluxes will be identical and the net flux will be zero.

Osmotic water flow across a membrane is also driven by the water chemical potential, in this case conferred by differences in hydrostatic pressure and/or osmotic pressure, the latter dependent on the water chemical activities on both sides of the membrane. Water flows by filtration (hydrostatic pressure difference) or osmosis (osmotic pressure difference). A difference in osmotic pressure across a membrane or at the mouth of a membrane pore indicates that the solution with higher solute concentration has a lower water concentration, and hence a lower water chemical potential ('fewer' water molecules can 'collide' with the membrane). For an insightful discussion of the hydrostatic equivalent of osmotic pressure, see Mauro (1957).

(a) Reflection coefficients and solution osmolality versus tonicity

A classic experiment in biology is to transfer cells (e.g. red blood cells) from normal plasma or serum to solutions of identical osmolalities but different composition, for instance NaCl, urea or sucrose solutions, all 300 $mosmol \cdot kg^{-1}$. The result of the experiment is that there is no appreciable change in cell volume in the NaCl solution, cell swelling in the urea solution and cell shrinkage in the sucrose solution. The explanation is that the solute reflection coefficients (σ_s) (which denote the relationship of water permeability and solute permeability for a given membrane) are different for the three solutes. Sucrose is impermeable, NaCl is moderately permeable and urea is very permeable. In practice, the value of σ can range between 0 (solute permeability = water permeability) to 1 (solute permeability = 0); σ can be less than zero if the solute permeability exceeds the water permeability. The equation describing osmotic volume flow can thus be written (for a single solute) as:

$$J_v = L_p\,(\Delta P - \sigma_s \cdot \Delta\pi) \tag{8.1}$$

where J_v = volume flow ($cm^3 \cdot cm^{-2} \cdot s^{-1} = cm \cdot s^{-1}$), L_p = hydraulic permeability coefficient ($cm^3 \cdot cm^{-2} \cdot s^{-1} \cdot atm^{-1} = cm \cdot s^{-1} \cdot atm^{-1}$), ΔP = difference in hydrostatic pressure (atm), σ_s = solute reflection coefficient (adimensional), and $\Delta\pi$ = difference in osmotic pressure (atm). The L_p is related to the osmotic water permeability (P_f or P_{os}) according to $P_f = L_p(RT/\bar{V}_w)$, where R and T have their usual meanings. (The comparison of P_f and P_d is an important biophysical criterion to demonstrate water transport via membrane pores (Whittembury and Carpi-Medina, 1988).) The term $\sigma_s.\Delta\pi$ is sometimes referred to as the **effective osmotic pressure difference**. From the above example, the NaCl solution is **isotonic** with the cell, whereas the sucrose solution is **hypertonic** and the urea solution is **hypotonic**, although all three solutions are **isosmotic**.

(b) *Volume flow and water flow*

Water flow (J_w) and volume flow (J_v) across a membrane are not necessarily the same. Consider the case of a membrane separating two solutions containing a single solute, at the same concentration (C_s) on both sides. Now C_s is suddenly increased on one side (side 1). If the membrane is impermeable to the solute ($\sigma_s = 1$), there will be a net water flux from side 2 to side 1 and $J_w = J_v$, since there is no solute flux. However, if $\sigma_s = 0$, then after adding solute to side 1 there will be:

1. a reduction in the unidirectional water flux from side 1 to side 2 [$J_w(1,2)$], because of the decrease in water concentration brought about by solute addition; and
2. an increase in solute flux in the same direction [$J_s(1,2)$].

Hence, there will be **net fluxes** of water towards compartment 1 and of solute towards compartment 2, minimizing the net volume flow between the compartments. Clearly, J_w and J_v will be different. This notion, i.e. that volume flow and water flow are not necessarily the same, is very important in designing correct experiments to study water transport, cell volume and its regulation.

(c) *Basics of water permeability of epithelial cell membranes*

The osmotic water permeability (P_f or P_{os}) of epithelial cell membranes varies greatly, from near zero (apical membranes of thick-ascending loop of Henle and of distal and collecting tubule cells in absence of antidiuretic hormone) to 400–600 $\mu m \cdot s^{-1}$ in proximal renal tubules. In unmodified lipid bilayers, P_f values range from < 1 to ca. 100 $\mu m \cdot s^{-1}$. Membranes of epithelial cells usually have P_f values near the high range of lipid-bilayer P_f, but in some instances they deviate considerably from this range (Tripathi and Boulpaep, 1989). Recent studies indicate that the very high P_f values of some epithelial cell membranes is due to pores (aquaporin family: Agre *et al.*, 1993; Engel *et al.*, 1994; and Chapter 2). These pores are highly selective for water. The molecular bases for the very low P_f values of other epithelial cell membranes are unknown.

8.1.2 Cell water and cell volume: basic concepts

The physical state of intracellular water and ions is not fully understood and remains an important problem in cell physiology. Evidence accumulated in the last few decades, from nuclear magnetic resonance spectroscopy, intracellular microelectrode techniques and quantitative fluorescent microscopy, supports the general view that cell water behaves like bulk water and that the activity coefficients of intracellular monovalent ions are close to the values predicted from the ionic strength, i.e. that these ions are quite mobile in the cytoplasm and distribute in total cell water volume. In contrast, divalent cations are compartmentalized and/or bound with relatively high affinity to intracellular molecules.

Regardless of the fact that intracellular water largely behaves as free-solution water, the cell cannot be modeled as a bag filled with a dilute solution. The cytoskeleton and the high concentration of macromolecules confer to the cytoplasm properties that are closer to those of gels than to those of fluids such as dilute salt solutions (e.g. Lechène, 1985). A recent hypothesis to explain cell volume regulation is based on the idea of macromolecular crowding, i.e. that the chemical activity of specific macromolecules is highly dependent on the total concentration of all macromolecules, via spatial restrictions (Minton, 1992; Minton *et al.*, 1992; Parker, 1993). Our discussion is based on the tenet that the cell membrane is more important than the binding properties of cytoplasmic components in determining fluxes and differences in composition between intra- and extracellular fluids.

(a) Cell-water volume and cell volume are not exactly the same

All cells contain a finite amount of solids, i.e. the amount of matter remaining after the cells are completely dehydrated. Cell solids amount on average to about 20% of cell mass or of cell volume, i.e. the remaining 80% or so is cell water, which in most cases is exchangeable with extracellular water. Some of the methods discussed below measure cell volume, while others measure cell-water volume. It is essential to keep this distinction in mind.

(b) The ideal response of cells to changes in external osmolality is osmometric

Consider the case of a cell initially at osmotic equilibrium with the extracellular solution (i.e. the driving force for osmotic water flow is zero). The solutes present are impermeant ($\sigma_s = 1$), there are no net solute fluxes during the experiment and the cell cannot regulate its volume. Now, we suddenly change the osmolality of the external solution and observe the change in cell volume until a new steady state is established. At this time, the cell-water volume (V_w(f)) will be related to the initial cell-water volume (V_w(i)) and to the initial and final extracellular osmolalities (π(i) and π(f), respectively) according to:

$$V_w(\mathrm{f}) = \frac{V_w(\mathrm{i}) \cdot \pi(\mathrm{i})}{\pi(\mathrm{f})} \tag{8.2}$$

which is a rearrangement of the statement that the total amount of intracellular solute (given by volume times osmolality) is constant. Equation (8.2) can be transformed to denote total cell volume (V_t) by adding to both sides the volume of cell solids (V_s), which is constant, and defining $V_t = V_w + V_s$:

$$V_t(\mathrm{f}) = V_s + \frac{V_w(\mathrm{i}) \cdot \pi(\mathrm{i})}{\pi(\mathrm{f})} \tag{8.3}$$

Equations (8.2) and (8.3) are represented graphically in Figure 8.1, which describes the ideal ('osmometric') response of the cell to changes in external osmolality. These equations can be made more practical by expanding the osmolality terms to include all solutes present in real experiments and their reflection coefficients. Inasmuch as cell volume regulatory events are in general slow when compared with the changes in cell volume following abrupt changes in external osmolality, equations (8.2) and (8.3) can be used to determine σ_s; for instance, by comparing the predicted and the observed change in cell volume when a solute with σ_s = 1 is replaced with another one with unknown σ_s'. Using equation (8.2), it can be shown that $V_w(\mathrm{f})'/V_w(\mathrm{f}) = \sigma_s'$.

Changes in the osmolality of one of the solutions bathing an epithelium will not necessarily result in an osmometric response. If both membranes are water permeable, there will be transepithelial water flow and the magnitude of the change in cell water volume will depend on the relative

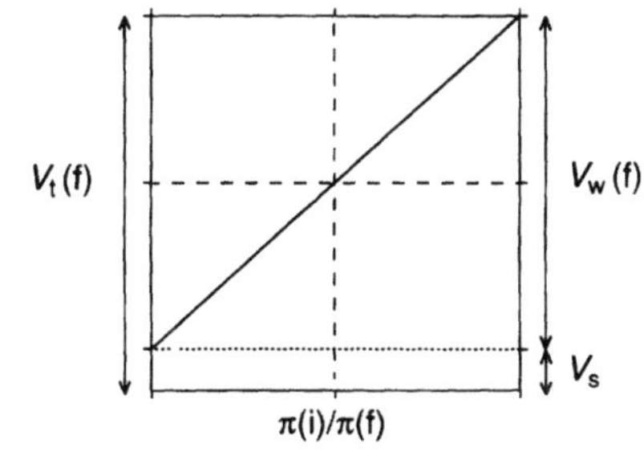

Figure 8.1 Relationship between steady-state total cell volume (V_t(f)) and cell-water volume (V_w), plotted in arbitrary units in the ordinates, and ratio of initial and final extracellular osmolalities (π(i)/π(f)). Relationship is a straight line depicting 'ideal' osmometric response of cell to changes in external effective osmolality. At infinite π(f): $V_t = V_s$ (volume of cell solids) and V_w = 0. See text for additional details.

osmotic water permeabilities of the two membrane domains (Cotton *et al.*, 1989).

8.2 METHODS OF MEASUREMENT OF CELL (OR CELL-WATER) VOLUME AND VOLUME CHANGES

The experimental techniques used to measure cell volume or cell-water volume are numerous (Table 8.1; and see MacKnight and Leader, 1989). Here we emphasize non-destructive methods used in living epithelia. These methods are based on one of three principles: measurements of cell dimensions (these methods assess total cell volume); measurements of the concentration of a known amount of a cell-water marker (these methods assess cell-water volume); and measurements of light-scattering by cell populations (these methods assess distribution of cell sizes).

The choices for the investigator are diverse and have distinct advantages and disadvantages. The main criteria for selection of a method are sensitivity, accuracy, time resolution, non-destructiveness, ease of execution and calibration, independence of the cell or tissue geometry, applicability to living tissues, and cost.

Table 8.1 Classification of non-destructive methods of measurement of cell (or cell-water) volume

In individual cells	*In cell populations*
1. Microscopic assessment of cell dimensions: three dimensions, two dimensions, one dimension	1. Measurement of total volume of the cells contained in a suspension upon centrifugation
2. Measurements of intracellular concentration of cell-water marker Electrophysiologic methods Optical methods	2. Assesement of the size distribution of cells in suspension (Coulter counter, light scattering) or attached to substratum (light scattering)

8.2.1 Measurements of cell volume from microscopic assessments of cell dimensions

(a) Three dimensions

The volume of any cell can be calculated by numerical integration techniques if the entire cell can be optically reconstructed, i.e. reconstructed in three dimensions. This approach became possible with the development of differential interference contrast microscopy (DIC or Nomarski microscopy). This methodology has two features that facilitate the task of three-dimensional reconstruction: it permits focusing on a thin optical slice in a thick translucent preparation; and it yields a much better edge resolution than possible with other microscopic methods, thus improving the identification of cell borders. Spring and associates (Spring and Hope, 1978; Spring and Hope, 1979; Persson and Spring, 1982; Spring, 1985) applied this method extensively to *Necturus* gallbladder epithelium. In addition to the use of DIC microscopy, their studies needed two additional technical developments: a miniaturized tissue chamber that permitted optical studies with short working-distance lenses while still allowing for superfusion of apical and basolateral surfaces of the epithelial sheet; and hardware and software that allowed for rapid acquisition and storage of large amounts of data. The experimental apparatus is depicted in Figure 8.2. The procedure was to use an objective driven by a

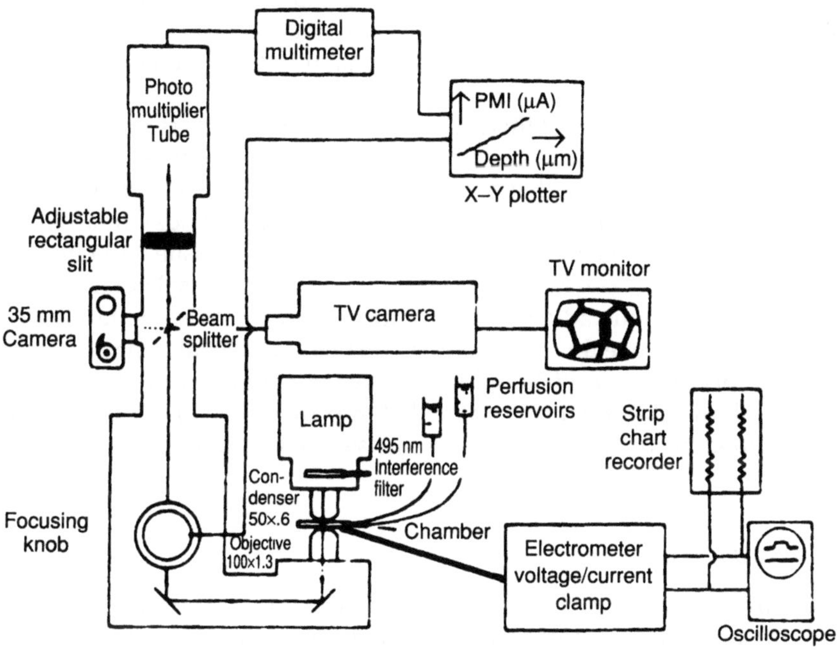

Figure 8.2 Experimental set-up for measurements of epithelial cell volume. The chamber in which the epithelium is mounted as a flat sheet is placed on the stage of an inverted microscope equipped with specialized optics (see text). The image is split and viewed on a television monitor while measurements are made with the photomultiplier tube. The output is displayed as the *y*-axis on an *x*–*y* plotter. The *x*-axis is the read-out of a potentiometer attached to the fine-focusing knob of the microscope. Voltage and current readings are amplified and controlled by a combination electrometer. The voltage and current signals are monitored on an oscilloscope and recorded on a strip-chart recorder. (Reproduced from Spring and Hope, 1978, with permission.)

step motor to focus on the tissue at adjacent planes, e.g. 1 μm apart. For instance, starting at the basal pole of the cell, a section was recorded, the objective was moved up and the overlying section was recorded, and so on, until reaching the apical pole of the cell. The surface area of each section was computed after tracing the edges. The volume of each section is given by surface area × height (1 μm). The volume of the cell is the sum of the volumes of all optical sections.

This elegant technique was applied to studies of water transport (Persson and Spring, 1982), cell volume regulation (Fisher *et al.*, 1981; Ericson and Spring, 1982b; Spring and Ericson, 1982; Larson and Spring, 1984; Furlong and Spring, 1990) and ion transport (Ericson and Spring, 1982a; Dausch and Spring, 1994) in *Necturus* gallbladder epithelium, and has also been used in amphibian urinary bladder (Kachadorian *et al.*, 1985; Davis and Finn, 1987) and frog skin (Spring and Ussing, 1986). Additional technical innovations permitted application of the optical sectioning technique to isolated, perfused mammalian renal tubules (Strange and Spring, 1987a,b).

The main advantages of this method are that it permits studies in living tissues, it is non-destructive and, in principle, it has a high resolution. The main disadvantages are that it is technically demanding, expensive and relatively slow (compared with indicator dilution methods). In addition, its accuracy depends on the optical properties of the tissue and objective detection of lateral cell borders, as well as apical and basolateral poles of the cell. Continuous advances in computerized microscopic techniques are likely to improve this methodology further, in particular in identifying cell borders. The use of lipophilic fluorescent probes can also help to detect cell membranes. Efforts aimed at improving the time resolution of the method resulted in the two-dimensional measurements described below.

(b) Two dimensions

Measurements of two dimensions permit accurate determination of cell volume under two sets of conditions: that the third dimension is known and does not change; or that the cell is fully described by two dimensions (e.g. cylinder, prolate spheroid) or one dimension (sphere). Because of the time necessary to complete the three-dimensional reconstruction of an entire *Necturus* gallbladder cell, Persson and Spring (1982) measured the dimensions of a 'volume element' of the cell, i.e. a slice 6 μm thick, improving considerably the time resolution. The procedure was validated by showing good correlation of the changes in cell volume with the changes in the dimensions of the volume element upon increasing the osmolality of the apical bathing solution. However, this procedure might not be adequate for cell-swelling experiments. Since the epithelial cells are closely packed together in monolayered epithelia, there is little if any space for increase in cell dimensions in the horizontal plane upon osmotic swelling.

(c) One dimension

Transporting epithelia are sheets of closely packed cells and the volume of the sheet is largely cell volume. If we neglect the volume of the lateral intercellular spaces, in a planar epithelium of uniform thickness, the cell volume will equal the cross sectional area × the mean height (or thickness). This simple idea was first exploited by MacRobbie and Ussing (1961) using frog-skin epithelium. With a conventional optical microscope, they assessed the thickness of the epithelium by successively focusing on the surface of the epidermis and on a pigmented cell just under the epithelium. Changes in the distance denoted changes in the thickness of the entire epithelium and were assumed to be linearly correlated with changes in epithelial cell volume. The reproducibility of the measurements was about 1 μm, which is quite appropriate for an epithelium 50 μm thick.

Fluorescent microbeads as markers of epithelial-cell poles in planar epithelia
Application of the method to thinner, monolayered epithelia required major improvements in reproducibility of the measurements. This was achieved by Crowe and Wills (1991) by using fluorescent microbeads to facilitate focusing on the surface of the epithelial cells and the underlying substratum. They used A6 cells grown on collagen-coated filters. Fluorescent microbeads (fluorescent polystyrene latex spheres 1 μm in diameter) were embedded in the collagen layer prior to plating the cells and added to the apical bathing solution for ca. 15 min prior to the experiment to allow for attachment to the apical surface. The distance between the beads at the apical surface and the beads embedded in collagen below the basal membrane was measured using a potentiometer coupled to the fine-focus knob of the microscope. Accuracy of the measurements was improved to 0.5 μm by averaging consecutive determinations (Crowe *et al.*, 1995b). The experimental set-up is illustrated in Figure 8.3 (Crowe and Wills, 1991). This method was simplified further by van Driessche *et al.* (1993): the focusing on the fluorescent microbeads was automated using a piezoelectric drive to move the objective and an algorithm based on light intensity in the center of the bead for precise location. This improves the accuracy to better than 0.1 μm. Technical problems described by Van Driessche *et al.* (1993) are random displacement of the apical microbeads, vertical displacements of the preparation, and limited time resolution, ranging from 6 to 12 s when tracking one to three beads.

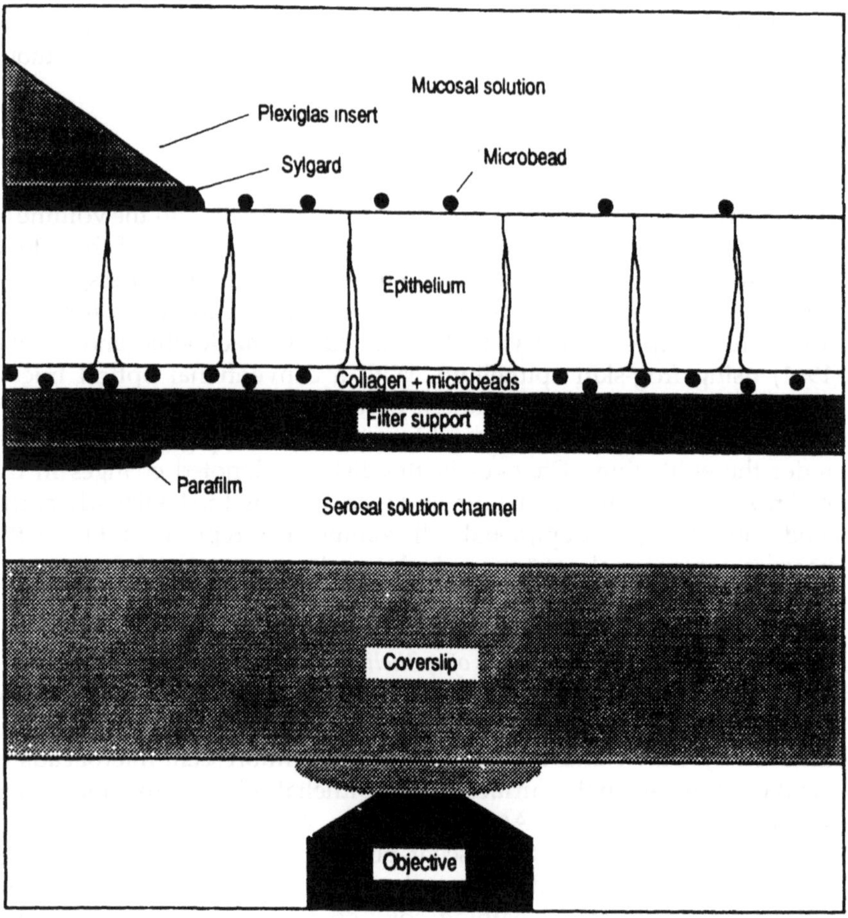

Figure 8.3 Diagram of experimental measurement of epithelial cell height. The epithelium (on collagen-coated filter support) is mounted between the apical and basolateral compartments, using a Sylgard ring to minimize edge damage. Approximate heights: epithelium 20 μm; collagen 4 μm; filter 50 μm; basolateral solution 100 μm; coverslip 170 μm. See text for details. (Reproduced from Crowe and Wills, 1991, with permission.)

Measurements of cell height in renal tubules

Cell volume of isolated perfused or collapsed renal tubules can be estimated from the mean cell height (tubule radius minus lumen radius). This has been done in numerous studies using a variety of optical systems. Major progress came from the use of video recording, which allows for off-line single-frame analysis with simultaneous measurements of outer and inner diameters and cell height at several places in the tubule segment (Welling *et al.*, 1983), and also from the use of DIC microscopy, first applied to isolated, perfused rabbit cortical-collecting duct (Kirk *et al.*, 1984a,b). A detailed description of geometric procedures that take into consideration cell shape in the calculation of volumes has been provided by Guggino *et al.* (1990).

Further improvements in measurements of cell volume in renal tubule cells were made in perfusion techniques, optical resolution and computerized data acquisition by Strange and Spring (1987a,b). In addition to the optical reconstructions described above, these investigators could assess cell height by focusing on the lateral walls of the perfused tubule segments. The time resolution of the measurements was thus increased to about 70 ms.

Whittembury and his coworkers have developed interesting technical approaches to measure cell volume in isolated renal tubules, both perfused and collapsed (for review, see Whittembury *et al.*, 1986). First, the use of an optical image splitter represented a major improvement in resolution. However, the method requires manual adjustment of the splitter and is limited in terms of time resolution (Whittembury *et al.*, 1986). Spatial resolution was improved by signal averaging of video frames (Carpi-Medina *et al.*, 1984; Lindemann, 1984; Whittembury *et al.*, 1986). For instance, 225 parallel scans perpendicular to the axis of a collapsed isolated tubule are averaged. The tubule is stained to improve contrast and the edges (solution–tubule) are identified from the first derivative of light intensity vs. distance (Figure 8.4). The technique can be applied to perfused tubules and to tubules with collapsed lumina. The nominal precision and time resolution of a measurement of diameter are 0.03 μm and about 20 ms, respectively. The method has been successfully used to measure apical and basolateral membrane P_f in mammalian tubule segments (Whittembury *et al.*, 1986).

Measurements in one dimension have major advantages, compared with three-dimensional reconstruction, in that they are simple and potentially easy to implement in most laboratories. In addition, they have much better time resolution than measurements of two or three dimensions. A disadvantage is that with the most sophisticated procedures the equipment is expensive. In addition, there is an uncertainty in that experiments in which the volume of the lateral intercellular spaces increases will yield a significant error in the calculated change in cell volume. If the spaces were to swell, for example during osmotic water flow from apical to basolateral surface (basolateral solution rendered hyperosmotic), then there will be a reduction in average cross-sectional area of the cells, and hence the true change in cell volume (in this case the cells shrink) will be greater than the value calculated from the change in cell height assuming a constant cross-sectional area. If these pitfalls and those mentioned in the previous section are taken care of, the potential usefulness of the methods based on cell-height measurements is great.

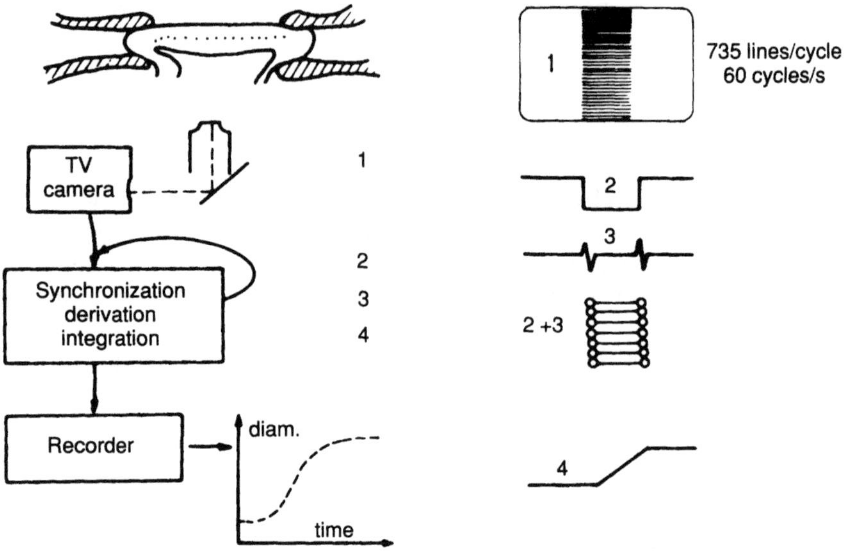

Figure 8.4 Experimental apparatus used to measure dimensions of isolated renal tubules. The tubule image is examined with an inverted microscope and a TV camera and displayed vertically on the TV monitor. The signal is analyzed (synchronization, comparison and derivation) to increase contrast at the edges of the image (circles in 2 + 3). The length of all dark segments of a chosen number of lines is summed for each frame. Signals are taken at ca. 20 ms intervals. (Reproduced from Whittembury *et al.*, 1986, with permission.)

8.2.2 Measurements of cell-water volume changes using intracellular indicators

Electrometric and fluorometric detection techniques have been used to measure the intracellular concentration of the indicator in a non-destructive fashion. Both are based on the use of the indicator-dilution principle. The volume of a solution (V_s) can be determined precisely if the amount (Q_i) and concentration (C_i) of one of its solutes can be measured (subscript s denotes solution; subscript i denotes indicator). The principle is contained in the definition of concentration:

$$C_i = \frac{Q_i}{V_s} \tag{8.4}$$

To apply this principle to the measurement of changes in cell-water volume, several properties of the solute (indicator) must hold: homogeneous distribution in total cell water; chemical stability; lack of binding, compartmentalization, metabolism or transport; inertness (lack of toxic or functional effects); and existence of methods for sensitive and accurate measurement of its concentration.

(a) Electrophysiologic methods

If an ion has the properties of a true cell-water indicator and can be measured with ion-sensitive microelectrodes, then measurements of cell-water volume can be carried out with great accuracy and excellent time resolution. The latter is a major advantage. It is routinely possible to record concentration changes with a time resolution of about 100 ms, and it is possible to improve this further.

Electrophysiologic methods based on the detection of native ions, such as Na^+ or Cl^- (Zeuthen, 1982; Tripathi and Boulpaep, 1988) or extraneous ions such as tetramethylammonium (TMA^+) (Reuss, 1985) have been used to measure changes in cell-water volume. The latter technique involves cell loading with the impermeant, non-transported ion TMA^+ and measurements of the intracellular $[TMA^+]$ ($[TMA^+]_i$) with intracellular microelectrodes. Once the cells are loaded, $[TMA^+]_i$ remains constant for several hours. Changes in $[TMA^+]_i$ are inversely proportional to cell water volume. Although under certain conditions the use of native permeant ions as cell-water markers can be validated, the fact that these ions are transported by the cell membranes and in many cases are involved in cell volume regulation makes them less useful.

Cell loading and measurement of $[TMA^+]_i$

Cells can be loaded with TMA^+ (or a similar quaternary-ammonium compound) by transient permeabilization with the pore-forming polyene antibiotic Nystatin (Reuss, 1985; Cotton *et al.*, 1989), by allowing spontaneous uptake (Alvarez-Leefmans *et al.*, 1992, but see below), or by simple diffusion from a microelectrode barrel (Serve *et al.*, 1988). To measure $[TMA^+]_i$, a double-barrel microelectrode is generally used, with one barrel TMA^+-sensitive and the other one to measure membrane voltage. For details, including calibration, see Chapter 7 and Altenberg *et al.* (1990). The TMA^+/K^+ selectivity is about 800, so that K^+ interference is small, i.e. 120 mM K^+ gives a signal equivalent to 0.15 mM TMA^+. Thus, the electrode's response deviates from the Nernst equation only at very low $[TMA^+]_i$. If $[TMA^+]_i > 1$ mM, its concentration can be simply calculated from:

$$[TMA^+]_i' = [TMA^+]_i \exp(\Delta V^*/S) \quad (8.5)$$

where $[TMA^+]_{i'}$ and $[TMA^+]_i$ denote two TMA^+ concentrations, ΔV^* is the difference in electrode voltage and S is the slope.

Calculation of changes in cell-water volume

Relative cell-water volumes can be calculated from the $[TMA^+]_i$ values according to:

$$V_c' = V_c \cdot ([TMA^+]_i/[TMA^+]_i') \quad (8.6)$$

where V_c' and V_c denote final and initial cell-water volumes, respectively, and $[TMA^+]_i'$ and $[TMA^+]_i$ denote final and initial TMA^+ concentrations, respectively. Equation (8.6) is a rearrangement of the equation indicating constant intracellular TMA^+ content.

Experimental applications: changes in cell-water volume in Necturus *gall-bladder epithelial cells*

Figure 8.5 depicts the changes in cell-water volume in an epithelial cell of *Necturus* gallbladder preloaded with TMA^+ and impaled with a double-barrel, TMA^+-sensitive microelectrode (Cotton *et al.*, 1989). The top records depict the estimated change in apical solution osmolality upon rapid change in superfusate from a control solution to one in which the osmolality was increased by addition of 40 mM sucrose. These records were obtained with an extracellular tetrabutylammonium (TBA^+)-sensitive electrode ($[TBA^+]$ was 1 mM in the control solution and 2 mM in the

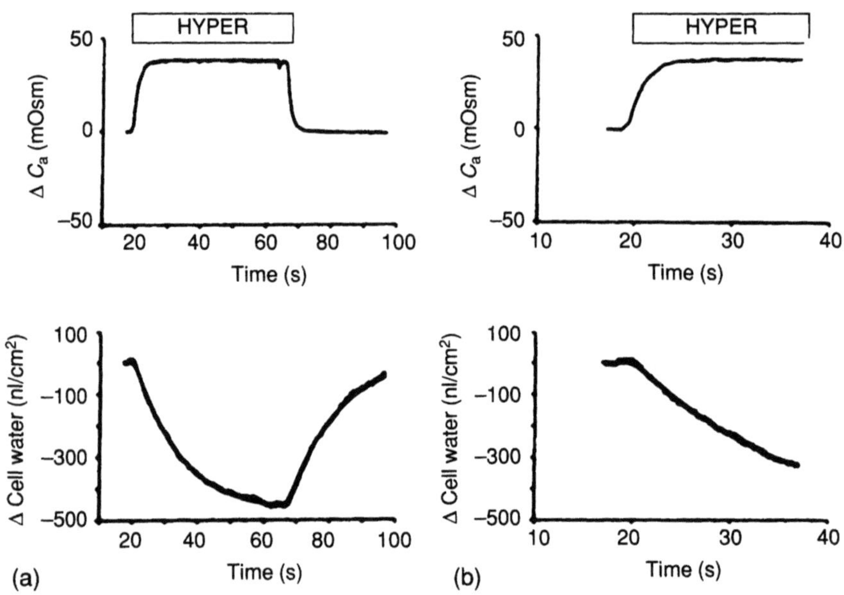

Figure 8.5 Changes in apical solution osmolality (ΔC_a), cell-water volume, and fitted change in cell-water volume (measured with tetramethylammonium technique) in response to exposure to hyperosmotic apical solution. (a) The apical solution osmolality was increased from 206 to 244 mosmol·(kg water)$^{-1}$. Initial cell-water volume was estimated at 2870 nl/cm^{-2} of epithelium, from mean cell height. The data for the complete experiment (ca. 70 s) are fitted to a mathematical four-compartment model (two bathing solutions, cell interior and lateral-intercellular space). Cell membrane hydraulic coefficients were calculated from the fit. (b) Same experiment as above except that the fit spans only 20 s. (Reproduced from Cotton *et al.*, 1989, with permission.)

hyperosmotic solution), taking advantage of the fact that TBACl and sucrose have similar diffusion coefficients; hence, the time course of the TBA^+ activity near the cell surface is an excellent indication of the time course of the sucrose concentration at the same site (for details see Cotton and Reuss, 1989; Reuss and Cotton, 1991). The bottom records denote the time course of the change in cell-water volume calculated from the change in $[TMA^+]_i$. The absolute cell volume was estimated from the mean cell height. The raw data in the bottom records are shown superimposed with the best fit of the equation describing the change in cell volume according to a four-compartment mathematical model (Cotton *et al.*, 1989). From this fit, the osmotic water permeability of the cell membranes was calculated. Note the time resolution in this experiment; the cell volume data were collected at a rate of 10 points per second.

Another example of the use of the TMA^+ technique to measure changes in cell-water volume is presented in Figure 8.6. Here we compare the effects of reducing apical solution $[Cl^-]$ from about 100 to 10 mM under control conditions (Figure 8.6a) and after elevation of cAMP levels (Figure 8.6b). The records in this figure denote, from top to bottom: transepithelial voltage, apical membrane voltage, basolateral membrane voltage and changes in cell-water volume, respectively. Under control conditions, the reduction in external $[Cl^-]$ produces minimal changes in voltages and a slow decline in $[Cl^-]_i$ (not shown), due to change in fluxes via the apical membrane Cl^-/HCO_3^- exchanger, and no change in cell volume. After elevating cAMP, which activates an apical membrane Cl^- channel, the reduction in external $[Cl^-]$ causes large changes in membrane voltages (due to the cAMP-activated Cl^- conductance), a rapid fall in $[Cl^-]_i$ (not shown) and a rapid decline in cell-water volume, due to net loss of KCl. For further details, see Cotton and Reuss (1991).

Advantages, disadvantages and use in other systems

The TMA^+ method has been used in *Necturus* gallbladder epithelial cells (Reuss, 1985; Cotton *et al.*, 1989; Cotton and Reuss, 1991), hepatocytes (Khalbuss and Wondergem, 1990) and neurons (Serve *et al.*, 1988; Ballanyi and Grafe, 1988; Alvarez-Leefmans *et al.*, 1992). Zeuthen and associates have used choline instead of TMA^+ in experiments in choroid plexus (Zeuthen, 1991) and retinal pigment epithelium (La Cour and Zeuthen, 1993).

The greatest advantages of this method are its sensitivity and time resolution: changes of 1–2% can be measured with a dwell time of 100 ms or less. Hence, it is the method of choice to measure changes in cell-water volume that occur rapidly and/or require curve fitting for quantitative analysis.

The disadvantages are several. First, it is necessary to validate that TMA^+ is a 'true' cell-water indicator in the cells under study, or that if

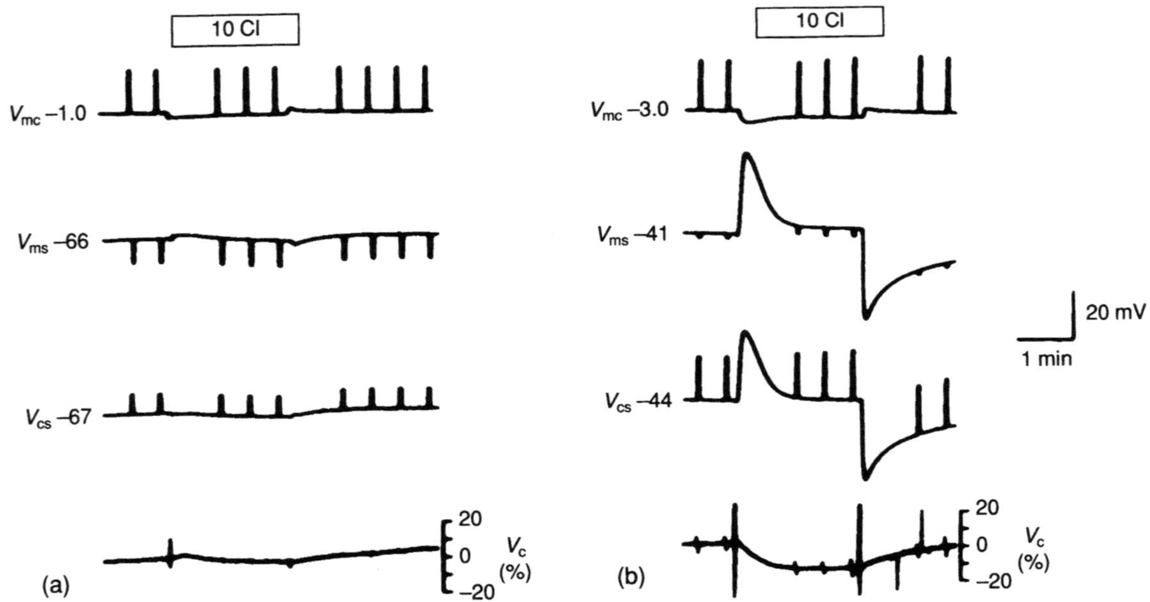

Figure 8.6 Changes in cell volume under isosmotic conditions. Effects of reducing luminal solution [Cl⁻] on voltages, resistances, and cell-water volume in *Necturus* gallbladder epithelium. V_{ms}, V_{mc} and V_{cs} denote transepithelial, apical membrane and basolateral membrane voltages, respectively. Upward changes in membrane voltages denote mucosa-positive (V_{ms}) or cell-positive (V_{mc}, V_{cs}) voltage changes. Voltage deflections result from transepithelial constant current pulses. Changes in cell-water volume (ΔV_c) were measured with the tetramethylammonium technique. (a) Luminal solution [Cl^-] was reduced from 98 to 10 mM (Cl^- replaced with cyclamate) during the indicated interval under control conditions (apical membrane has active K^+ channels, but not Cl^- channels). There is a minimal change in cell-water volume, reflecting reversal of apical membrane Cl^-/HCO_3^- exchange. (b) Same experimental perturbation, after elevating cAMP levels. Cell volume decreases because cAMP activates apical membrane Cl^- channels; upon lowering external [Cl^-], the cell looses both K^+ and Cl^-. (Reproduced from Cotton and Reuss, 1991, with permission.)

transport does occur, the rate of extrusion after completion of the loading phase is slow (Alvarez-Leefmans *et al.*, 1992) or can be minimized by maintaining a low level of indicator in the extracellular solution (Zeuthen, 1991). Second, under some conditions the $[TMA^+]_i$ (of a few mmoles/l) may have undesirable pharmacological or osmotic effects. Ballanyi *et al.* (1992) have proposed the use of a low (0.1–1 mM) concentration of decamethonium instead of TMA^+ (the selectivity is 10-fold greater). Third, correct use of this method requires a preparation that can be impaled with double-barrel microelectrodes (tip diameters ca. 0.2–0.5 μm), i.e. relatively large, low input-resistance cells, or cells that are coupled via gap junctions. Fourth, the method requires basic electrophysiologic and micromanipulation equipment and the investigator must possess skills in the use of intracellular microelectrode techniques.

An interesting variation of the method described above is to use TMA^+ as an extracellular fluid marker. The $[TMA^+]$ can be measured with a large-tip ion-selective microelectrode. Suspended cells are incubated in a constant-volume, closed chamber. If the initial ratio of intra- to extracellular volumes is large, then measurable signals can be obtained when there are net water fluxes across the cell membrane (for instance, by changing extracellular osmolality: Blumenfeld *et al.*, 1988).

(b) Fluorescence methods

These methods are also based in the indicator dilution principle, but the cell-water indicator is a fluorescent molecule and the method of detection is fluorometric. The technique has been applied to proximal renal tubules (Tauc *et al.*, 1990), to cancer cells of epithelial origin (Altenberg *et al.*, 1994) and to nonepithelial cells (Muallem *et al.*, 1992; Crowe *et al.*, 1995a).

In human breast cancer cells, changes in cell volume were estimated from the change in intracellular concentration of CMAC (glutathione-transferase substrate), BCECF (pH-sensitive dye) or Fura-2 (Ca^{2+}-sensitive dye) (Altenberg *et al.*, 1994). Single-cell dye fluorescence was measured excluding the edges, with a ×100, 1.3 NA oil-immersion objective. For the ion-sensitive dyes BCECF and Fura-2, excitation light was near the isosbestic points, i.e., at wavelengths insensitive to pH_i and $[Ca^{2+}]_i$, respectively. To reduce photobleaching, total light exposure was limited to 40–200 ms per second. The output was digitized and stored in a computer at a rate of 0.2–1 Hz. Fluorescence records were corrected for photobleaching.

Figure 8.7a shows the changes in fluorescence in a human breast cancer cell loaded with BCECF and exposed to several solutions of different osmolalities. The changes in fluorescence elicited by the anisosmotic solutions are superimposed on the fluorescence decay due to photobleaching. Figure 8.7b depicts the fluorescence after correction for photobleaching

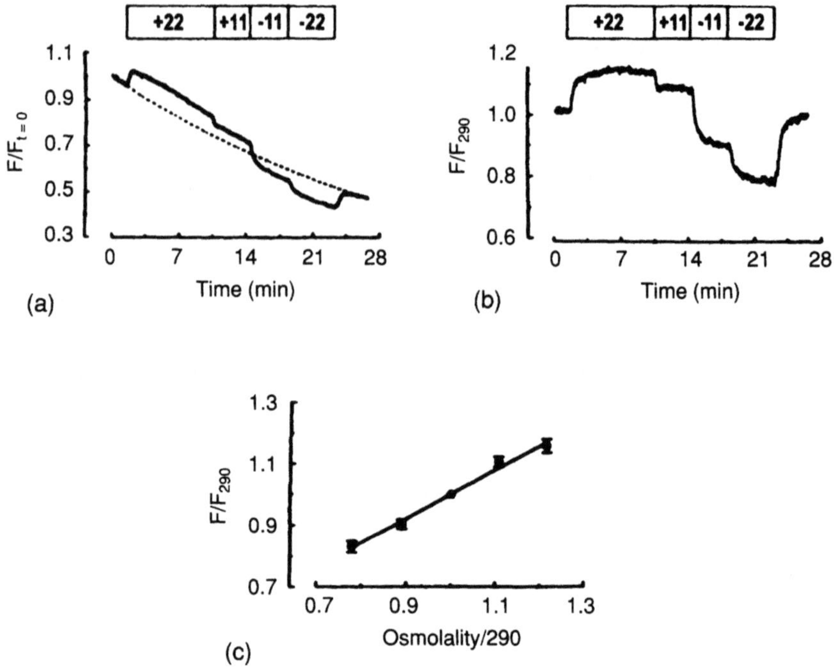

Figure 8.7 Effects of changes in bath osmolality on relative fluorescence of cell volume markers. (a) Typical experiment in a human breast-cancer cell (MCF-7) loaded with BCECF. $F/F_{t=0}$ is relative fluorescence normalized to the value at $t = 0$. Numbers in the top bars indicate percentage change in bath osmolality. The dotted line is a fit of a single-exponential to the data points. The continuous decay in $F/F_{t=0}$ is due to photobleaching. (b) The data shown in (a) were corrected for fluorescence decay and normalized to the value in isosmotic solution (F/F_{290}). (c) Relationship between F/F_{290} and relative bath osmolality normalized to the isosmotic solution (osmolality/290). The basic solution was HCO_3^-/CO_2-buffered containing 100 instead of 140 mM NaCl; the osmolality was changed with sucrose. Values are means ± SEM of 6 to 12 experiments. The line is the best fit of a straight line to the data. (Reproduced from Altenberg *et al.*, 1994, with permission.)

(Altenberg *et al.*, 1994). A calibration curve (Figure 8.7c) can be constructed from this kind of experiment. The fluorescence changes were ca. 80% of the ideal osmotic response of the cell, likely because of underestimation of the compartmentalized dye. These cells do not appreciably regulate their volume after osmotic swelling (Altenberg *et al.*, 1994).

Major improvements in this methodology can be expected from the use of confocal fluorescence microscopy which potentially allows for estimations of dye concentration in the cytoplasm, i.e. excluding dye-binding compartments, and because of the optical advantages involved in detecting from a thin layer of cytoplasm.

8.2.3 Measurements of cell volume changes from light scattering

Measurements of light scattering have been used extensively to measure cell volume in suspensions (Terwilliger and Solomon, 1981; Mlekoday *et al.*, 1983). Recent improvements in methodology have also permitted its application to groups of cells plated on glass coverslips, either isolated or confluent (McManus *et al.*, 1993). The physical bases of light scattering by cells are not entirely understood. They are thought to include large-particle effects such as interference and diffraction and small-particle effects; the former result in near-forward scattering and the latter in large-angle scattering (for further details, see McManus *et al.*, 1993 and references therein).

The method described by McManus *et al.* (1993) measures scattering of laser light provided by a 633 nm He–Ne laser (30° light-incidence angle, 8 mm diameter light spot) by cells plated on a glass coverslip positioned in an *ad hoc* cuvette with continuous superfusion. The light was measured with a photomultiplier tube (axis at 90° with respect to the incident light beam) and provided a continuous output proportional to cell volume (Figure 8.8). This technique has been used with red blood cells, glioma and

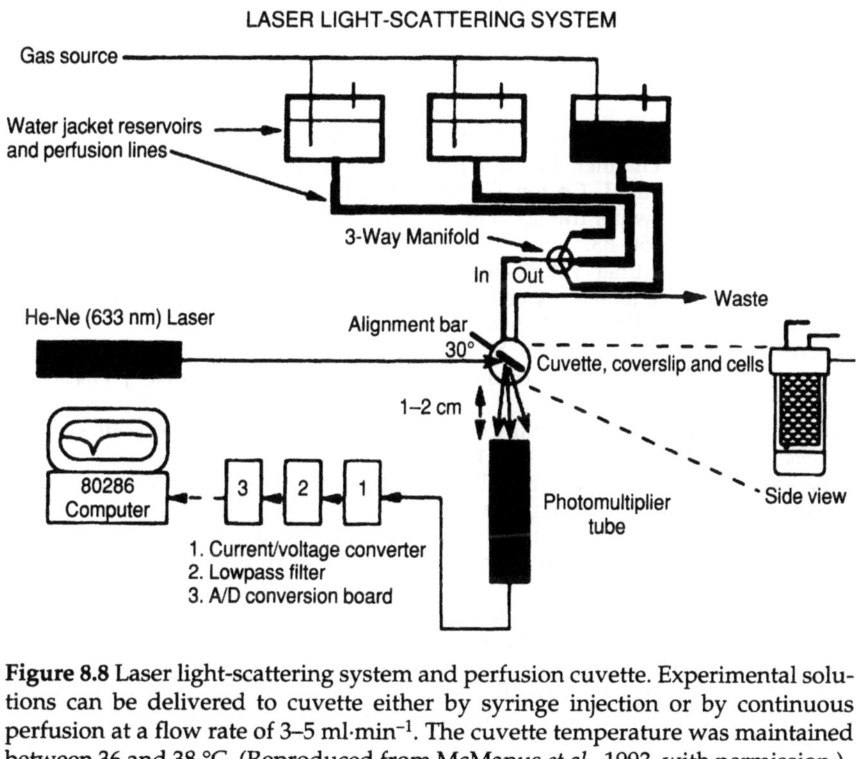

Figure 8.8 Laser light-scattering system and perfusion cuvette. Experimental solutions can be delivered to cuvette either by syringe injection or by continuous perfusion at a flow rate of 3–5 ml·min^{-1}. The cuvette temperature was maintained between 36 and 38 °C. (Reproduced from McManus *et al.*, 1993, with permission.)

Ehrlich ascites cells and cultures of thick-ascending limb of Henle's loop. Validation of the light-scattering signals was obtained from comparison with known volume regulatory responses (McManus *et al.*, 1993).

The main advantages of this method are its relative technical simplicity and the good sensitivity and time resolution. In dense cultures, rapid volume changes of the order of 2% or less can be detected with a time resolution of 1 s or less (Strange and Morrison, 1992; Strange, personal communication). The main disadvantages are the difficulties in obtaining absolute values of cell volume or of cell volume changes and the necessity, dictated by the current technology, to study cell populations. This prevents its use with epithelia consisting of more than a single cell type. In addition, slow changes in cell volume are more difficult to measure accurately because of laser-induced fluctuations in the output of the photomultiplier tube. These and other shortcomings are discussed by McManus *et al.* (1993). The method described by these investigators represents a major improvement over previous attempts and will probably be used extensively in the near future. It is amenable to simultaneous measurements with fluorescent dyes. Chamber modification and use of a rigid permeable support for the epithelial cell culture may allow for measurement of transepithelial electrical properties (McManus *et al.*, 1993).

REFERENCES

Agre, P., Preston, G.M., Smith, B.L. *et al.* (1993) Aquaporin CHIP: the archetypal molecular water channel. *American Journal of Physiology*, **265**:F463–F474.

Altenberg, G., Copello, J., Cotton, C. *et al.* (1990) Electrophysiological methods for studying ion and water transport in *Necturus* gallbladder epithelium. *Methods in Enzymology*, **192**:650–83.

Altenberg, G.A., Deitmer, J.W., Glass, D.C. *et al.* (1994) P-glycoprotein-associated Cl^- currents are activated by cell swelling but do not contribute to cell volume regulation. *Cancer Research* **54**:618–22.

Alvarez-Leefmans, F.J., Gamino, S.M. and Reuss, L. (1992) Cell volume changes upon sodium pump inhibition in *Helix aspersa* neurons. *Journal of Physiology* (London), **458**:603–19.

Ballanyi, K. and Grafe, P. (1988) Cell volume regulation in the nervous system. *Renal Physiology and Biochemistry*, **1**:114–41.

Ballanyi, K., Strupp, M., and Grafe, P. (1992) Electrophysiological measurement of volume changes in neurons, glial cells, and muscle fibers *in situ*, in *Practical Electrophysiological Methods: A Guide for* In Vitro *Studies in Vertebrate Neurobiology*, (eds H. Kettenmann and R. Grantyn), Wiley-Liss, New York, pp. 363–6.

Blumenfeld, J.D., Gullans, S.R. and Hebert, S.C. (1988) A simple method for continuous measurement of the volume of cells in suspension. *American Journal of Physiology*, **254**:C192–C199.

Carpi-Medina, P., Lindemann, B., González, E. *et al.* (1984) The continuous measurement of tubular volume changes in response to step changes in contraluminal osmolality. *Pflügers Archiv*, **400**:343–8.

Cotton, C.U. and Reuss, L. (1989) Measurement of the effective thickness of the mucosal unstirred layer in *Necturus* gallbladder epithelium. *Journal of General Physiology*, **93**:631–47.

Cotton, C.U. and Reuss, L. (1991) Effects of changes in mucosal solution Cl^- or K^+ concentration on cell water volume of *Necturus* gallbladder epithelium. *Journal of General Physiology*, **97**:667–86.

Cotton, C.U., Weinstein, A.M. and Reuss, L. (1989) Osmotic water permeability of *Necturus* gallbladder. *Journal of General Physiology*, **93**:649–79.

Crowe, W.E. and Wills, N.K. (1991) A simple method for monitoring changes in cell height using fluorescent microbeads and an Ussing-type chamber for the inverted microscope. *Pflügers Archiv*, **419**:349–57.

Crowe, W.E., Altamirano, J., Huerto, L. and Alvarez-Leefmans, J. (1995a) Calcium-dependent volume changes in single N1E-115 neuroblastoma cells measured with a fluorescent probe. *Neuroscience*, **69**:283–96.

Crowe, W.E., Ehrenfeld, J., Brochiero, E. and Wills, N. K. (1995b) Apical membrane sodium and chloride entry during osmotic swelling of renal (A6) epithelial cells. *Journal of Membrane Biology* **144**:81–91.

Dausch, R. and Spring, K.R. (1994) Regulation of NaCl entry into *Necturus* gallbladder epithelium by protein kinase C. *American Journal of Physiology*, **266**:C531–C535.

Davis, C.W. and Finn, A.L. (1987) Interactions of sodium transport, cell volume, and calcium in frog urinary bladder. *Journal of General Physiology*, **89**:687-702.

Engel, A., Walz, T. and Agre, P. (1994) The aquaporin family of membrane water channels. *Current Opinion in Structural Biology*, **4**:545–53.

Ericson, A.-C. and Spring, K.R. (1982a) Coupled NaCl entry into *Necturus* gallbladder epithelial cells. *American Journal of Physiology*, **243**:C140–C145.

Ericson, A.-C. and Spring, K.R. (1982b) Volume regulation by *Necturus* gallbladder: apical Na^+-H^+ and Cl^--HCO_3 exchange. *American Journal of Physiology*, **243**:C146-C150.

Fisher, R.S., Persson, B.-E., and Spring, K.R. (1981) Epithelial cell volume regulation: bicarbonate dependence. *Science*, **214**:1357–1359.

Furlong, T.J. and Spring, K.R. (1990) Mechanisms underlying volume regulatory decrease by *Necturus* gallbladder epithelium. *American Journal of Physiology*, **258**:C1016–C1024.

Guggino, W.B., Markakis, D. and Amzel, L.M. (1990) Measurements of volume and shape changes in isolated tubules. *Methods in Enzymology*, **191**:371–9.

Kachadorian, W.A., Sariban-Sohraby, S. and Spring, K.R. (1985) Regulation of water permeability in toad urinary bladder at two barriers. *American Journal of Physiology*, **248**:F260–F265.

Khalbuss, W.E. and Wondergem, R. (1990) An electrophysiological technique to measure change in hepatocyte water volume. *Biochemistry and Biophysical Acta*, **1029**:51–60.

Kirk, K.L., DiBona, D.R. and Schafer, J.A. (1984a) Morphologic response of the rabbit cortical collecting tubule to peritubular hypotonicity: quantitative examination with differential interference contrast microscopy. *Journal of Membrane Biology*, **79**:53–64.

Kirk, K.L., Schafer, J.A. and DiBona, D.R. (1984b) Quantitative analysis of the structural events associated with antidiuretic hormone-induced volume reabsorption in the rabbit cortical collecting tubule. *Journal of Membrane Biology*, **79**:65–74.

La Cour, M. and Zeuthen, T. (1993) Osmotic properties of the frog retinal pigment epithelia. *Experimental Eye Research*, **56**:521–530.

Larson, M. and Spring, K.R. (1984) Volume regulation by *Necturus* gallbladder: basolateral KCl exit. *Journal of Membrane Biology*, **81**:219–232.

Lechène, C. (1985) Cellular volume and cytoplasmic gel. *Biology of the Cell*, **55**:177–180.

Lindemann, B. (1984) Real-time area-tracker records cellular volume changes from video images. *Review of Scientific Instruments*, **55**:1788–90.

MacKnight, A.D.C. and Leader, J.P. (1989) Volume regulation in epithelia: experimental approaches, in *Methods in Enzymology*, Vol. 17, *Biomembranes (Part R: Transport Theory: Cells and Model Membranes*, (eds S. Fleischer and B. Fleischer), Academic Press, Inc., San Diego, pp. 744–90.

MacRobbie, E.A.C. and Ussing, H.H. (1961). Osmotic behaviour of the epithelial cells of frog skin. *Acta Physiologica Scandinavica*, **53**:348–65.

Mauro, A. (1957) Nature of solvent transfer in osmosis. *Science*, **126**:252–3.

McManus, M., Fischbarg, J., Sun, A., Hebert, S. and Strange, K. (1993) Laser light-scattering system for studying cell volume regulation and membrane transport processes. *American Journal of Physiology*, **265**:C562–C570.

Minton, A.P. (1992) Confinement as a determinant of macromolecular structure and reactivity. *Biophysical Journal*, **63**:1090-100.

Minton, A.P., Colclasure, G.C. and Parker, J.C. (1992) Model for the role of macromolecular crowding in regulation of cellular volume. *Proceedings of the National Academy of Sciences*, **89**:10504–6.

Mlekoday, H.J., Moore, R. and Levitt, D.G. (1983) Osmotic water permeability of human red cells. *Journal of General Physiology*, **81**:213–30.

Muallem, S., Zhang, B.-X., Loessberg, P.A. *et al.* (1992) Simultaneous recording of cell volume changes and intracellular pH or Ca^{2+} concentration in single osteosarcoma cells UMR-106-01. *Journal of Biological Chemistry*, **267**:17658–64.

Parker, J.C. (1993) In defense of cell volume? *American Journal of Physiology*, **34**:C1191–C1200.

Persson, B.-E. and Spring, K.R. (1982) Gallbladder epithelial cell hydraulic water permeability and volume regulation. *Journal of General Physiology*, **79**: 481–505.

Reuss, L. (1985) Changes in cell volume measured with an electrophysiologic technique. Proceedings of the *National Academy of Sciences USA*, **82**:6014–18.

Reuss, L. and Cotton, C.U. (1991) Electrophysiological methods in the study of water transport across cell membranes, in *Cell Membrane Transport: Experimental Approaches and Methodologies*, (eds D.L. Yudilevich, R. Devés, S. Perán *et al.*), Plenum Publishing, London, pp. 239–54.

Serve, G., Endres, W. and Grafe, P. (1988) Continuous electrophysiological measurements of changes in cell volume of motoneurons in the isolated frog spinal cord. *Pflügers Archiv*, **411**:414–15.

Spring, K.R. (1985) The study of epithelial function by quantitative light microscopy. *Pflügers Archiv*, **405**:S23–S27.

Spring, K.R. and Ericson, A.-C. (1982) Epithelial cell volume modulation and regulation. *Journal of Membrane Biology*, **69**:167–76.

Spring, K.R. and Hope, A. (1978) Size and shape of the lateral intercellular spaces in a living epithelium. *Science*, **200**:54–8.

Spring, K.R. and Hope, A. (1979) Fluid transport and the dimensions of cells and interspaces of living *Necturus* gallbladder. *Journal of General Physiology*, **73**:287–305.

Spring, K.R. and Ussing, H.H. (1986) The volume of mitochondria-rich cells in frog skin epithelium. *Journal of Membrane Biology*, **91**:21–6.

Strange, K. and Morrison, R. (1992) Volume regulation during recovery from chronic hypertonicity in brain glial cells. *American Journal of Physiology*, **263**:C412–C419.

Strange, K. and Spring, K.R. (1987a) Absence of significant cellular dilution during ADH-stimulated water reabsorption. *Science*, **235**:1068–70.

Strange, K. and Spring, K.R. (1987b) Cell membrane water permeability of rabbit cortical collecting duct. *Journal of Membrane Biology*, **96**:27–43.

Tauc, M., Le Maout, S. and Poujeol, P. (1990) Fluorescent video-microscopy study of regulatory volume decrease in primary culture of rabbit proximal convoluted tubule. *Biochimica et Biophysica Acta*, **1052**:278–84.

Terwilliger, T.C. and Solomon, A.K. (1981) Osmotic permeability of human red cells. *Journal of General Physiology*, **77**:549–70.

Tripathi, S. and Boulpaep, E.L. (1988) Cell membrane water permeabilities and streaming currents in *Ambystoma* proximal tubule. *American Journal of Physiology*, **254**:F188–F203.

Tripathi, S. and Boulpaep, E.L. (1989) Mechanisms of water transport by epithelial cells. *Quarterly Journal of Experimental Biology*, **74**:385–417.

Van Driessche, W., De Smet, P. and Raskin, G. (1993) An automatic monitoring system for epithelial cell height. *Pflügers Archiv*, **425**:164–71.

Welling, L.W., Welling, D.J. and Ochs, T.J. (1983) Video measurement of basolateral membrane hydraulic conductivity in the proximal tubule. *American Journal of Physiology*, **245**:F123–F129.

Whittembury, G. and Carpi-Medina, P. (1988) Renal reabsorption of water: are there pores in proximal tubule cells? *News in Physiological Sciences*, **3**:61–5.

Whittembury, G., Lindemann, B., Carpi-Medina, P. *et al.* (1986) Continuous measurements of cell volume changes in single kidney tubules. *Kidney International* **30**:187–91.

Zeuthen, T. (1982) Relations between intracellular ion activities and extracellular osmolarity in *Necturus* gallbladder epithelium. *Journal of Membrane Biology*, **66**:109–21.

Zeuthen, T. (1991) Water permeability of ventricular cell membrane in choroid plexus epithelium from *Necturus* maculosus. *Journal of Physiology*, **44**:133–51.

9

Methods and experimental analysis of isolated epithelial cell membranes

Austin K. Mircheff

The techniques of subcellular fractionation analysis make it possible to characterize transport processes in isolated membrane samples. With this ability, one can learn how transporters are distributed between the apical and basolateral surfaces of an epithelium and study transport activities under a wider range of experimental conditions than can be sustained by intact cells. One also has an avenue for studying the genesis of epithelial polarity and for elucidating the sometimes surprising methods that epithelial cells employ to regulate their transport activities.

The goal of isolating a sample of apical or basolateral membrane vesicles in order to characterize transport activities would seem to involve solving the essentially 'preparative' problem of obtaining a workable

Epithelial Transport: A guide to methods and experimental analysis.
Edited by Nancy K. Wills, Luis Reuss and Simon A. Lewis.
Published in 1996 by Chapman & Hall, London. ISBN 978-94-010-7168-0

yield of the membranes of interest, sufficiently free of contaminating membranes from other organelles. The broader goal of learning how some transport activity is distributed among the various organelles is more obviously an analytical problem. The distinction between the preparative and analytical approaches is important, as DePierre and Karnovsky (1973) argued forcefully and eloquently, and one should always recognize the crucial analytical dimensions of what may appear to be simple preparative problems. This chapter summarizes the tools that are available for isolating subcellular membranes, and illustrates how these can be combined in an analytically sound fashion. We begin by reviewing the basic biological facts which make it imperative that membrane isolation be approached as an analytical problem.

9.1 THE CELL LYSATE AS A MULTICOMPONENT MIXTURE

When a cell is lysed, a few of its organelles remain largely intact. These typically include the nucleus, mitochondria, peroxisomes, lysosomes, secretory vesicles and, in a few epithelial cell types, the brush border. Some organelles exist as small vesicles. These include the coated and uncoated transport vesicles which ferry membrane constituents and other products from one subcellular site to another. Other organelles are quite extensive and upon cell lysis they fragment into smaller particles, which generally take the form of sealed membrane vesicles. These include the basolateral plasma membranes, endoplasmic reticulum, Golgi complex and, for most epithelial cell types, the apical plasma membranes. The endosomal system, which functional and immunocytochemical studies indicate comprises at least three compartments (an apical early endosome, a basolateral early endosome and a late endosome), is at the moment enigmatic; some workers argue that the endosomes are tubular reticula, while other see them as dynamic systems of vesicles. Most, if not all, of the spatially extensive organelles are organized into distinct microdomains. The most complex of these is the Golgi complex, which is now seen to consist of a cis-reticular network, cis-, medial- and trans- saccules, a trans-reticular network and a system of peripheral vesicles. As has been particularly well documented for the renal proximal tubule, the brush border surfaces of epithelial cells are organized into distinct microvillar and intermicrovillar microdomains. Some workers have reported distinctions between the basal and lateral domains of the basolateral plasma membranes of certain epithelial cell types, and it is still not possible to exclude the possibility that each of these domains is organized into microdomains.

9.1.1 Subcellular membrane populations

As will be discussed in more detail later, the researcher has available a number of methods for separating the components of the mixture of organelles and organellar fragments in a cell lysate. Several factors make the use of these methods somewhat less straightforward than we might like. As DeDuve (1964) recognized in the early development of subcellular fractionation, most organelles which remain intact, and certainly all organellar membrane fragments, are represented in the cell lysate by populations of particles that frequently are quite diverse or heterodisperse with respect to the physical properties detected by the available separation procedures. Even a perfectly uniform membrane might yield a heterodisperse population of particles upon cell lysis, the dispersity reflecting randomness in the size of the particles that result and randomness in the presence or absence of sparsely distributed membrane constituents. If an organelle is organized into microdomains, additional heterogeneity may result from the varying degree to which the two domains are captured in the various membrane fragments.

In addition to the challenge created by the unavoidable heterodispersity of subcellular membrane populations is the problem that membranes from one organelle may closely resemble membranes from other organelles with respect to any given physical characteristic. For example, the membrane vesicles derived from the plasma membranes, endoplasmic reticulum, Golgi complex and endosomes all have similar, broad size distributions, so their sedimentation coefficients do not differ from each other. It also often happens that Golgi membranes and endoplasmic reticulum membranes have similar equilibrium densities, so that cell fractionation schemes may have to employ a combination of three or more separation procedures. Before embarking on a discussion of the various membrane separation procedures, we will digress to examine how one recognizes different populations of subcellular membranes.

9.1.2 Membrane markers

The intact organelles in a cell lysate retain distinct morphologies which make them easy to recognize, although quantitation of morphological data requires expertise outside the scope of this volume; collecting such data can be time-consuming even for the experienced practitioner. Many of the suborganellar membrane fragments are morphologically nondescript: plasma membrane vesicles do not look much, or at all, different from Golgi vesicles, vesiculated lysosomes, endosomal vesicles or any of the variety of uncoated membrane vesicles.

Traditionally, membrane constituents which have catalytic or ligand binding activities that are relatively easy to measure have been used as intrinsic markers for identifying membranes, and a standard repertoire of biochemical markers has evolved over the years. The primary subcellular locations of several of these markers are summarized in Figure 9.1.

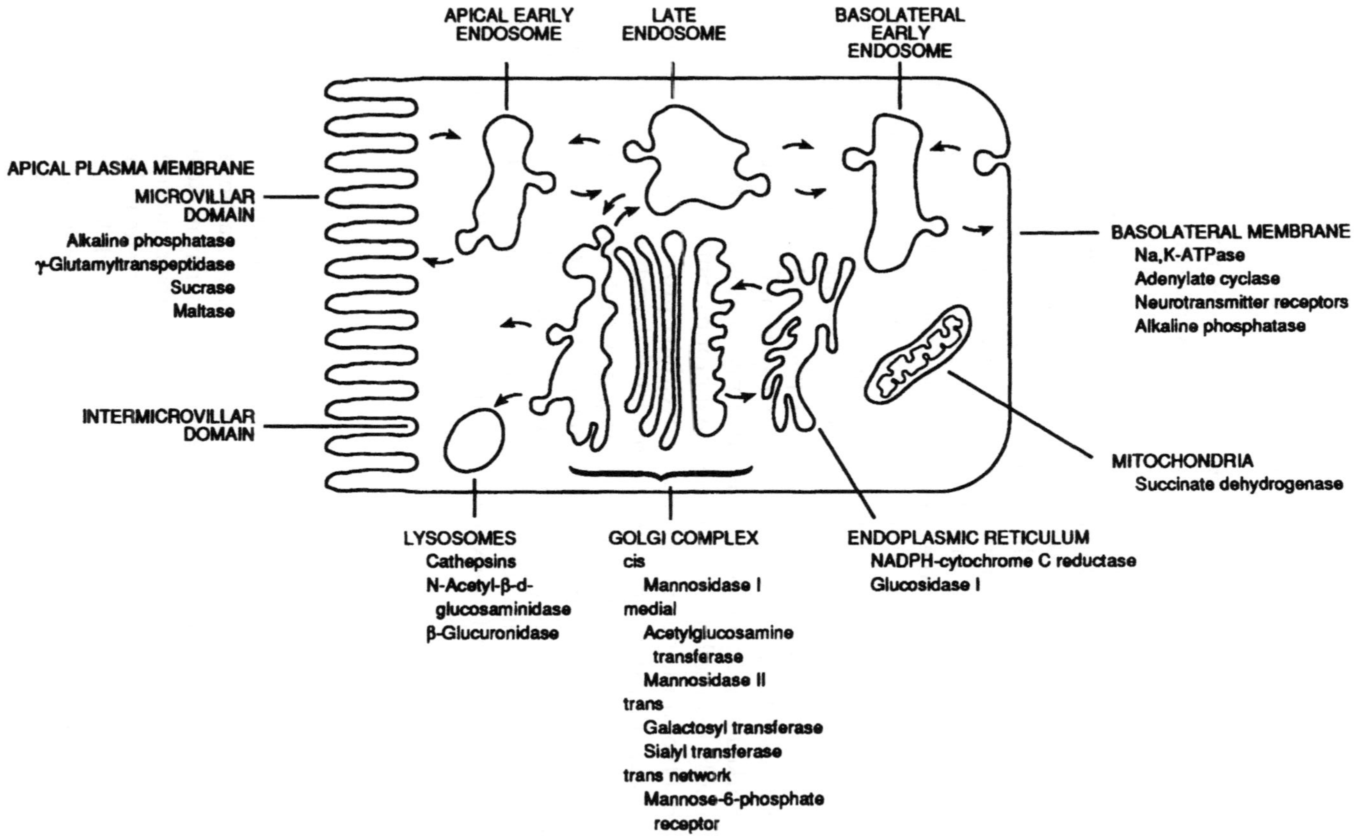

Figure 9.1 Important locations of some useful membrane markers.

In using intrinsic membrane markers one must always be alert to the possibility that a marker may be associated with a number of different organelles. This is particularly clear for plasma membrane constituents. It is now well established that plasma membrane constituents may turn over many times during the life of a single cell. This means that each constituent must exist at some steady-state level in each stage of the membrane assembly and degradation cycle, from the endoplasmic reticulum through the Golgi complex and endosomes to the lysosomes. At least in certain cells, the turn-over fluxes of plasma membrane constituents are superimposed on much larger recycling fluxes between the plasma membrane and the endosomal compartments. For these reasons, it is necessary to expand the repertoire of membrane markers to include functional and extrinsic markers. Metabolically incorporated labels, such as [^{35}S]-methionine, may be used in pulse-chase protocols to distinguish the compartments of the membrane assembly pathway. Internalized receptor ligands or fluid phase markers may be used to discern membranes from the endosomal system. Plasma membrane may be marked by reagents which adsorb firmly, or attach covalently, to extracellularly exposed surfaces or reactive groups. Sulfo-N-hydroxysuccinimidyl-biotin and other biotin derivatives have proven to be useful for this purpose, since different avidin-conjugated ligands can be employed to confirm the specificity of the labeling, quantitate the amount of biotin present in isolated membrane samples and precipitate biotinylated membrane constituents. Recent studies using these strategies suggest that, in some exocrine acinar cell types, plasma membrane constituents recycle between small surface-expressed pools and large intracellular pools (Lambert *et al.*, 1993a).

9.2 CELL DISRUPTION

9.2.1 Media

The medium in which cells or tissue fragments are suspended at the time of lysis will influence the design of the subsequent fractionation scheme. The cell swelling caused by a hypotonic medium may make the cells more fragile and influence the lines along which their surface membranes rupture. Early procedures for isolation of intestinal brush borders involved homogenization in 5 mM Na-EDTA, which favored cleavage at the plane of the terminal web, leaving brush borders intact. On the other hand, this hypotonic medium will also damage organelles, such as mitochondria, secretory vesicles and lysosomes, which the investigator may prefer to remain intact. The authors have found that an approximately isotonic medium containing 5% (w/v) sorbitol, 0.5 mM Na-EDTA and 5 mM histidine-imidazole buffer (pH 7.5) is useful for a wide range of tissue fragments and of freshly isolated and cultured epithelial cell types.

9.2.2 Devices

Various devices are available for breaking cells open. Probably the mildest and most likely to leave the largest number of organelles intact is the hand-held Dounce apparatus, in which sheer is generated as a glass pestle is moved into and out of a closely fitting glass mortar containing the sample. The magnitude of the sheer is determined, in part, by the rate of movements and the clearance between the mortar and pestle. The apparatus is usually marketed with 'tight' and 'loose' pestles. Larger sheer rates are generated in the similar Potter–Elvehjem apparatus, in which the pestle is constructed of Teflon® and rotated by an electric motor as the mortar is manually raised and lowered so that the pestle passes through the sample.

Very large sheer rates are generated in the OmniMixer® (Sorvall Instruments), Polytron® (Brinkmann Instruments) and Tissumizer® (IKA-Works) devices. In this type of device a motor-driven propeller drives fluid through narrow apertures in a shaft which is immersed in the sample. Varying sized shafts are available to accommodate a range of sample volumes. The propellers are designed to rotate at rates which cause cavitation as well as sheer. One potential problem with cavitation is that some proteins may denature at the gas–liquid interface. With tissues that contain a good deal of mucus, cavitation generates a thick froth which entraps membranes and which may require considerable time to dissipate. In many cases this may be tolerable, but in some it may be necessary to decrease the motor speed to eliminate cavitation.

Cavitation can also be exploited as the disrupting principle. Commercially available cavitation bombs (Parr) can be pressurized from laboratory compressed N_2 cylinders. Samples are allowed to equilibrate at high pressure inside the bomb. The cavitation that occurs as the sample is released to atmospheric pressure is sufficient to disrupt most cell types. This release is done through a small orifice, limiting the use of cavitation to cell suspensions and tissue dispersions that will pass readily through the orifice.

9.3 SEPARATION METHODS

9.3.1 Centrifugation

Virtually all membrane isolation schemes employ differential centrifugation (differential sedimentation), which separates membranes on the basis of their sedimentation coefficient (the rate at which they move through a centrifugal field). Nuclei and secretory vesicles can usually be sedimented with forces of 1000 × g for 10 min. Mitochondria and lysosomes typically sediment at 10 000 × g for 20 min. Plasma membrane vesicle samples can frequently be obtained with centrifugation for as low as 20 000 × g for 30 min, although the yields in such samples will be small, with most of the plasma membrane vesicle population(s) remaining in the supernatant fraction. The entire population of plasma membrane vesicles can usually be sedimented by centrifugation at 100 000 × g for 60 min.

It is never possible to obtain a highly purified membrane or organelle sample with a single differential sedimentation step, for the simple reason that the volume occupied by the membrane pellet will entrap any membranes or soluble components that occupied this space at the start of the centrifugation run. Additional slowly sedimenting membranes will inevitably migrate into this space during the time necessary for all of the more rapidly sedimenting membranes to reach the pellet. The simplest solution to this problem is to resuspend the pellet and repeat the centrifugation procedure several times more, essentially diluting out the more slowly sedimenting membrane populations.

An alternative to repeated cycles of centrifugation and resuspension is the technique of rate-zonal centrifugation, which is typically performed in a swinging bucket rotor. The sample to be analyzed is layered over a density gradient or a column of medium, and then centrifuged at a force–time combination sufficient for complete sedimentation of the rapidly sedimenting components. One requirement for this technique is that the medium must be less dense than the equilibrium density of the sedimenting membrane population. If nuclei have been removed by low-speed centrifugation, it is frequently possible to obtain mitochondrial samples virtually free of contamination by suborganellar membranes with a single rate-zonal centrifugation run.

As already noted, most cell lysis procedures fragment the endoplasmic reticulum, Golgi complex, endosomes and plasma membranes into populations of small vesicles, or microsomes, which have broad, overlapping size distributions. It is rare, indeed, that a highly purified, representative sample of one of these populations can be obtained with only differential centrifugation methods. The classic solution to this problem has been to subject microsomal or mixed microsomal–mitochondrial fractions to centrifugal fields until the particles reach their equilibrium positions on density gradients (isopycnic centrifugation). Density gradient centrifugation is usually done with swinging bucket rotors. Vertical rotors have been introduced which allow gradients to re-orient in a centrifugal field, markedly decreasing the path length that membranes must traverse to reach their equilibrium positions. Very large-scale analyses can be performed in zonal rotors, which are essentially bowls that can accommodate a single sample and gradient with total volumes between 250 ml and 1.5 l.

Density gradients can be constructed with various solutes. The most common solute for density gradients used in isolation of subcellular membranes is sucrose, although other highly water-soluble molecules, including glycerol and sorbitol, have also been used. While these low molecular weight polyols provide a benign medium for soluble proteins and membranes, to achieve solutions with large enough densities they must be present in such high concentrations that they perturb osmotically reactive vesicles and organelles. One strategy for avoiding extreme hypertonicity has been to use polymers, such as dextran (Pharmacia) and ficoll, low molar concentrations of which can be sufficiently dense to allow for a useful separation range without untoward osmotic impact. A more recent strategy has been to use colloidal suspensions of silica particles (Percoll®) or solutions of a tri-iodinated benzamide derivative of glucose (Metrizamide®).

Percoll is sufficiently massive that density gradients form under the same centrifugal forces needed for resolution of microsomal membrane populations. Thus, one will almost always arrange the Percoll concentration in a fashion that the self-formed gradient covers the desired density range. Glycerol, sorbitol and sucrose gradients, on the other hand, are always constructed in advance of the membrane separation procedure. The simplest way to construct a density gradient is to use a syringe or transfer pipet to layer a sequence of solutions of decreasing density into the centrifuge tube, yielding a discontinuous, or step, density gradient. Most membranes in microsomal fractions will accumulate in discrete bands at the steps. These bands are easy to see and to collect with a transfer pipet or syringe. However, they may not necessarily correspond precisely to the distributions of the membrane population of interest, and their discrete appearance may belie a good deal of heterogeneity.

It is often advisable to construct continuous rather than discontinuous density gradients. These are easily constructed with a conventional two-chambered gradient maker comprising a reservoir chamber connected to a mixing chamber which, in turn, empties into the centrifuge tube. In the simplest arrangement, fluid with the density desired at the high density limit of the gradient is placed in the mixing chamber, and fluid corresponding to the low density end of the gradient is placed in the reservoir chamber. The fluid emerging from the mixing chamber is allowed to flow down the side of the centrifuge tube so that each successively less dense incremental volume will layer, with little turbulence and mixing, over the fluid already in the tube. If the fluid levels are identical to begin with, if the fluid levels fall at the same rate (i.e. if fluid flows from the reservoir chamber to the mixing chamber at half the rate that it flows from the mixing chamber into the centrifuge tube) and if the mixing chamber is adequately mixed, then the density of the fluid in the centrifuge tube will decrease linearly with position from bottom to top.

Several embellishments on this simple theme give one much greater control over the shape and reproducibility of the density gradient. One can carefully control the emptying rate by using flexible tubing and a peristaltic pump, rather than gravity, to drive fluid from the mixing chamber to the centrifuge tube. The AutoDensiFlow® (Litton Industries), has been designed to deliver fluid to, or remove it from, a surface, so that it is useful both in constructing and in collecting fractions from density gradients. The shape of the density gradient can be altered to optimize the spacial separations between populations of interest. By applying a piston to the surface of the fluid in the mixing chamber to keep its volume constant, a gradient is obtained in which the composition varies as a rectangular hyperbola as a function of volume emerging from the mixing chamber. If the high density fluid is placed in the mixing chamber, and the increasingly less dense fluid emerging from the gradient maker is layered on top of the fluid already in the tube, then the gradient will have a concave shape. If the high density solution is placed in the reservoir chamber and the low density solution placed in the mixing chamber, the gradient will have a convex shape, asymptotically approaching the density of the fluid in the reservoir chamber. With this arrangement, fluid emerging from the gradient maker will have to be delivered through a length of rigid tubing to the bottom of the centrifuge tube, rather than layered on top of the fluid already in the tube.

9.3.2 Density perturbation

The density at which certain membrane populations equilibrate may be influenced by their history of exposure to hypertonic media. This is often true for endoplasmic reticulum membranes. If they are originally suspended in a low density medium and layered on top of a density gradient such that they sediment toward their equilibrium position, they may equilibrate at a significantly lower density than if they had been suspended in a high osmolarity medium and layered underneath a density gradient such that they float toward their equilibrium position. This phenomenon probably reflects an interaction between the membrane's permeability to the major solute in the density gradient medium and the forces which influence its tendency to resume its original shape after it has been shrunken osmotically. The different responses of membranes after exposure to a hypertonic medium can be exploited in cell fractionation schemes.

Other methods have been used to alter selectively the equilibrium densities of certain membrane populations. Digitonin, which binds with high affinity to cholesterol, increases a membrane's density depending on how much cholesterol it contains, either by extracting cholesterol from the membranes, leaving behind other more dense constituents, or by adding its own mass to the mass of the membrane's other constituents. Digitonin has some detergent activity, so that it cannot be used in schemes to isolate sealed membrane vesicles for transport studies.

Horseradish peroxidase (HRP) can be introduced into the endosomal system as a fluid phase component or covalently attached to a ligand that binds to receptors which are then internalized from the plasma membranes. When crude membrane samples are subsequently incubated with H_2O_2 and diaminobenzide, the dense reaction product forming within vesicles that contain HRP significantly increases their net density. By a judicious selection of the incubation time, one can preferentially load the early endosomal compartments with HRP. In at least some cell types, traffic between the early and late endosomes stops when the temperature is reduced below 18°C, allowing for greater selectivity in loading the early endosomal compartment. It may also be possible to load the late endosomal compartment selectively by allowing cells to take up HRP at 37°C, then transferring cells to 18°C medium and allowing HRP to wash out of the early endosome. One difficulty that should be anticipated in the use of this technique is that the oxidizing reaction conditions may inactivate membrane marker enzymes and transporters.

9.3.3 Selective membrane aggregation

Most subcellular membranes aggregate in the presence of millimolar concentrations of divalent cations. The apical membranes of intestinal and proximal tubular epithelial cells notably do not participate in such aggregates. These surface membranes can then be harvested by differential centrifugation after other membranes have been precipitated in the presence of Mg^{2+} or Ca^{2+}. Most recent studies have employed Mg^{2+} to avoid Ca^{+2}-dependent activation of proteases and phospholipases. Both total cell lysates and post-nuclear supernatant fractions have been used as the starting points for such separations. Mg^{2+} is added to a final concentration of 10 mM, the sample is incubated in the cold for 12–20 min, and the membrane aggregates are sedimented by centrifugation at 3000 × g. The resulting supernatant fraction is then centrifuged at 30 000 × g for 30 min to sediment the apical membrane vesicles.

9.3.4 Selective adsorption of plasma membranes

A number of schemes for isolation of plasma membranes from cells in suspension or of the apical membranes of cells from intact epithelia or epithelia-like culture sheets have been based on the membranes' selective adsorption to glass beads, charged polyacrylamide beads and lectin-coated beads. Basolateral membranes adhering to plastic culture dishes have been isolated after cells were lysed by osmotic swelling.

9.3.5 Free-flow electrophoresis

Free-flow electrophoresis separates membranes on the basis of their surface charges. The apparatus is designed so that a uniform curtain of buffer, contained within glass plates, flows at a right angle to the field between two axial electrodes. The sample to be analyzed is pumped, with minimal turbulence, into the flowing buffer through a port located near the cathode. Thus, while the electric field causes membranes to move in the X dimension at a rate proportional to their electrophoretic mobility, determined by their net surface charges and hydrodynamic properties, the flowing buffer transports them in the Y dimension. The buffer emerging from the electrophoresis chamber is collected in 120 fractions. The capacity of this apparatus is essentially unlimited, as long as the buffer flow rate is uniform and constant and the electric field can be kept constant. Unfortunately, free-flow electrophoresis is not yet very widely employed, and the apparatus, which is rather expensive, is not available in most departments. On the other hand, electrophoretic separations of subcellular membranes can also be achieved in preparative, density-gradient electrophoresis apparatus.

9.3.6 Phase partitioning

Phase partitioning is an especially versatile separation procedure in that it operates on the basis of a variety of membrane surface properties, including properties determined by membrane phospholipid composition and carbohydrate content that are independent of surface charge. This means that phase partitioning can be used to resolve membrane samples which may appear indistinguishable with respect to the properties detected by density gradient centrifugation and electrophoresis. The countervailing limitations of this separation procedure are related to limitations in the amount of material that can be analyzed and the length of time required for each analysis. Because this procedure is so broadly applicable, at least for analytical purposes, but not yet widely employed, we will discuss it in some detail.

When water and two different water-soluble polymers, most frequently dextran and polyethyleneglycol (Carbowax®, Union Carbide), are mixed together at polymer concentrations above a critical value, the system separates into two immiscible phases. In a typical dextran–polyethyleneglycol–water system, the dextran-rich phase is the more dense and so forms the lower phase of a standard two-phase solvent system. Different membranes partition into the two phases and the interface in ratios that depend on the polymer concentrations, pH, ionic strength and temperature. Conditions can frequently be found that will cause the different membrane populations in a subcellular fraction to partition differently. These differences are almost always quantitative rather than qualitative; it seldom happens that one population partitions entirely into the lower phase and interface, and the other entirely into the upper phase. However, even small quantitative differences in partitioning behavior can be amplified and exploited by the principle of counter-current distribution that is employed in chromatographic processes.

An apparatus that automatically performs counter-current distribution separations can be obtained from the instrumentation shops in the Department of Chemistry at the University of Lund, in Sweden, and the Department of Biology, University of Sheffield, United Kingdom. Both operate according to the same principle. As illustrated in Figure 9.2, two lucite disks, each with 60 or 120 radially oriented cavities machined into its surface, are placed together on the horizontal plane so that the cavities combine to form chambers. In the upper disk there are holes which allow access to each of the chambers. The chambers are individually loaded with aliquots of the lower phase. Because most useful partitioning is between the upper phase and the lower phase plus interface, the volume of lower phase is limited to the portion of the chamber formed by the lower disk. Aliquots of the upper phase are then added such that they are contained in the portion of each chamber formed by the upper disk.

The membrane sample to be analyzed is sedimented, resuspended in a volume of the upper phase, and placed into several adjacent chambers, usually equivalent to 10% of the total number that will be used for the analysis. One can devote all the chambers to a single separation, or one can analyze several samples simultaneously by devoting a series of chambers to each. The largest number that can be analyzed in a 120-chamber apparatus with reasonable resolution is six, each using 20 chambers.

After loading, the apparatus is programmed to mix the phases, rest while the phases separate, and then rotate the upper disk for one cavity's distance with respect to the lower disk. In a 120-chamber analysis in which the starting sample has been placed in chambers 1–12, the first transfer carries membranes in the upper phase of chamber 12 to chamber 13, and it transfers a fresh upper phase on to the membranes remaining in the lower phase of chamber 1. This cycle is repeated for as many transfers as desired. A surprisingly small number of transfers produces an approximately normal distribution for each membrane population in the sample. The position of the mode of each depends on the coefficient of partitioning between the upper phase, which corresponds to the mobile phase of a conventional chromatography system, and the lower phase plus interface, which correspond to the stationary phase.

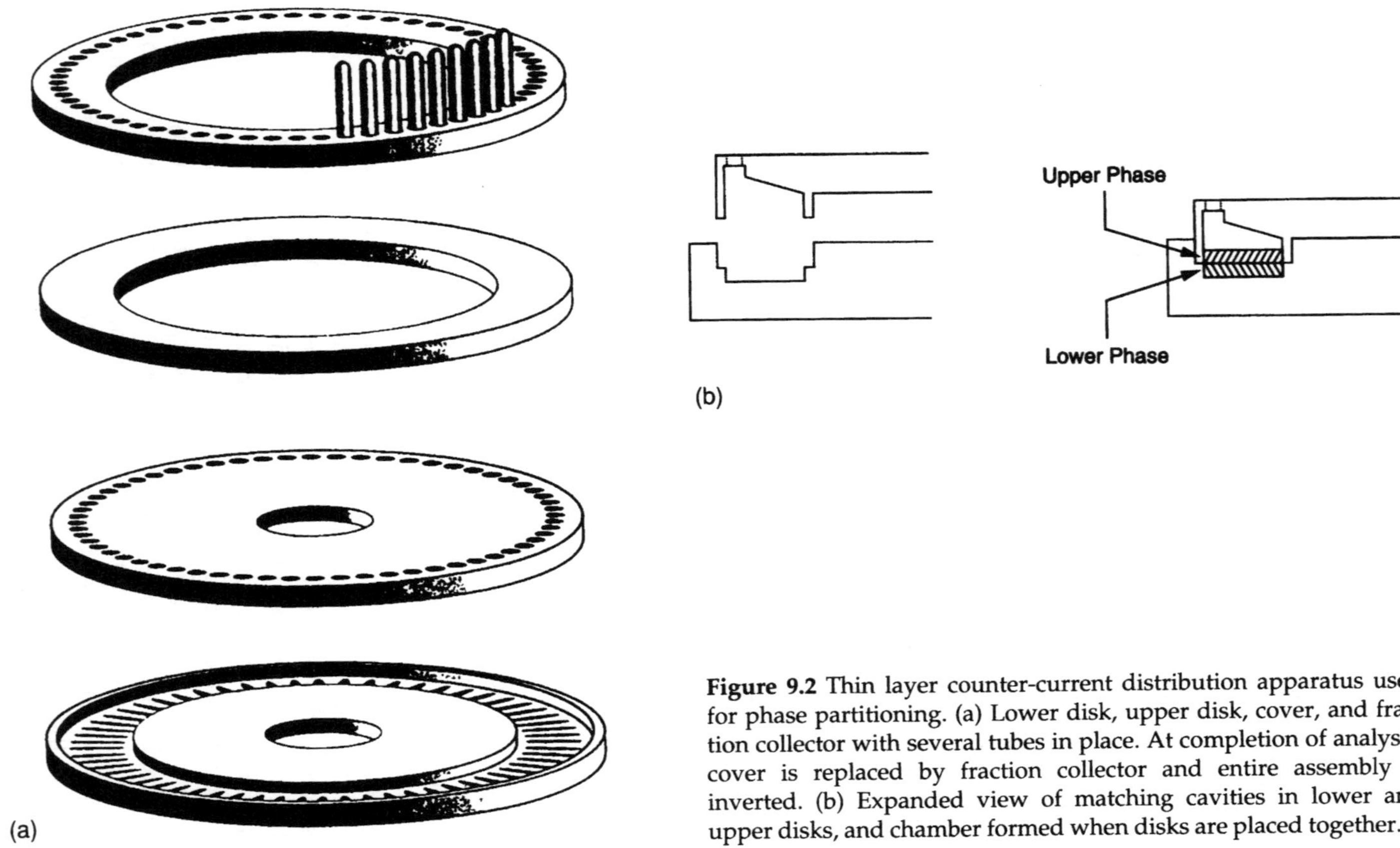

Figure 9.2 Thin layer counter-current distribution apparatus used for phase partitioning. (a) Lower disk, upper disk, cover, and fraction collector with several tubes in place. At completion of analysis, cover is replaced by fraction collector and entire assembly is inverted. (b) Expanded view of matching cavities in lower and upper disks, and chamber formed when disks are placed together.

In some cases it will be of interest to examine how variations in the phase system composition will influence membrane partitioning behavior. With a 120-chamber apparatus, one can analyze samples in as many as three different phase systems with reasonable resolution. Chambers 1–40 are prepared with the first system, 41–80 with the second and 81–120 with the third. Samples, in the appropriate upper phases, are placed in chambers 21–22, 61–62 and 101–102, and subjected to 18 transfers. Like free-flow electrophoresis, phase partitioning is not yet very widely employed. However, a regular series of international conferences have focused on this technology, and several volumes on the topic have been published (e.g. Fisher and Sutherland, 1989; Walter and Johansson, 1994). If one is contemplating establishing a cell fractionation facility, it is worth noting that the counter-current distribution apparatus is less expensive than an ultracentrifuge.

9.4 CHARACTERIZATION OF ISOLATED MEMBRANES

During the design of a membrane isolation scheme, and also once a final product has been isolated, it will be of interest to characterize the various subcellular fractions in a variety of different ways. In addition to the catalytic activities illustrated in Figure 9.1, one can quantitate selected immunoreactivities and transport activities.

9.4.1 Transport activities of isolated membrane vesicles

Methods employing radiotracers and potential-, pH- and ion-sensitive dyes (a particularly broad selection of dyes is offered by Molecular Probes, Inc., Eugene, Oregon, USA) are available for studying transport phenomena once membrane vesicle samples have been isolated. Tracer flux methods are certainly the most commonly employed means of studying transport in isolated membrane vesicles. The intravesicular space can be loaded with a desired medium by resuspension, usually in a small volume, with membrane protein concentrations in the neighborhood of 10 mg/ml, a cycle of quick freezing (e.g. by immersion in a liquid N_2 bath), thawing, and trituration through a narrow-gauge syringe needle. Equilibration at room temperature or 37°C for up to 30 min may enhance vesicle membrane resealing. An aliquot of the sample is mixed with an aliquot containing the radiolabelled substrate, often with a 5–10-fold dilution. The uptake process is stopped by dilution (again, 5–10-fold) with an ice-cold quenching medium, which is usually similar to the reaction medium but with the radiolabelled transport substrate removed, then quickly filtered to separate membranes (and the substrate they have taken up) from the substrate remaining free in the reaction medium. Nitrocellulose filters with 0.2–0.45 μm pore diameters are suitable for most membrane samples and transport substrates. The filter is then rinsed with two additional aliquots of the quench buffer.

For many purposes it will be necessary to measure the uptake time-course over small intervals. In such cases it is usually best to use a separate reaction tube for each time point, rather than withdrawing a series samples from an ongoing reaction over time. One of the best systems is to place the reaction medium containing the transport substrate in a discrete droplet at the bottom of a polystyrene tube. If the tube is tilted to a 45° angle, a small droplet of membrane sample can be held on the side of the tube, above the reaction medium. The two are mixed when the tube is turned upright and touched to a vortex mixer. If one measures time by counting beats on a metronome, it is possible, with practice, to run reproducible reactions for as little as 0.5 sec. It is critical to obtain accurate unidirectional flux measurements. With tracer uptake studies these are approximated by initial rate measurements. Thus, one should take care to assure that the uptake process is proceeding linearly over the time interval that will be standard for a particular study.

Some studies of Na^+/H^+ antiport and ATPase-driven H^+ transport have used pH-sensitive dyes. Most work has been done with acridine orange, the visible light absorbance spectrum and fluorescence intensity of which are sensitive to pH. A number of ion-sensitive dyes have been synthesized which, in principle, should also prove highly useful for measurements of transport in membrane vesicles.

Potential-sensitive dyes can be used to measure the permeation of charged species for which selective dyes are not available. The Na^+ coupled transport of neutral solutes, such as glucose or amino acids, can be measured in this way. Potential-sensitive dyes can also be used to measure the relative permeabilities of pairs of ions with so-called bi-ionic potential measurements. If only one of the ionic species in a system is very highly permeant (e.g. because the membranes have been treated with a selective ionophore), a gradient of this species will charge the membrane potential to the species' Nernst potential. If another permeant species is added to the system, it dissipates the membrane potential at a rate proportional to its concentration and permeability coefficient.

9.4.2 Immunoblotting and immunoprecipitation

A final means of characterizing isolated membrane vesicles is with regard to their contents of immunoreactive constituents. This is done with dot- or slot-immunoblotting, in which a template allows one to apply subcellular samples in a regular array to filter papers, which may be nylon, nitrocellulose or paper that has been treated in such a fashion that proteins attach to it covalently. In Western blotting, samples are first resolved by sodium dodecylsulfate–polyacrylamide gel electrophoresis, then transferred to the paper. The papers are then incubated, with appropriate washing procedures, with the primary antibody and with reagents that will detect the primary antibody (e.g. ^{125}I-labeled protein A) or a secondary antibody conjugated with horseradish peroxidase or alkaline phosphatase. Labels, such as [^{35}S]-methionine or [^{3}H]-fucose, that have been metabolically incorporated into specific proteins or glycoproteins can be detected by autoradiography after the subcellular sample has been solubilized and the molecules of interest precipitated and resolved by SDS-polyacrylamide gel electrophoresis. This is most often accomplished with a primary antibody followed by a secondary antibody or protein A conjugated to Sepharose beads.

9.5 MULTIDIMENSIONAL SUBCELLULAR FRACTIONATION

With the repertoire of separation and characterization methods outlined above, one may contemplate isolating membranes from any cell type that can be lysed. We will illustrate by showing how these tools are combined in a subcellular fractionation scheme that has already been generalized for a wide variety of tissues and cell types.

9.5.1 Keeping track of marker distributions

A crucial aspect of the analytical approach is a balance sheet of the marker contents in each of the terminal fractions that arise during an isolation procedure. It is often useful to know the marker contents of key intermediate fractions as well. Thus, the volumes of the fractions of interest must be measured, and samples must be set aside for marker determinations. The resulting marker distribution data will be useful in several ways. The overall purification factor for a membrane sample is traditionally calculated from the enrichment of a characteristic marker's specific activity relative to the specific activity in the initial homogenate. Purification of the desired membrane sample with respect to potentially contaminating membranes can be evaluated from decreases in the specific activities of the relevant markers. With a balance sheet, it is possible to adjust the calculated purification factors to account for any activation or inactivation of the markers that occur during the course of the isolation procedure.

It will often happen that the selected marker for the population of interest is associated with additional membrane populations so that, in order to estimate more accurately the extent to which the membrane population of interest has been purified, it will be necessary to adjust the calculated purification factor for the percentage of the cell's total marker content that is localized to that population. A sample spread-sheet for marker distribution data is presented in Appendix C.

9.5.2 A general subcellular fractionation scheme

As in most membrane isolation schemes, the initial cell lysate is centrifuged at a speed great enough to sediment nuclei, intact secretory vesicles and connective tissue fragments, i.e. 1000 × g for 10 min. The pellet should be resuspended, homogenized and centrifuged at least once more to obtain reasonable yields of plasma membrane vesicles and other suborganellar membrane populations in the low-speed supernatant. One can evaluate the number of cycles that suffice by measuring the distributions of plasma membrane markers between the combined low-speed supernatant fraction and the low-speed pellet. With reference to the spread-sheet in Appendix C, the final low-speed pellet and combined supernatant fractions are designated P_o and S_o. The mitochondria and the plasma membranes and other microsomal vesicle populations are then resolved from S_o. An intermediate cycle of differential centrifugation at 10 000 × g for 20 min may be performed to sediment the mitochondria and, possibly, lysosomes, but this is likely to diminish the yields of the plasma membranes and other microsomal populations, since the more massive elements of these populations will sediment with the mitochondria.

S_o is then subjected to isopycnic density gradient centrifugation. If one can manage to keep the volume of the initial cell lysis suspension and combined low-speed supernatants reasonably small, virtually the entire S_o fraction can be applied to the density gradient, saving the time that would be consumed by an intermediate differential centrifugation step for harvesting and concentrating the membranes. This strategy also has the advantage of decreasing the likelihood that membranes might combine to form aggregates that are difficult to disperse.

When working on a new membrane isolation problem, it is advisable to employ continuous, rather than step, density gradients and, after the centrifugation run, to collect the gradient in a series of equal-volume fractions. The membranes in each of the density gradient fractions can be harvested by high-speed centrifugation (100 000 to 250 000 × g for 60 min) after the density gradient solute has been diluted with buffer. The series of pellets, collectively designated ΣP_i, is resuspended in buffer and samples are set aside for marker determinations. A sample of the pooled supernatants (ΣS_i) should also be set aside for marker determinations.

The principal reason for emphasizing the use of continuous density gradients is that the distributions of membrane markers among the resulting fractions very often allow one to perceive the distributions of different populations of membrane and to evaluate, at least in a provisional fashion, the extent to which neighboring populations overlap each other's distributions. One may find that a marker which had been selected because of its presumed specific localization to a particular membrane

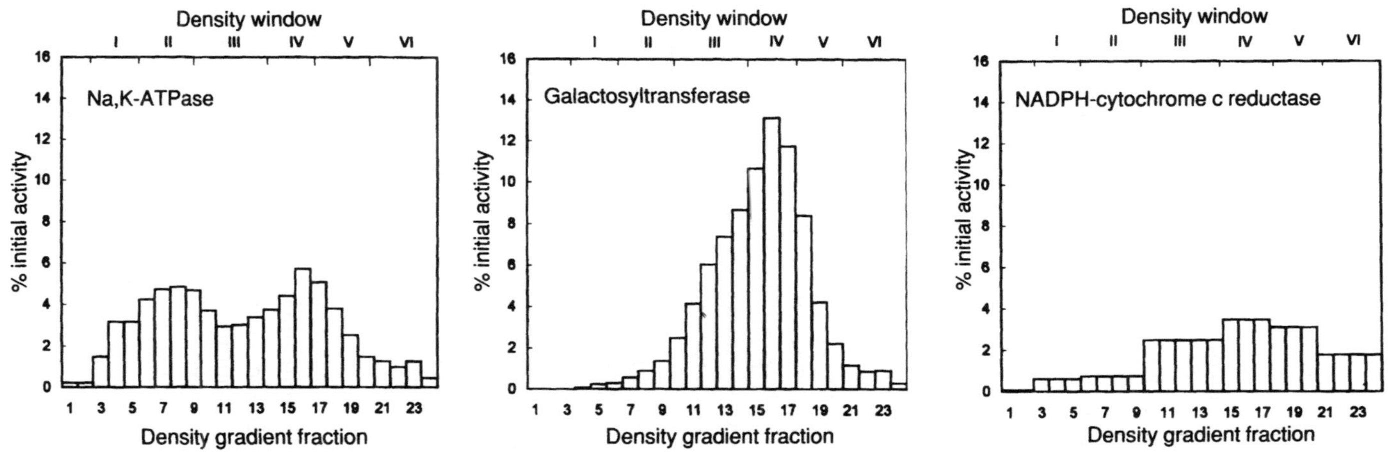

Figure 9.3 Density distributions of the Na,K-ATPase, galactosyltransferase, and the NADPH-cytochrome c reductase activities in the ΣP_1 fraction from rat exorbital lacrimal gland fragments. Fractions are pooled into density windows prior to determination of NADPH-cytochrome c reductase and subsequent separation procedures. (Redrawn from Bradley *et al.*, 1994.)

population is in fact associated with one or more distinct populations of vesicles, as evident from a multi-modal density distribution. This will frequently be the case for Na,K-ATPase and other plasma membrane markers. This point is illustrated in Figure 9.3, which summarizes the density gradient distributions of Na,K-ATPase, galactosyltransferase and NADPH-cytochrome c reductase from rat exorbital lacrimal glands (e.g. Bradley *et al.*, 1994). The Na,K-ATPase density distribution reveals three distinct modes, indicating at least three different populations of Na,K-ATPase-containing membranes, in fractions 8, 16 and 23. A shoulder in fractions 4 and 5 suggests the presence of an additional Na,K-ATPase-containing membrane population.

It would be difficult to know, *a priori*, whether an observed multiplicity of Na,K-ATPase-containing membrane populations reflects a microdomain organization of the plasma membranes, varying association of remnant cytoskeletal components with otherwise identical plasma membrane fragments, or localization of the plasma membrane marker to a variety of distinct organelles. Additional marker distribution data are informative, although not necessarily decisive, in this regard. The peak centered in fraction 16 closely parallels the major peak of galactosyltransferase, a parallel which would suggest that a significant component of the total Na,K-ATPase is associated with Golgi membranes. The parallel peaks of Na,K-ATPase and galactosyltransferase are overlapped by the distributions of NADPH-cytochrome c reductase and succinate dehydrogenase. Because succinate dehydrogenase is well established as a mitochondrial inner membrane marker and NADPH-cytochrome c reductase as an endoplasmic reticulum marker, one would infer that these distributions reflect overlap of mitochondrial, endoplasmic reticulum membrane and Golgi populations. Apart from the practical matter of resolving these three presumptive membrane populations, it would remain necessary to address the question of whether the Na,K-ATPase in this region of the density gradient was associated with the Golgi membranes or with some other membrane population.

The mitochondria can be easily separated from the other membrane populations by rate-zonal centrifugation, and they can be seen to contain little of the plasma membrane, Golgi or endoplasmic reticulum marker. To separate microsomal membranes which equilibrate at the same densities but which appear to be derived from different organelles, and to learn which populations account for the Na,K-ATPase activity, one might employ a density perturbation method, free-flow electrophoresis or phase partitioning. Each of these methods has been used in such situations during analysis of one or another cell type. Phase partitioning in aqueous dextran-polyethyleneglycol two-phase systems has proven to be extremely effective in most of the instances in which it has been attempted.

When each region of the density gradient is analyzed in the same two-phase system, the result is a two-dimensional fractionation, such as depicted in Figure 9.4, which provides a basis for systematically mapping the cell's biochemical organization. The membrane populations discerned from the two-dimensional distributions of Na,K-ATPase and galactosyltransferase, along with their presumed subcellular origins, are illustrated in Figure 9.4. This map emphasizes the diversity of membrane populations derived from the Golgi complex and presumably reflects the complexity of its microdomain organization. It also emphasizes the diversity of subcellular compartments which contain Na,K-ATPase activity, including the Golgi complex and endosomes.

The location of the endoplasmic reticulum population in Figure 9.4 was deduced from the two-dimensional distribution of NADPH-cytochrome c reductase (not shown here). The endoplasmic reticulum population is overlapped by several Golgi populations in the two-phase system employed for this particular analysis. Additional membrane populations can be discerned from the two-dimensional distributions of other markers, including most notably acid phosphatase. These tend to be centered about partitioning fractions 6–10 from density windows II–VI, and they appear to reflect additional compartments along the endosome–lysosome and secretory vesicle assembly pathways.

While the complete map from which Figure 9.4 is abstracted is remarkably crowded with sometimes overlapping membrane populations from diverse subcellular origins, it has proven feasible to obtain purified samples of virtually all of the populations which have been delineated. This has been done with systematic variations in the composition of the two-phase system. Decreasing the pH and increasing the polymer concentration both have the effect of decreasing mobility of the Golgi and plasma membrane populations, yielding purified samples of the endosomal membranes. Increasing the pH and the initial sample loading concentrations tends to increase the mobility of the Golgi and plasma membrane populations, yielding purified samples of the endoplasmic reticulum population.

9.5.3 Some applications of the general fractionation scheme

The analytical approach we describe here has been applied to a variety of questions in epithelial cell physiology. These include:

- the assembly and intracellular transit of such plasma membrane components as aminooligopeptidase (Ahnen *et al.*, 1983) and the Na,K-ATPase α and β subunits (Mircheff *et al.*, 1992);
- the existence of intracellular pools of Na/H antiporters (Hensley *et al.*, 1990) and amino acid transporters (Hensley and Mircheff, 1994) in renal proximal tubule cells;

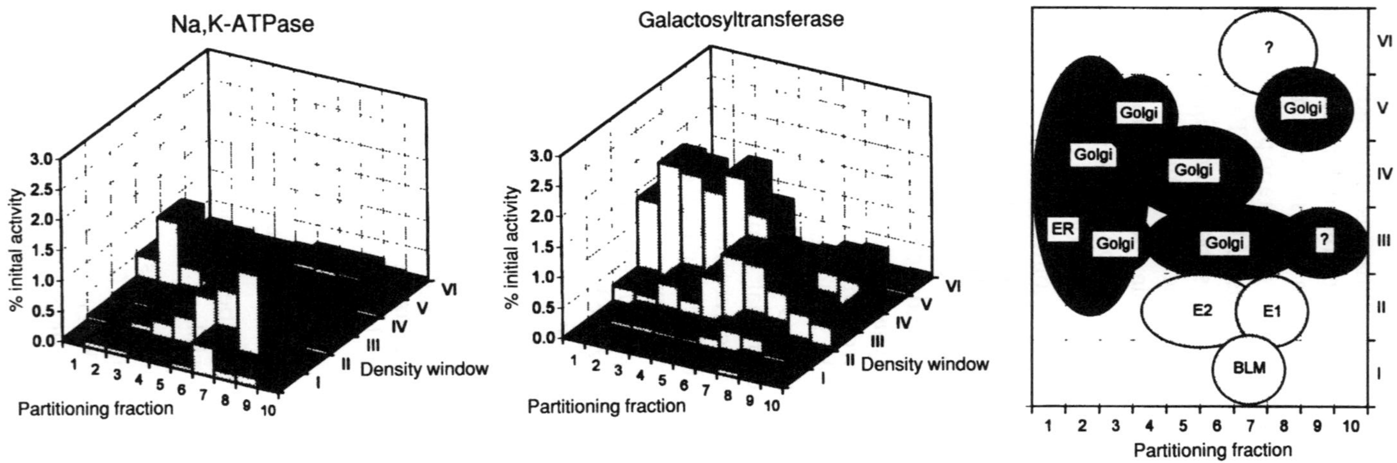

Figure 9.4 Two-dimensional distributions of Na,K-ATPase and galactosyltransferase generated when each of the six density windows illustrated in Figure 9.3 is analyzed by partitioning in a Dextran-polyethylenegylcol two-phase system with pH 8.0. The right-hand panel depicts the positions of the major endoplasmic reticulum membrane population (ER) and of the membrane populations inferred from the depicted marker distributions (BLM = basolateral membranes; G = Golgi-derived membranes; E = endosomal membranes; ? = subcellular origin at present unknown). (Redrawn from Bradley *et al.*, 1994.)

- the ability of secretagogues to stimulate recruitment of Na,K-ATPase from intracellular compartments of exocrine acinar cells (Lambert *et al.*, 1993b);
- the ability of parathyroid hormone to stimulate internalization and subsequent inactivation or degradation of Na/H antiporters in isolated renal proximal tubules (Hensley *et al.*, 1989);
- the ability of acute renal arterial hypertension to induce internalization of Na,K-ATPase in the proximal tubule (Zhang *et al.*, 1994).

REFERENCES

Ahnen, D.J., Mircheff, A.K., Santiago, N.A. *et al.* (1983) Intestinal amino-oligopeptidase: distinct molecular forms during assembly on intracellular membranes *in vivo. J. Biol. Chem.* **258**:5960–6.

Bradley, M.E., Lambert, R.W. and Mircheff, A.K. (1994) Isolation and identification of plasma membrane populations. *Meth. Enzymol.*, **228**:432–48.

DeDuve, C. (1964) Principles of tissue fractionation. *J. Theor. Biol.* **6**:33–59.

DePierre, J.W. and Karnovsky, M.L. (1973) Plasma membranes of mammalian cells. A review of methods for their characterization and isolation. *J. Cell Biol.* **56**:275–303.

Fisher, D. and Sutherland, I.A. (1989) *Separations Using Aqueous Phase Systems*, Plenum Publishing, London.

Hensley, C.B. and Mircheff, A.K. (1994) Complex subcellular distributions of sodium-dependent amino acid transport systems in kidney cortex and LLC-PK_1/Cl_4 cells. *Kidney Int.* **45**:110–22.

Hensley, C.B., Bradley, M.E. and Mircheff, A.K. (1989) Parathyroid hormone-induced translocation of Na/H antiporters in rat proximal tubule. *Am. J. Physiol.* **257**:C637–C645.

Hensley, C.B., Bradley, M.E. and Mircheff, A.K. (1990) Subcellular distribution of Na/H antiporter activity in rat renal cortex. *Kidney Int.* **37**:707–16.

Lambert, R.W., Maves, C.A. and Mircheff, A.K. (1993a) Carbachol-induced increase of Na^+/H^+ antiport and recruitment of Na^+/K^+-ATPase in rabbit lacrimal acini. *Curr. Eye Res.* **12**:539–51.

Lambert, R.W., Maves, C.A., Gierow, J.P. *et al.* (1993b) Plasma membrane internalization and recycling in rabbit lacrimal acinar cells. *Invest. Ophthalmol. Vis. Sci.* **34**:305–16.

Mircheff, A.K. (1989) Isolation of plasma membranes from polar cells and tissues: apical/basolateral separation, purity, and function. *Meth. Enzymol.* **172**:18–34.

Walter, H. and Johansson, G. (1994) *Meth. Enzymol.* 228.

Zhang, Y., Mircheff, A.K., Balkovetz, D.F. *et al.* (1994) Rapid redistribution of renal sodium transporters during acute hypertension. *Hypertension* **24**(abstract):408.

10

Methods and experimental analysis of single ion channels

Simon A. Lewis

10.1 PATCH CLAMP METHODOLOGY

The patch clamp technique introducted by Neher and colleagues (Neher, Sakmann and Steinbach, 1978; Hamill *et al.*, 1981) has been used to study ion channels at both the single-channel and the whole-cell level in animal, plant and bacterial cells. The successful application of the technique is critically dependent on the ability to form a high resistance (10–100 GΩ) seal between the patch pipette and the plasma membrane of the cell under study. The formation of a giga-seal effectively isolates the membrane patch both electrically and chemically. Electrical isolation of the patch allows the current flowing through a single channel to be resolved and the patch to be voltage clamped by simply applying a voltage to the pipette. Chemical isolation of a patch of membrane from a cell allows the normal ionic environment of the patch to be manipulated. The giga-seal is also mechanically very stable and enables the patch to be either excised from the cell or ruptured, thus creating a number of different and useful recording configurations.

Formation of a giga-seal requires particular care in the fabrication of

Epithelial Transport: A guide to methods and experimental analysis.
Edited by Nancy K. Wills, Luis Reuss and Simon A. Lewis.
Published in 1996 by Chapman & Hall, London. ISBN 978-94-010-7168-0

patch pipettes as well as in tissue preparation. The following section considers some of the principles involved in patch pipette fabrication and tissue preparation which lead to the formation of a giga-seal.

10.1.1 Patch pipette fabrication

The required instrumentation, along with the techniques of fire polishing and coating and electrode geometries involved in making a patch clamp electrode, have been described in detail by Hamill *et al.* (1981), Corey and Stevens (1983) and Sakmann and Neher (1983). For a detailed discussion of these aspects of pipette fabrication, the novice is advised to read the above publications for in-depth and detailed step-by-step instructions. The principles involved in the fabrication of a patch pipette will be discussed here because (when one is just starting out in patch clamping and has no insight into the channel size) careful preparation of the pipettes will increase the probability of finding channels.

There is no rigorous set of rules for producing a pipette that will form giga-seals on all cells and thus one is forced to proceed using a regimented, logical and empirical approach for individual cell types. The following series of rules have been suggested by Corey and Stevens (1983) when determining the species of glass to use and in some cases the subspecies. The ideal pipette should have the following properties:

- low series resistance
- low noise characteristics
- chemical inertness
- high yield of giga-ohm seals
- capacity for reuse (to form a number of seals).

Series resistance of a pipette is located from the tip of the pipette and extends towards the shaft. Thus rapidly tapering (bullet-shaped) pipettes will have a lower series resistance than more gently tapering (wedge-shaped) pipettes. Series resistance (1–10 MΩ) is negligible for excised or intact patch recording where the seal/membrane resistance is at least 1000 times higher than the series resistance (thus the command voltage does occur across the membrane); however, in the whole-cell configuration membrane/seal resistance is much lower and significant voltage drops can occur along the pipette, leading to an incorrect estimate of the voltage drop at the membrane. The shape of the glass pipette and hence its series resistance depend on the type of glass used. Bullet-shaped tips are more easily fabricated from soft glass with low melting points than from harder glass with a higher melting point.

Another property in the choice of glass is its **noise characteristics**. Hard glass (borosilicate glass such as boralex, Kimax, and Corning 7040 and 7052; aluminosilicate glass such as Corning 1723) has low intrinsic noise characteristics, while soft glass has higher intrinsic noise characteristics (e.g. soda glass or flint glass available as blue-tip hematocrit tubing, Fisher and Kimble #R-6).

The properties of low series resistance and low noise characteristics in a pipette seem to be mutually exclusive, but a survey of capillaries by Rae (1985) discovered a glass with a melting point lower than that of soft glass and a noise level lower than that of the hard glasses. Thus Corning 8161 (Potash Lead glass) has been described as the best general-purpose glass available for patch pipette fabrication. Two other recommended Corning glasses (both borosilicate) are 7040 because of its low noise characteristics and 7052 which is described as the glass of last resort (of these three) since it has the highest melting point and highest noise level.

One must also be concerned about the **chemical inertness** of the glass to be used in fabrication of a patch pipette. Cota and Armstrong (1987) demonstrated that K-currents, recorded in the whole-cell mode from a primary cell culture of rat pituitary gland, exhibit fast inactivation when either soda (VWR hematocrit tubing, blue brand) or Corning 8161 (potash lead) glasses are used in the fabrication of patch pipettes. No inactivation occurred when hard borosilicate glass (Kimax-51) was used, or when high concentrations of EGTA were used. The authors concluded that the current inactivation observed was due to a block of K^+ channels by divalent or multivalent cations released from the soft glasses. Copello *et al.* (1991) determined that the offending divalent ion was barium.

Borosilicate glass is used more frequently than soda glass, and in one case (Richards and Dawson, 1986) Corning 7052 produced seals in 80% of the attempts while Kimax (Boralex) yielded seals in less than 10% of the attempts on freshly isolated turtle colon enterocytes.

The noise levels of a pipette fabricated from soft glass can be reduced by applying a layer of Sylgard #184 to within 100 μm of the tip. Also the chemical inertness of soft glass can be cirumvented by using high concentrations of EGTA to buffer the release of multivalent cations from the glass.

The first criterion in choosing a glass is **frequency of seal formation** as well as mechanical stability of the seal. Next is noise level and geometry and last is **reusability** of the pipette. It has been suggested that pipettes once used cannot be reused. However, Rae and Levis (1984) were able to obtain seven sequential seals using Corning 7052.

10.1.2 Preparation of epithelial cells

The formation of a giga-seal requires that the membrane be relatively free from any substance (e.g. glycocalyx, basement membrane, mucus etc.) which would restrict the pipette from making close contact with the membrane. The enzymes hyaluronidase and collagenase have been used to improve the success rate of acquiring high resistance seals. However, care should be taken when using an enzyme treatment since the enzyme activity might modify channel kinetics, channel conductance or selectivity. Although channels might be stable to such enzymes as collagenase and hyaluronidase, other enzymes (e.g. trypsin) have been shown to hydrolyze epithelial Na^+ channels (Lewis and Alles, 1986). Given this observation for some serine proteases, such enzymes must be used with a degree of caution. If such enzymes are used then the investigator must determine the effect of the enzyme on the *in vitro* preparation. Otherwise the results should be interpreted with prudence.

Three types of epithelial preparation have been studied: tubular preparation, flat sheet epithelia and tissue cultured cells. Epithelia perform vectorial (directed) transport due to the two cell membranes which are in series with each other and which contain differing transport properties. It has been demonstrated that, upon dissociation of an epithelium into individual cells, the integral membrane proteins are redistributed over the entire membrane surface (Ziomek *et al.*, 1980; Dragsten *et al.*, 1981). Thus for studies of ion channels in the apical membrane, one must leave the epithelium intact and select cells away from the epithelial edges. This is particularly critical for tissue cultured cells.

Two approaches can be used to gain access to the apical surface of renal tubules. The apical membrane can be exposed by ripping the tubule lengthwise with a sharp needle. The tubule is then secured to a coverslip and transferred to the experimental setup (Hunter *et al.*, 1984). Alternatively, the tissue can be perfused from one end and the apical membrane approached from the opposing end (Gogelein and Greger, 1984). In both methods seals were formed without the need for enzymatic cleaning of the apical surface.

Patch clamping the basolateral membrane is more difficult since the basement membrane and basolateral membrane are intimately linked. Because the basement membrane acts as a mechanical support for the epithelial cells, removal of this structure normally leads to dissociation of the cells. It was recently demonstrated that removal of the basement membrane resulted in the appearance of an apical membrane potassium channel in the basolateral membrane (Copello *et al.*, 1993). A novel approach has been used to circumvent the problem of channel redistribution. Briefly, the lateral membrane at the open end of a single-ended perfused tubule can be directly patch clamped (Gogelein and Greger, 1984). Even under these conditions there is a possibility of channel mixing between the two membranes because of the disrupted tight junctional complex.

The bathing solution used in the patch clamp chamber must be carefully chosen. Many studies have been performed at room temperature (which is appropriate for amphibian cells) with a NaCl solution usually buffered to pH 7.2–7.4 with 10 mM HEPE. In two tissues – *Necturus* proximal tubule (Lopes and Guggino, 1987) and rabbit urinary bladder (Donaldson and Lewis, 1990) – removal of bathing solution HCO_3/CO_2 causes a 15 mV and 10 mV depolarization, respectively. In the rabbit urinary bladder this depolarization is a result of a decrease in basolateral membrane K^+ permeability (decrease in number of K^+ channels). It is best then to incubate the cells in the same solutions and at the same temperature in which the *in vitro* epithelium was first studied.

10.1.3 Formation of a giga-seal

To monitor the formation of a giga-seal, voltage pulses of appropriate size are applied to the pipette and the resultant current output, which is proportional to pipette-seal resistance, is observed on an oscilloscope screen. Hence any increase in seal resistance will cause a decrease in current output. When the pipette is pushed up against the cell, the current pulses will become slightly smaller, reflecting an increase in seal resistance. At this point, application of gentle suction (20–30 cm H_2O) should result in a large and relatively rapid increase in resistance, signaling the formation a giga-seal which should persist even after the release of suction. Verification of giga-seal formation can be obtained by simply increasing the gain of the patch clamp and monitoring the oscilloscope trace. Typically the trace should be flat except for capacitive spikes at the start and end of the voltage pulse.

The success rate for the establishment of giga-seals is critically dependent on the size and shape of the pipette to be used and the relative cleanness of the membrane surface to be patched. In addition, the success rate can be improved by following some general rules of procedure as outlined by Hamill *et al* (1981):

1. Pipettes should always be moved through the air–water interface with a slight positive pressure (10 cm of H_2O) to the pipette so that pipette solution streams out of the pipette. The solution stream helps to keep the pipette tip clean and free from obstruction by cell debris until contact with a cell is made.
2. Do not reuse a pipette (unless properties of glass permit multiple attempts).
3. Remove debris from the air–water interface and filter all solutions to be used.
4. Avoid using PO_4^{2-} buffered Ringers when Ca^{2+} is present since small crystals may form at pipette tip by precipitation. HEPES appears to be an effective alternative buffer, although its use may not be physiologically appropriate and may interfere with some channels (see discussion above).

5. Slightly hypo-osmotic (5–10%) pipette filling solutions improve the possibility of giga-seal formation.

Although the nature of the seal between the membrane and the glass is unknown, it is known that application of suction to the pipette produces an omega-shaped deformation of the cell membrane, giving an increase in the area of glass–membrane contact (Sigworth and Neher, 1980). Thus the highest resistance seals should be obtained in patches where the membrane is drawn far into the pipette, producing longer sealing regions (Auerbach and Sachs, 1984).

Sakmann and Neher (1984) recognize four possible causes of altered channel function (caused by membrane deformation) which must be guarded against:

- local leaks in the membrane;
- altered receptor or channel mobility caused by detachment from cytoskeleton elements;
- narrowing of the pipette tip by the formation of a cytoplasmic bridge between the membrane in the pipette tip and the cell surface membrane (bottleneck);
- ill-defined voltage gradients in the area of membrane where the high resistance seal is formed (rim effects).

To ensure that giga-seal formation is not altering channel properties, the results should be compared with conventional current measurements. In studies where this has been done, channels appear to function in a normal manner.

10.1.4 Patch configurations

The different patch configurations that can be obtained are displayed diagrammatically in Figure 10.1. Formation of each configuration starts in the cell-attached mode with the formation of a giga-seal. From this position the combination of either rupturing the membrane patch with suction and/or withdrawal of the pipette will produce one of three well-defined configurations. For a detailed review of configuration formation refer to the seminal paper of Hamill *et al.* (1981) and to Sakmann and Neher (1984). For more details of experimental methods, refer to the examples given with each configuration.

(a) Cell-attached patch

A patch of membrane is isolated (electrically) from the rest of the cell by the formation of a giga-seal enabling the investigator to study channels within the patch in their normal ionic environment. This configuration

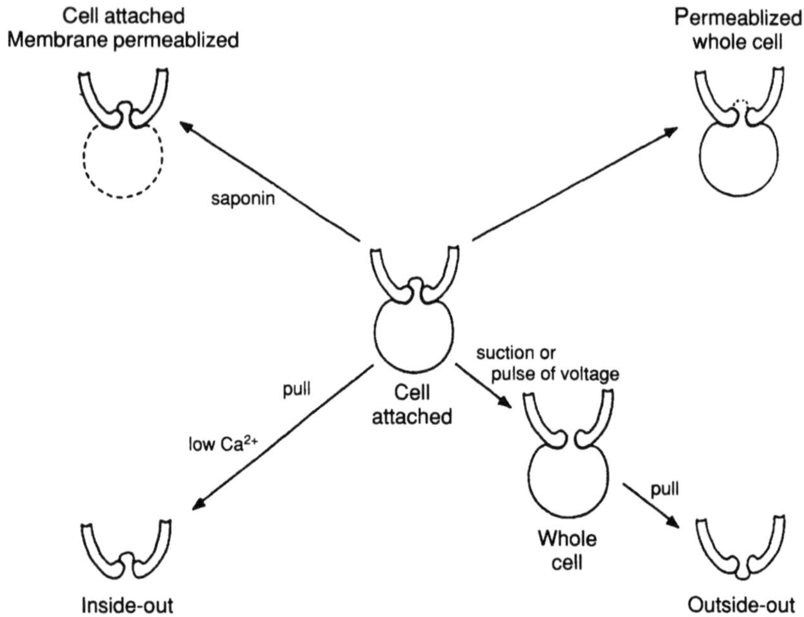

Figure 10.1 The different patch configurations available.

has a number of specific uses. The small area occupied by the patch allows experiments to be conducted using ions (in the pipette) which are advantageous for the study of channel selectivity but would be harmful if applied to the whole cell. In addition, the measurement of current is limited to a small area of the cell membrane which is chemically and physically isolated from the bathing medium, thus the addition of a substance to the bathing medium which produces changes in the activity of channels within the patch must be acting via intracellular signals. The cell-attached mode can be used to study the mechanisms underlying hormone and transmitter activation of channel activity.

When using the cell-attached patch recording configuration, one has to be aware that small cells may have input resistances in the range of 10 GΩ and that because of this high resistance the applied potential may only partially appear across the patch membrane. This uncertainty in the absolute value to which the membrane is being clamped, when using the cell-attached configuration, will introduce errors into the construction of current–voltage relationships, limiting the usefulness of this configuration.

Another limitation with the cell-attached configuration is that it is not easy to change the cytoplasmic or pipette solutions. To circumvent these problems a number of investigators have developed a perfusion system which allows for the rapid exchange of pipette filling solutions (e.g. Lapointe and Szabo, 1987). In addition cytoplasmic composition can be changed in the cell-attached mode by permeabilizing the cell membrane by the rapid bath exposure of the membrane to saponin, thereby allowing rapid equilibration between the bath solution and the cell interior (Dunne and Petersen, 1986).

(b) Inside-out patch

After the formation of a giga-seal, the pipette may be drawn away from the cell surface to form an inside-out patch. Sometimes this procedure causes a vesicle to be formed in the membrane tip. The probability of vesicle formation may be reduced by using a low concentration of Ca^{2+} in the bath. If a vesicle does appear, subsequent exposure of the vesicle to air, or alternatively touching the vesicle against an oil droplet or Sylgard surface within the bath, will cause the outer shell of the vesicle membrane to rupture, producing an inside-out patch.

The inside-out configuration is the one of choice for the construction of I-V relationships since the voltage across the patch can be quickly and accurately clamped. In this configuration the cytoplasmic face of the membrane is exposed to the bath solution. This allows the composition of the solutions on either side of the membrane to be controlled. Typically the solution in contact with the intracellular surface of the membrane is changed repeatedly (for methods see Yellen, 1982; Fenwick *et al.*, 1982; Kakei and Ashcroft, 1987) in order to study its influence on membrane currents and channel selectivity.

(c) Whole-cell configuration

To obtain the whole-cell recording mode, the patch of membrane under the seal is broken by either high voltage or suction. If the giga-seal remains intact the patch clamp electrode will now be in contact with the cytoplasm. Electrical access to the cell interior is indicated by a sudden increase in capacitive transients from an applied test pulse and, depending on the cell input resistance, a shift in current level. The capacitive transients are due to the cell capacitance and can be reduced by employing the capacitance cancellation circuitry of the patch clamp amplifier. If there is a high level of Ca^{2+} buffering capacity (e.g. 10 mM EGTA) in the pipette solution, resealing of the patch membrane will not occur and whole-cell recordings can be made.

Whole-cell recording can be done under voltage or current clamp conditions. In the current clamp mode, and in the absence of current injection, the potential recorded will be the cell membrane potential. This approach has been used to measure membrane potential in small cells such as red blood cells where conventional microelectrode techniques have failed to give reliable measurements of membrane potential (Hamill, 1983). Voltage clamping the membrane potential of a cell in the whole-cell configuration is simply a matter of applying a potential to the pipette electrode. This allows conventional voltage clamp experiments to be carried out on cells not normally amenable to the approach using the more traditional microelectrode technique.

In the whole-cell configuration the cytoplasm of the cell is in direct contact with the pipette solution and the composition of the pipette solution can be used to dialyze the intracellular environment of the cell, enabling intracellular contents to be changed as pipette solution is changed. The ability to alter the intracellular environment has obvious advantages but in some cases dialyzing the cell may result in the loss of essential cytoplasmic components. To prevent the loss of such components a new recording configuration has been developed. In this configuration a giga-seal is first formed but then, instead of gaining access to the interior of the cell by disrupting the membrane patch by suction, the membrane patch is permeabilized by the addition of either ATP^{4-}(Lindau and Fernandez, 1986) or the polyene antibiotic nystatin (Horn and Marty, 1988) to the patch pipette. The permeabilized patch provides electrical access to the cell and allows for the exchange of small ions but prevents the rapid diffusion of large molecules from the cell. Electrically the only difference between this approach and the standard whole-cell configuration is that the time constant of the capacitive transient is longer for the permeabilized patch.

In using the whole-cell technique one must be aware that the quality of the voltage clamp depends on the size of the series resistance which in the whole-cell mode arises mainly from the residual resistance of the broken patch membrane that provides electrical access to the cell interior. Series resistance has two detrimental effects. First it slows the charging of the cell membrane capacitance. Secondly, it yields errors in membrane potential when large currents flow. To offset these effects, series resistance compensation can be used to alter pipette potential in such a way as to compensate for the potential drop across the series resistance.

(d) Outside-out patches

This configuration is obtained from the whole-cell mode. Once the patch has been ruptured, the pipette is withdrawn from the cell surface causing a patch of membrane to reform across the pipette tip, but now with the extracellular membrane facing the bathing solution. This configuration is analogous to the inside-out configuration and has similar advantages and disadvantages. The outside-out patch is the one of choice for examining ionic channels controlled by externally located receptors. The extracellular solution can be easily exchanged, allowing the effects of different agonists and permeating ions to be tested. However, results should be interrupted with care since the formation of the outside-out patch may cause major structural rearrangement of the membrane and cytoskeletal elements and, as a consequence, channel properties (Trautmann and Siegelbaum, 1983).

10.2 DATA COLLECTION

A typical set-up for recording data from single channel recordings contains certain basic components. A high-gain current-to-voltage converter is used to record single-channel currents and a number of such patch clamp amplifiers are commercially available (Axon Instruments Inc., 1429 Rollins Rd., Burlingame, CA 94010; Dagan Inc., 2855 Park Ave., Minneapolis, MN 55407; List Electronics, Pfungstaedter Strasse 18-20, D-6100 Darmstadt/Eberstadt, Germany; Warner Instruments Hamden, CT 06510). Voltage pulses for testing electrode resistance and for stimuli are provided by a stimulator. The outputs of the patch clamp amplifier are displayed on an oscilloscope and are stored on a data recorder (e.g. an FM tape-recorder or a PCM adapter–video recorder system). A variable filter is usually included in the set-up, enabling filtering frequency to be varied. The polarity of the amplifier output and the filter setting and sampling rates will be discussed further.

10.2.1 Polarity conventions of patch-clamping

Patch clamp users have defined their polarity convention to coincide with microelectrode voltage clamp techniques. First, all voltages are refered to the outside of the cell membrane as ground (i.e. V_{cell} minus V_{bath}); and second, the current flow (movement of positive charge) from the cell interior to the outside bath is positive.

The most basic information on the possible type of channel in the patch can be obtained by monitoring the polarity of current flow and relating it to the imposed ion gradients across the patch. For example, if an isolated patch of membrane was held at zero potential and bathed on the cytoplasmic side by a KCl solution and with a NaCl solution of equal concentration filling the patch pipette, then upward channel openings (positive current flow) would indicate that a K^+ channel is in the patch.

For the inside-out and cell-attached configurations, the voltage and current polarities equal the cell membrane potential and direction of current flow as per the above convention. For an outside-out patch (or whole-cell patch), the voltage and current polarities are inverted so that they are consistent with the above convention. Thus a measured negative voltage and the corresponding currents are both inverted.

10.2.2 Filter settings and sampling rate

A question that must be addressed while collecting data, but more important while analyzing data (particularly when studying cation channels), is what filter settings one should use. Unfortunately there is no unique answer, but certain guidelines can be followed. First, the maximum filter setting used for recording data should be equal to or greater than the inherent frequency response of the data recorder. In addition, the amplitude of the input signal must not exceed the maximum allowable input of the recorder since in many cases there is an inherent relaxation time for a saturated amplifier. Modulation of the patch clamp gain in conjunction with low-pass filtering can be used to achieve maximum recording frequency with maximum allowable gain.

When filtering the signal for data analysis, one should use a Bessel filter since it rounds the leading and falling edges of a square pulse, as opposed to a Butterworth filter which has a characteristic response to the leading edge of a square pulse of a lag followed by an overshooting but rapidly dampened sinusoid. The **corner frequency** of a filter is equal to the –3 dB (decibel) point (the frequency at which the signal is reduced to 70.8% of its DC value) and the **roll-off** of the filter (the steepness by which the filter attenuates the signal as the signal frequency exceeds the corner frequency) is given in decibels per octave (an octave equals the doubling of the frequency). Typically, the steepness of the roll-off is expressed as either the number of poles (the filter rolls off at –6 dB per octave per pole beyond the corner frequency) or more simply as xdB per octave. Thus an 8-pole filter (48 dB per octave) will decrease the amplitude by 99.6% in an octave (e.g. from 1 to 2 kHz or 2 to 4 kHz). A **decibel** is defined as a unit for expressing the ratio of the voltage or current and is equal to 20 times the common logarithm of the ratio of voltage or current (e.g. a voltage ratio of 10 is equal to 20 dB and a voltage ratio of 3.16 is equal to 10 dB).

The first rule of filtering is that the current amplitude of the single-channel opening should exceed the peak-to-peak background noise level. As stated above, the higher the signal-to-noise ratio, the greater the accuracy with which current flow through a channel may be determined. In order to maximize the ratio of channel current to background noise, the filter setting can be adjusted for every voltage at which single channel currents are measured. However, the cut-off frequency of the filter must always be greater than the frequency with which the channel is opening and closing. If one filters at a frequency which is lower than the frequency of channel opening and closing, then the amplitude of the opening is decreased in size, yielding an underestimate of channel magnitude (see Yellen, 1984). The best approach is to measure the amplitude of a channel which is open for a period of time that exceeds the filter cut-off frequency.

10.3 DATA ANALYSIS

Once an adequate seal has been formed and channel activity has been observed and recorded, the next challenge is to identify the channel. It must be remembered that the channels isolated by the patch pipette are integral membrane proteins which facilitate the diffusion of ions across cell membranes. Conformational changes in these channels will be resolved electrically as pulses of current with a fixed amplitude which are typically rectangular in shape and rise and return to a set baseline. This baseline value is a function of the seal resistance, voltage clamp potential and solution composition. Analysis of these ideal transitions will yield information on single-channel conductance, channel selectivity and channel kinetics. In addition, the dependence of the single-channel current on voltage, concentration and species of permeant ion can also yield information about the conduction process, such as size of the channel pore and the number of affinity of ion-binding sites within the channel.

Before proceeding with the analysis we must ask ourselves a very straightforward question. To what detail or depth do we want to analyze this channel? The answer depends on the quality and quantity of the collected data, the assurance that there is only one channel in the patch of membrane being studied, and the level of the computer software analysis programs.

Whatever the quality of the data, some basic information concerning channel identity can be obtained. Basic information on channel selectivity and conductance can be obtained by measuring single-channel current amplitudes. Information concerning the kinetics of channel opening and closing is more difficult to obtain, requiring high quality single-channel recordings which exhibit long periods of channel activity.

Identification of channel type can be performed in the cell-attached mode but is most convincingly demonstrated on excised patches, since these configurations allow easy alterations of the solution composition which bathes the exposed membrane face and rapid and accurate alternation of the clamp potential.

10.3.1 Single-channel current amplitude

Measurement of current can be performed manually by displaying the current (actually a voltage) on a storage oscilloscope or a paper chart recorder. Knowing the gain of the patch clamp (i.e. volts/picoampere) one calculates the current flow through a single channel for specific ionic conditions and clamp voltages.

A number of computer software packages are available for automated data analysis, including automated generation of amplitude histograms.

Once the data have been stored in the computer memory, a computer program counts the number of times a current value falls within a current window or bin of specified width; it then increments this bin by the specified width and again counts the number of events that fall within that current bin. This process continues, starting at a specified lower current limit and ending at a specified upper current limit. The total number of bins is limited by computer memory and display capabilities. The data are displayed as a histogram with the current on the abscissa and number of occurrences per bin on the ordinate. In a simple case two current peaks are observed: one representing the mean closed current level, the other the mean open current level. The difference in current between these two levels equals the current flow through the channel.

It is from these amplitude histograms, generated at different clamp voltages and under varying ionic conditions, that one achieves initial estimates of single-channel conductance versus voltage relationships, channel selectivity and the open time probability of the channel and its possible modulation by agonists and antagonists.

(a) Current–voltage relationship

To construct a current (*I*) vs. voltage (*V*) relationship, one plots the value for single-channel current obtained from the amplitude histogram at different clamp potentials against the clamp voltage. The shape, slope and intercept of the plot with the voltage axis all contain valuable information about the channel's conduction process.

The shape of the plot contains information on the type of conduction process. A linear plot is obtained if the channel is acting as an ohmic conductor, while rectification is indicated by a curvilinear plot. (See Chapter 2 for a partial list of ion channels measured with the patch clamp technique.)

The slope of the *I–V* relationship gives the conductance, which may depend on the voltage and the ionic conditions. When symmetrical solutions are used to bath the membrane patch the *I–V* curve passes through the origin. Imposition of an ionic gradient across the patch should simply displace the plot to either the left or the right, depending on the direction of the imposed ion gradient and channel selectivity. The point where the plot intersects the voltage axis is the reversal potential. The value of the reversal potential is used in the calculation of channel selectivity.

(b) Channel selectivity

The **reversal potential** is defined as the voltage at which there is no net flow of ions through the channel and it is obtained by either extrapolation or interpolation of the measured points on the *I–V* curve. Note that the measurement of the reversal potential by extrapolation assumes that the *I–V* relationship is linear, which may not be true in the case of channels that display rectification.

Thus, if the ion activities on both sides of the patch are known, the reversal potential can be used to estimate channel selectivity. For a channel that is perfectly selective for one ion species, the reversal potential will equal the Nernst potential for that ion. Commonly, however, there is a deviation from the ideal Nernst potential which reflects the channels selectivity to other ions. For example, to determine the relative permeabilities of Cl^- and Na^+ in a K^+ channel, *I–V* curves are constructed in the presence of a number of different ion gradients. The first would be a KCl gradient to determine the K^+ to Cl^- selectivity of the channel, and the next a mixed solution of NaCl and KCl to determine the K^+ to Na^+ selectivity. The observed reversal potentials are then inserted in the Goldman equation (Goldman, 1943) and used to determine the relative permeabilities.

(c) Open channel probability

In addition to determining channel selectivity and voltage-conductance relationships, amplitude histograms can be used to calculate the fraction of time that the channel spends in the open state, i.e. the probability that the channel is open (P_o). The amplitude histogram is ideally composed of two current peaks, one representing the closed state of the channel and the other representing the open state of the channel. The amount of time the channel spends in the open state is equal to the total number of times the current is above a specified value, multiplied by the rate at which the data was sampled by the computer (time per point). Similarly the amount of time the channel spends in the closed state is equal to the total number of times the current is below a specified value, multiplied by the rate at which the data was sampled by the computer. The open probability is then the ratio of the amount of time spent in the open state divided by the sum of time in both the open and closed states.

Measurements of P_o can be used to quantify the effect that channel blockers have on channel activity. This approach has the added advantage that blockers can be used to quantify the contribution of the channel to the macroscopic current. Thus is it possible to compare microscopic and macroscopic currents, enabling one to determine whether channel function is altered by the mechanics of membrane patch isolation.

10.3.2 Channel kinetics

The current flowing through a single channel can be thought of as a series of rectangular current pulses that have a relatively fixed amplitude but that undergo time-independent transitions between any number of different conformational states. The time a channel resides (dwell-time) in any given conformational state is an exponentially distributed random variable.

Kinetic analysis is complicated by a number of factors. For an accurate analysis of channel kinetics, long records containing many channel openings and closings are required since the standard deviation of an exponential distribution is equal to its mean. It is also necessary that the statistical probabilities governing the channel fluctuations, recorded over these long time periods, do not change. This does not always hold true and alterations in channel environment (like channel run-down) or drug desensitization can alter open and close times.

Even if the above conditions are met, kinetic analysis can be greatly complicated if there is more than one channel in the patch. The approaches to determining the kinetics of single and multiple channel patches are discussed separately.

(a) Single channel patches

Consider the simplest possible situation: a two-state model having an open and a closed state, the distribution of which can be described by a single exponential:

$$\text{CLOSED} \underset{k_{oc}}{\overset{k_{co}}{\longleftrightarrow}} \text{OPEN} \qquad (10.1)$$

where k_{co} and k_{oc} are respectively the rate constants of channel opening and closing given in units of s^{-1}.

The mean duration of a given state will be equal to the inverse of the sum of all possible exit rates from that state. Then the mean time spent in the open state (τ_o) is given by:

$$\tau_o = 1/k_{oc} \qquad (10.2)$$

while the mean closed time (τ_c) will be given by:

$$\tau_c = 1/k_{co} \qquad (10.3)$$

and the probability of the channel being open is:

$$P_o = \tau_o/(\tau_o + \tau_c) \qquad (10.4)$$

Unfortunately, most channels have multiple open and closed states which, unlike different conductance levels, cannot be distinguished from each other in the current record. However, multiple conformational states can be detected by examining the distribution of open and closed times. The time spent in each state is described by a sum of multiple exponentials, the number of exponentials indicating the number of open or closed states.

Technically the challenge is to be able to resolve as many of the actual channel transitions as possible, including the briefest openings and closings, and to perform this task as quickly and as accurately as possible. Obviously the use of computer-aided analysis is necessary. For a full discussion of the theory behind the process of kinetic analysis, refer to Colquhoun and Sigworth (1983).

For construction of open time histograms, the time taken for the opening and subsequent closing of a channel (termed an event) is first recorded and then placed in an appropriate bin, the width of which is a prespecified time interval. The number of events that occupy a specific bin is then plotted against the time interval represented by the bin. The process is similar to construction of a closed histogram but the event measured is the time between closing and subsequent opening of a channel. The histograms are then fitted with an exponential distribution, and the fitted distribution is usually tested to determine the goodness of the fit. This fitted exponential distribution can be used to formulate a model to account for the observed number of conformational states. Thus, deriving kinetic parameters from multistate channels requires the investigator to choose a kinetic model. For the case of the channel that has one open state and two closed states, the number of possible kinetic models one can choose from is three (Dionne, 1981), which means that the derived kinetic parameters are not unique but are model dependent.

(b) Multiple-channel patches

So far we have tacitly assumed that each patch of membrane contains only one channel. Such a prediction might be expected if channel density was low and if channels were evenly distributed over the membrane surface. It is becoming increasing evident that channels appear in clusters, with more than one channel in the membrane patch. This occurrence of more than one channel in a patch is serious since it will yield overestimates for the open probability of a single channel and incorrect estimates for calculated kinetic parameters independent of how simple or complex the model used. Unfortunately the empirical observation of only two peaks in an amplitude histogram (one open and the other closed) does not guarantee that only one channel exists in the patch. A simple example is if we have two identical and independent channels, each having a probability of being open of 0.01 (1% of the time). It is easily demonstrated that the probability of both channels being open is extremely low (0.0001). The rule of thumb that one can use is that the minimal number of channels in a given patch of membrane is equal to the maximal number of channels observed.

The probability of finding r identical and independent channels open (P_r) in a patch of membrane containing N channels is given by:

$$P_r = \frac{N!}{r!(N-r)!} \cdot P_o^{\,r}(1 - P_o)^{N-r} \tag{10.5}$$

where P_o is the probability of a single channel opening (Labarca *et al.*, 1980; Colquhoun and Hawkes, 1983). When $r = N$ then $P_r = P_o^{\,r}$. This equation has been used on epithelial Na^+ channels to determine the number of channels in an obviously multiple channel patch (Palmer and Frindt, 1986).

Several approaches have been used to assess channel number. First and most obvious is to record for long periods of time (i.e. observe many closing transitions), an approach that increases the probability of observing the opening of a second channel before closure of the first. Next, if the channel is voltage gated, then by setting the voltage to increase P_o to a maximum value will also increase the probability of observing simultaneous channel openings. The last approach is to calculate (Colquhoun and Hawkes, 1983) the probability that, in a given record length with known channel kinetics, no double openings will be observed. It is important to stress that for patches in which channel activity is high (e.g. a P_o of 50%) the lack of double openings is highly suggestive of only a single channel.

An alternative approach that is now being used for determining the P_o for suspect multiple channel patches (i.e. patches which demonstrate a low value for P_o) is simply to state that the reported open time probability for the channels in a patch is equal to N (number of channels in the patch) times P_o (open probability of a single channel) (Sackin and Palmer, 1987). Such a relationship, although qualitative in terms of the absolute P_o, nevertheless yields important information if for instance the channel is voltage or agonist gated.

Determination of kinetic parameters in obviously multichannel patches is extremely difficult, if not impossible, and requires not only picking what one hopes is an adequate model but also assuming that the channels are indeed identical and functionally independent.

An approach to obtaining qualitative information about multiple-channel patches is the use of noise (fluctuation) analysis. Constructing *power spectral density* (PSD) curves for single-channel data can yield insight into the complexity of the channel kinetics and in theory, for channels with simple kinetics (one open and one closed state), the number of channels in the membrane patch. The qualitative information available is briefly outlined below.

- A single Lorentzian in a PSD suggests that the channel kinetics are simple, i.e. one open and one closed state.
- Multiple Lorentzians in a PSD suggest that channel kinetics are more complex and consist of more than one open or closed states. Determination of kinetic parameters is model dependent; however, changes in corner frequency and plateau values as a function of voltage

or agonist concentration will yield some insight into the underlying kinetic processes.

- For the case of simple channel kinetics (e.g. equation (10.1)) there is enough information in the PSD to calculate the number of channels (N) in the patch. (See section 10.5 for analysis and interpretation of PSD.)

10.4 INTERPRETATION OF PATCH CLAMP DATA

In this section, the principles discussed above are applied to the study of ion channels in epithelial tissues. First, we would like to stress the need to have a good macroscopic 'fingerprint' of the channel that one is looking for, before initiating patch clamp studies. Many of the preliminary reports of patch clamping epithelial cell membranes tended to be qualitative in nature, lacking extensive selectivity data, voltage dependence, kinetic analysis, etc. Other reports demonstrated the existence of a conductance which could not be localized to a particular membrane or was not found in the intact epithelium.

This fingerprint can consist of any one or all of the following points.

- Pharmacological blockers.
- Ion selectivity.
- Current–voltage relationship.
- Kinetic data from fluctuation analysis.
- Channel density from which rough estimate of single-channel conductance can be estimated (ligand binding studies).
- Single-channel currents (or conductance) from fluctuation analysis.
- Known regulators of channel activity (e.g. Ca^{2+}, cyclic nucleotides, protein kinases or phosphatases).
- Membrane location (i.e. apical or basolateral).
- Localization to a single cell type in an epithelium composed of multiple cell types.

By comparing this macroscopic fingerprint to the single-channel fingerprint, one can feel confident that indeed the channel measured is at least partially involved in determining the macroscopic conductive properties of that membrane.

In addition to identifying a particular channel's contribution to a known membrane conductance, other useful information can be obtained from interpretation of patch clamp data:

- Determination of the channel density in the membrane.
- For a basolateral membrane K^+ channel of known properties, the number of Na^+-K^+-ATPases required to counter the calculated K^+ flux through the channel can be estimated.
- A clustering of channels might suggest a cellular mechanism of channel turnover or regulation by insertion and withdrawal.

- The appearance or disappearance of channel activity in a patch might suggest the loss of a regulatory component.
- Microscopic selectivity of a channel (e.g. a finite Na^+ permeability) might account for an unexplained Na^+ conductance of the membrane.

This is only a partial list, which will grow as more epitheliologists employ patch clamp methodology to study epithelial transport.

10.5 CURRENT FLUCTUATION ANALYSIS: A METHOD FOR DETERMINING THE PROPERTIES OF SINGLE ION CHANNELS

The above section has assessed single-channel properties using the patch clamp technique. There are, however, limitations to patch clamping. The ability to observe a single-channel event is dependent upon:

- the single-channel current amplitude being larger than the background noise of the measuring system;
- the channel properties not being changed during the patching procedure;
- the density of channels in a single patch of membrane being low.

An alternative approach to patch clamping is to measure the fluctuations in the macroscopic current of an ensemble of channels (usually numbering millions of channels per square centimeter).

The equipment neeeded to perform current fluctuation is similar to that previously described for transepithelial measurements of epithelial resistance, short-circuit current and voltage. The most important piece of equipment is an ultra-low noise clamp (Lindemann and Van Driessche, 1977) which is used to continuously short-circuit the epithelium (Appendix A). The short-circuit current is then filtered by a high-pass RC filter (0.53 Hz). This high-pass filter removes the DC component of the signal and allows the current signal (noise) to be further amplified. Typically one amplifies the signal such that peak-to-peak signal is close to ±10 V. This amplified signal is then low-pass filtered to eliminate aliasing artifacts. Next, the output of the filter is digitized using a laboratory computer. The digitizing rate depends on the kinetic properties of the channel or the blocker kinetics. The digitized data are divided into blocks which typically range from 512 to 2048 data points and one typically collects at least 16 blocks of data. The lowest frequency point is equal to the inverse of the product of the number of data points per block and the digitizing rate. The maximum frequency is equal to the inverse of two times the digitizing rate (e.g. digitizing at 5 ms a point will yield a maximum frequency of 100 Hz).

Each block of digitized data is analyzed using a fast Fourier transform program which decomposes the current fluctuations into the sum of sine and cosine waves of different frequencies. One then computes the power spectra density (PSD), which graphically represents the data as the logarithm of power on the abscissa ($A^2 \cdot$sec) vs. the logarithm of frequency (Hz) on the ordinate. The shape of such a plot depends on the kinetic properties of the transport process being analysed. As an example, for a channel that has two states (equation (10.1), the PSD (in a log–log representation) is described by a Lorentzian function which has a plateau at low frequencies and then bends downward and has a slope of ca. –2 as the frequency increases (Figure 10.2a). The equation which describes this shape is:

$$S(\mathrm{f}) = S_o/[1+(f/f_c)^2] \qquad (10.6)$$

where $S(\mathrm{f})$ is the power as a function of frequency, S_o is the plateau value (power at 0 Hz or DC), f_c is the corner frequency (the frequency at which the power has decreased by one half) and f is the frequency (Hz). The corner frequency is equal to the sum of the opening and closing rate constants (k_{co} and k_{oc} respectively, with units of s^{-1}) and is described by the equation:

$$2\pi f_c = k_{oc} + k_{co} \qquad (10.7)$$

The plateau is described by the following equation:

$$S_o = 4Mi^2ak_{oc}k_{co}/(k_{oc} + k_{co})^3 \qquad (10.8)$$

where M is the number of channels in the membrane, i is the current flow through a single channel, and a is area of the membrane. The probability of the channel being open (P_o, which cannot be determined from fluctuation analysis) is given by:

$$P_o = k_{co}/(k_{co} + k_{oc}) \qquad (10.9)$$

and the macroscopic current (the short-circuit current, I_{sc}) is given by:

$$I_{sc} = MiP_o \qquad (10.10)$$

Inspection of equation (10.8) shows that we have four unknowns (M, i, k_{co} and k_{oc}) but we have only three equations with their respective data points. Thus one cannot solve for the four unknowns with the available data set. This problem was overcome for the amiloride-sensitive sodium transporting epithelia, since this channel did not demonstrate a measurable Lorentzian component. The lack of a Lorentzian component suggested that either the spontaneous rate of opening and closing for this channel was much slower than the lowest measured frequency of the PSD or that the probability the channel being open was unity. One can induce the channel to fluctuate spontaneously by using a reversible channel blocker – in this instance, amiloride. Thus the rate of going from open to closed is a concentration-dependent rate constant, i.e. $k_{oc} = k_{oc}'$[Amil], so

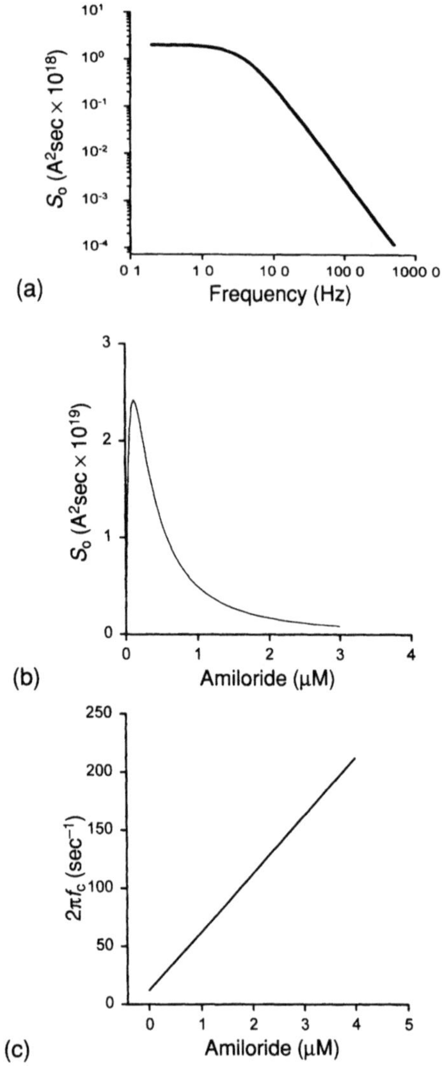

Figure 10.2 (a) Example of a power spectral density. Note that both axes have a logarithmic scale. This example of a PSD was calculated using equation (10.9), a S_o of 20×10^{-18} amp^2·s and a corner frequency (f_c) of 3.8 Hz. (b) Plateau (S_o) value as a function of the amiloride concentration. The curve was calculated using equation (10.8). The following values were used: $M = 10 \times 10^6$ channels per cm^2; $i = 0.7$ pA; $k_{oc} = 50$ s^{-1}μM^{-1}; $k_{co} = 12$ s^{-1}; and the amiloride concentration was allowed to change from 0 to 3 μM. The k_D for this example is 0.24 μM. Note that the S_o peaks at 0.12 μM amiloride, i.e. at an amiloride concentration equal to half the k_D. (c) Relationship between the amiloride concentration and the corner frequency (actually $2\pi f_c$). The zero amiloride intercept is equal to the rate at which amiloride dissociates from channel and the slope is the association rate constant.

that k_{oc}' has units of $s^{-1}M^{-1}$), where [Amil] is the amiloride concentration in the bath. The corner frequency at finite bath amiloride concentration is given by:

$$2\pi f_c = k_{co} + k_{oc}'[\text{Amil}] \tag{10.11}$$

In the absence of bath amiloride, the plateau value of the PSD (equation (10.8)) is zero. As the amiloride concentration in the bath is increased, the plateau value and corner frequency of the PSD increase (Figure 10.2b,c). When the concentration of amiloride equals one half of the dissociation constant for amiloride ($k_D = k_{co}/k_{oc}'$) the plateau value reaches a maximum. At higher concentrations of amiloride the plateau value of the PSD decreases and approaches zero as the amiloride concentration approaches infinity. In contrast, the corner frequency increases as a linear function of the amiloride concentration.

A linear regression analysis of a plot of the corner frequency and the amiloride concentration allows one to determine the two rate constants, where k_{co} is the intercept and k_{oc}' is the slope (Figure 10.2c). It is now possible to calculate the channel density, single-channel current and probability of the channel being open as a function of the amiloride concentration. The value of the I_{sc} used to determine these parameters is the component of the total amiloride-sensitive I_{sc} which is not inhibited by the submaximal concentration of amiloride.

For patch clamp data in which there are multiple channels (of the same molecular species) one does not have to resort to channel blockade as long as there are discrete channel open and closed events, since in this instance one has an independent estimate of one of the four unknowns, i.e. one can directly measure *i*. However, one still must use a kinetic model before one can estimate the number of channels in the patch.

REFERENCES

Auerbach, A. and Sachs, F. (1984) Patch clamp studies of single ionic channels. *Ann. Rev. Bioeng.* **13**:269–302.

Colquhoun, D. and Hawkes, A.G. (1983) The principles of the stochastic interpretation of ion-channel mechanisms, in *Single-Channel Recording*, (eds B. Sakmann and E. Neher), Plenum Press, New York, pp. 135–76.

Colquhoun, D. and Sigworth, F.J. (1983) Fitting and statistical analysis of single channel records, in *Single-Channel Recording*, (eds B. Sakmann and E. Neher), Plenum Press, New York, pp. 191–263.

Copello, J., Simon, B., Segal,Y. *et al.* (1991) Ba^{2+} release from soda glass modifies single maxi K^+ channel activity in patch clamp experiments. *Biophys. J.* **60**:931–41.

Copello, J., Wehner, F. and Reuss, L. (1993) Artifactual expression of a maxi-K^+ channels in basolateral membrane of gallbladder epithelia cells. *Am. J. Physiol.* **264**:C1128–C1136.

Corey, D.P. and Stevens, C.F. (1983) Science and technology of patch recording

electrodes, in *Single-Channel Recording*, (eds B. Sakmann and E. Nener), Plenum Press, New York, pp. 53–68.
Cota, G. and Armstrong, C.M. (1987) Potassium channel 'inactivation' induced by soft glass patch pipettes. *Biophys. J.* **53**:107–9.
Dionne, V.E. (1981) The kinetics of slow muscle acetylcholine-operated channels in the garter snake. *J. Physiol.* **310**:159–90.
Donaldson, P.J. and Lewis, S.A. (1990) The effect of serosal hypertonic challenge on basolateral membrane potential in the rabbit urinary bladder. *Am. J. Physiol.* **258**:C248–C257.
Dragsten, P.R., Blumenthal, R. and Handler, J.S. (1981) Membrane asymmetry in epithelia: is the tight junction a barrier to diffusion in the plasma membrane. *Nature* **294**:718–22.
Dunne, M.J. and Peterson, O.H. (1986) GTP and GDP activation of K^+ that can be inhibited by ATP. *Pflügers Arch.* **407**:564–5.
Fenwick, E.M., Marty, A. and Neher, E. (1982) A patch-clamp study of bovine chromaffin cells and their sensitivity to acetylcholine. *J. Physiol.* (Lond.) **331**:577–97.
Gogelein, H. and Greger, R. (1984) Single channel recordings from basolateral and apical membrane of renal proximal tubules. *Pflüger Arch.* **408**:282–90.
Goldman, D.E. (1943) Potential, impedance, and rectification in membranes. *J. Gen. Physiol.* **27**:37–60.
Hamill, O.P. (1983) Potassium and chloride channels in red blood cells, in *Single-channel Recording* (eds B. Sakmann and E. Neher), Plenum Press, New York, pp. 451–71.
Hamill, O.P., Marty, A., Neher, E. *et al.* (1981) Improved patch clamp techniques for high resolution current recording from cells and cell-free membrane patches. *Pflügers Arch.* **391**:85–100.
Horn, R. and Marty, A. (1988) Muscarinic activation of ionic currents measured by a new whole-cell recording method. *J. Gen. Physiol.* **92**:145–59.
Hunter, M., Lopes, A.G., Boulpaep, E.L. and Giebisch, G.H. (1984) Single channel recordings of calcium-activated potassium channels in the apical membrane of rabbit cortical collecting tubules. *PNAS* **81**:4237–9.
Kakei, M. and Ashcroft, F.M. (1987) A microflow superfusion system for use with excised membrane patches. *Pflügers Arch.* **409**:337–41.
Labarca, P., Coronado, R. and Miller, C. (1980) Thermodynamic and kinetic studies of the gating behavior of a K^+-selective channel from the sacroplasmic reticulum membrane. *J. Gen. Physiol.* **76**:397–424.
Lapointe, J.-Y. and Szabo, G. (1987) A novel holder for allowing internal perfusion of patch pipettes. *Pflügers Arch.* **410**:212–16.
Lewis, S.A. and Alles, W.P. (1986) Urinary kallikrein: a physiological regulator of epithelial Na^+ transport. *PNAS* **83**:5345–8.
Lindau, M. and Fernandez, J.M. (1986) IgE-mediated degranulation of mast cells does not require opening of ion channels. *Nature* **319**:150–3.
Lindemann, B. and Van Driessche, W. (1977) Sodium-specific membrane channels of frog skin are pores: current fluctuations reveal high turnover. *Science* **221**:292–4.
Lopes, A.G. and Guggino, W.B. (1987) Volume regulation in the early proximal tubule of the *Necturus* kidney. *J. Memb. Biol.* **97**:117–25.
Neher, E., Sakmann, B. and Steinbach, J.H. (1978) The extracellular patch clamp: a method for resolving currents through individual open channels in biological membranes. *Pflügers Arch.* **375**:219–26.
Palmer, L.G. and Frindt, G. (1986) Amiloride-sensitive Na^+ channels from the

apical membrane of the rat cortical collecting tubule. *PNAS* **83**:2767–70.
Rae, J.L. (1985) The application of patch clamp methods to ocular epithelia. *Curr. Eye Res.* **4**:409–20.
Rae, J.L. and Levis, R.A. (1984) Patch clamp recordings from the epithelium of the lens obtained using glasses selected for low noise and improved sealing properties. *Biophys. J.* **45**:144–6.
Richards, N.W. and Dawson, D.C. (1986) Single potassium channels blocked by lidocaine and quinidine in isolated turtle colon epithelial cells. *Am. J. Physiol.* **251**:C85–C89.
Sackin, H. and Palmer, L.G. (1987) Basolateral potassium channels in renal proximal tubules. *Am. J. Physiol.* **253**:F476–F487.
Sakmann, B. and Neher, E. (1983) Geometric parameters of pipettes and membrane patches, in *Single-Channel Recording*, (eds B. Sakmann and E. Neher), Plenum Press, New York, pp. 37–52.
Sakmann, B. and Neher, E. (1984) Patch clamp techniques for studying ionic channels in excitable membranes. *Ann. Rev. Physiol.* **46**:455–72.
Sigworth, F.J. and Neher, E. (1980) Single Na^+ channel currents observed in cultured rat muscle cells. *Nature* **287**:447–9.
Trautmann, A. and Siegelbaum, S.A. (1983) The influence of membrane patch isolation on single acetylcholine-channel current in rat myotubules, in *Single-channel Recording* (eds B. Sakmann and E. Neher), Plenum Press, New York, pp. 473–80.
Yellen, G. (1982) Single Ca^{2+} activated non-selective cation channels in neuroblastoma. *Nature* **296**:357–9.
Yellen, G. (1984) Ionic permeation and blockade in Ca^{2+}-activated K^+ channels of Bovine chromaffin cells. *J. Gen. Physiol.* **84**:157–86.
Ziomek, C.A., Schulman, S. and Edidin, M. (1980). Redistribution of membrane proteins in isolated mouse intestinal epithelial cells. *J. Cell Biol.* **86**:849–57.

11

Epithelial cell culture

Nancy K. Wills

The application of tissue culture techniques to epithelial cells has revolutionized the study of morphologically complex or inaccessible tissues such as endocrine and exocrine glands, renal tubules, airway epithelia and ocular epithelial cells. Epithelial physiologists can now examine cell differentiation and other regulatory processes under precisely controlled conditions. When used in conjunction with molecular biological and other genetic methods, these fundamental processes can be assessed at the molecular level. This chapter focuses mainly on techniques for growing differentiated epithelial monolayers on permeable supports. This approach can yield morphologically simple monolayer epithelia that are amenable to a variety of electrophysiological, molecular and optical methods for studies of epithelial cell function.

Although it is beyond the scope of this chapter to list all epithelial cell

Epithelial Transport: A guide to methods and experimental analysis.
Edited by Nancy K. Wills, Luis Reuss and Simon A. Lewis.
Published in 1996 by Chapman & Hall, London. ISBN 978-94-010-7168-0

lines, an impressive array is now commercially available from American Type Culture Collection which publishes a detailed catalogue. Other useful references concerning cultured epithelia include the excellent texts by Freshney (1987, 1992), and reviews by Pollard and Walker (1990), Jakoby and Pastan (1979), Matlin and Valentich (1989) and Traub (1985).

Before describing some basic terminology, it is useful to note the important role that tissue culture models have played in recent breakthroughs concerning genetic diseases. For example, cultured epithelial cells from cystic fibrosis patients have been used to identify the basic mechanisms of this disorder (Chapter 14). Cell culture techniques have also provided important insights into cell–cell interactions and protein trafficking. In particular, cultured epithelial cells have played a major role in our present knowledge of the factors which lead to the establishment of epithelial tight junctions and the polarization of the cell surface into apical and basolateral membrane domains (cf. Fleming, 1992; see also Chapter 3).

11.1 BRIEF HISTORY AND TERMINOLOGY OF EPITHELIAL CELL CULTURE

Tissue culture is not a new technique: it dates to the beginning of the twentieth century (cf. Freshney, 1987). The method was developed for studies of cell growth and behavior in the absence of other influences. Initial studies involved the use of fragments or **explants** of tissues from an organism that were then placed in nutrient solutions and incubated at the appropriate temperature and pH. New cells would then grow out of the tissue explant, leading to use of the term **tissue culture**.

For our purposes, it is important to distinguish different types of tissue culture. The term cell culture will refer to cultures derived from cells that are initially subjected to chemical, mechanical or enzymatic disaggregation, i.e. they are dispersed as single cells. The term **organ culture** means that a three-dimensional culture of an intact tissue has been attained (i.e., without dispersement into single cells). **Reconstituted** or **histotypic cultures** will refer to a system in which the dissociated cells have been recombined in some way to look at interactions between different cell types or to produce a three-dimensional structure. For example, in a histotypic culture, mammary gland epithelial cells form a three-dimensional network of tubules when grown in appropriate gel matrices (Freshney, 1987).

11.2 MAJOR TYPES OF EPITHELIAL CELL CULTURE

Two types of epithelial cell culture have been extensively used for studies of epithelial cell function: primary cell cultures and continuous cell lines. **Primary cell culture** was defined by Schaeffer (1978) as: 'a culture started from cells, tissues, or organs taken directly from organisms. A primary culture may be regarded as such until it is subcultured for the first time. It then becomes a cell line.' Each subculturing, i.e. re-suspension and plating of the cells, is referred to as a **passage**. A **continuous cell line** was defined as one which has an essentially infinite life span.

11.2.1 Primary cell culture methods

Primary cell culture involves the following four basic steps:

(a) The appropriate excision and isolation of the epithelium of interest.
(b) Disaggregation of the epithelial cells using mechanical or biochemical procedures.
(c) Selection or separation of the cell type of interest.
(d) Cell seeding (or plating) and promotion of cell growth.

(a) Isolation of the epithelium

Aseptic procedure is a crucial consideration in obtaining tissues for primary culture. Separate sets of sterile dissection instruments should be used for incising the skin, opening body cavities and removing the organ of interest. Also to maintain sterility, the dissection should take place in an area outside the tissue culture laboratory. Care should be taken to avoid puncturing the intestine or large blood vessels, in order to minimize the possibility of contaminating the specimen or exposing the cells to potentially harmful substances such as heme. The tissues should be stored in sterile isosmotic buffered saline solution (such as Ringer's solution or phosphate buffered saline) or in culture medium and immediately transported to the tissue culture laboratory for further processing of the epithelium, if necessary. Tissues may be stored in cold solutions (4°C) but should never be exposed to excessive heat or cold. The temperature tolerance of the tissue will depend on the species and specific epithelium.

(b) Disaggregation of epithelial cells

Figure 11.1 illustrates the main steps involved in the dissociation of epithelial cells (for details see Bashor, 1979). Generally, the epithelium is carefully isolated under sterile conditions, minced into 1–3 mm pieces and placed in a calcium-free solution (to disrupt the tight junctions; see Chapter 3) containing digestive enzymes such as trypsin, pronase, collagenase, hyaluronidase, elastase, pancreatin, papain or deoxyribonuclease. The action of these enzymes is then stopped by placing the specimen on ice, or by adding serum or specific inhibitors. These procedures require

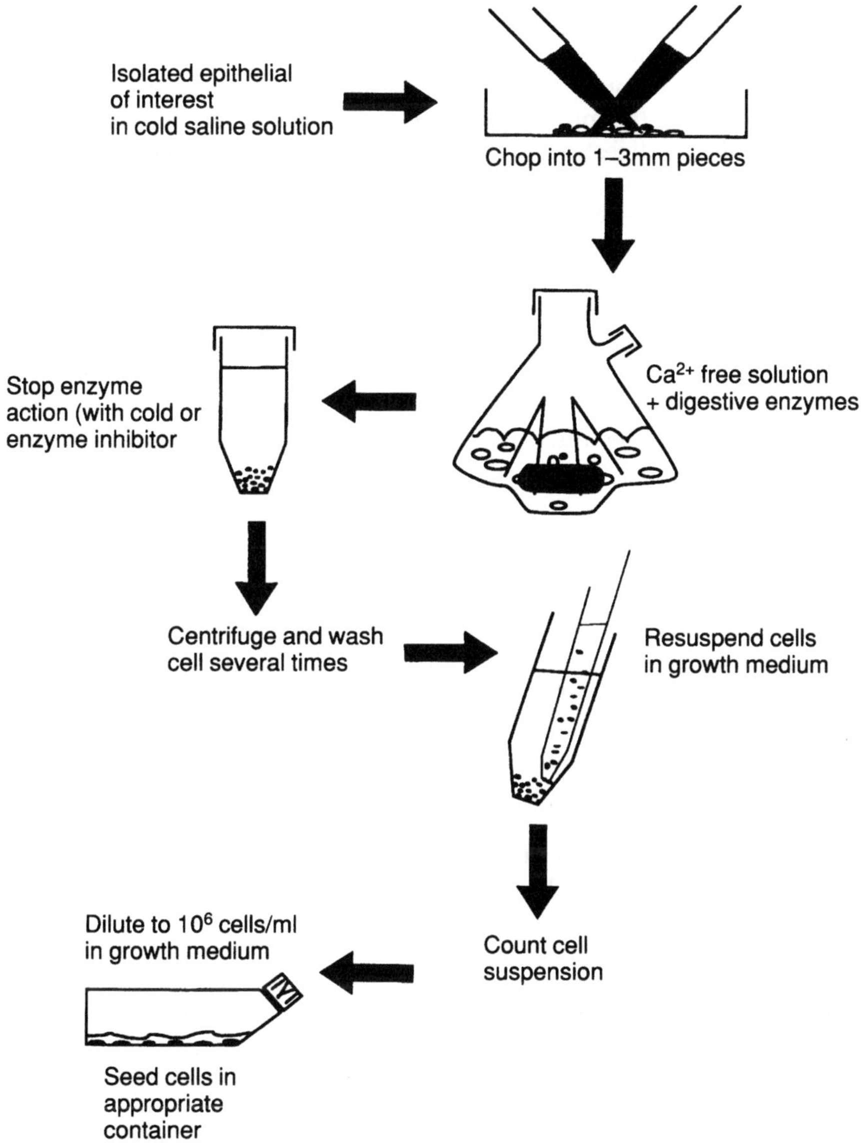

Figure 11.1 Steps involved in preparation of single cells by enzymatic dissociation for primary culture. Following chopping of the tissue into 1–3 mm pieces, the fragments are stirred in calcium-free solutions containing enzymes. Enzyme action is then stopped, the cells are washed, and then resuspended in culture medium as shown. (Adapted from R. I. Freshney, *Culture of Animal Cells: A Manual of Basic Technique*, copyright 1987 Alan R. Liss. Reprinted by permission of John Wiley & Sons, Inc.)

careful monitoring since membrane-bound proteins can be destroyed by these enzymes and cells can be damaged by prolonged exposure to calcium-free solutions.

There are numerous procedures for treating the minced tissues. Many involve agitating the minced fragments by pipetting back and forth, then allowing the undigested fragments to settle. The 'cloudy' solution is collected, centrifuged, and washed several times in enyzme-free solution. Alternatively, some cell types can be released mechanically by forcing the tissue through a fine sieve. Once the cells have been harvested, their viability can be assessed by using indicator dye methods such as trypan blue exclusion. Although trypan blue exclusion alone does not guarantee viability, one can use this approach to estimate the number of viable cells in a known volume of the final suspension. Cells can then be plated at an optimal density in sterile plastic flasks, petri dishes, carrier beads, or on permeable membrane filters (see below). A typical cell seeding density is $0.2–2 \times 10^5$ cells/cm^2.

(c) Selection or separation of the cell type of interest

Many epithelia consist of more than one cell type. In addition, adjacent connective tissue and smooth muscle may make it difficult to harvest the cells of interest without further processing. One means of separating dissociated epithelial cells is to sort, on the basis of specific gravity, using Percoll density gradients and centrifugation (e g. Bello-Reuss and Weber, 1987). In addition to these techniques, other biochemical approaches such as affinity chromatography, elutriation or unit gravity methods are sometimes used. Growth of unwanted cells such as fibroblasts can be inhibited by using growth selective media (containing L-amino acids instead of D-amino acids) or factors that are inhibitory to fibroblast growth (see Freshney, 1987 for further details of these separation and selection methods). More recently, immuno-separation methods or 'panning' techniques were developed to enrich or deplete cell populations of particular cell types. In this approach, cell suspensions are placed in petri dishes or containers that are coated with antibodies to a specific receptor or protein (Turksen and Aubin, 1994; Warrington *et al.* 1992; Gong *et al.* 1992). Cells that bind the antibody become attached to the dish. One can then collect the attached cells (enriched population) or those remaining in the fluid suspension (depleted population).

For primary cultures, it is necessary to identify specific cell types. Usually, this cannot be determined solely by light microscopic examination of cell morphology. For this reason, electron microscopy or biochemical methods such as antigenic determination or assays for enzymes, receptors or specific transport or uptake mechanisms are often employed.

For example, a useful method for identifying intercalated cells of the distal nephron is peanut-lectin binding (Bello-Reuss and Weber, 1987).

(d) Promotion of cell growth

Successful cell culture ultimately depends on the choice of an appropriate growth medium, which is often a trial and error process. The chemical composition of cell culture growth medium varies for different epithelia and species. Many companies now market standard prepared media or custom-made media. Two common sources are GIBCO Laboratories (Grand Island, New York) and Sigma Chemical Co. (St Louis, MO). In general, it is necessary to supplement the medium with antibiotics, vitamins, hormones, growth factors and/or serum.

11.2.2 Antibiotics and buffers

Media are usually supplemented with antibiotics such as penicillin and streptomycin (for a detailed list of tissue culture antibiotics see Bashor, 1979). To prevent fungal infections, nystatin or amphotericin B are sometimes employed. However, investigators interested in studying isolated ionic channels should be aware that these antibiotic compounds incorporate into the plasma membrane to form pores with channel-like activity. Therefore, it may be necessary to remove antibiotics prior to certain types of study. Some investigators also include buffers such as HEPES in the medium to maintain a constant extracellular pH. Again, investigators who wish to assess ion channel activity should be warned that this compound affects certain anion channels (Hanrahan and Tabcharani, 1990).

11.2.3 Defined media versus supplementation with serum

It is advantageous to use culture media with known concentrations of nutrients, growth factors, and hormones. For epithelial tissues, some frequently used hormones include transferrin, insulin and hydrocortisone, depending on the specific epithelium and species (cf. Traub, 1985). In many cases, the exact growth requirements are unknown and the culture medium is supplemented with serum. In some situations, heat inactivated serum is employed. Typical sera include fetal calf serum, horse serum, human placental serum and murine (mouse) serum. Several epithelia are extremely sensitive to serum quality and type, so any serum used in cell culture must be carefully evaluated for toxicity, and the promotion of cell growth and differentiation. In addition, the optimal amount of supplementation (usually 1, 5, 10 or 15% v/v) should be determined experimentally.

11.2.4 Subculturing and continuous cell lines

When cells are to be harvested for subculture, the cells should be actively growing and subconfluent to obtain the maximum number of healthy viable cells. Some epithelia detach from the anchoring substratum when exposed to calcium-free and magnesium-free buffered salt solutions. When using enzymes such as trypsin to release the cells, it is necessary first to wash the epithelium in salt solutions in order to remove residual serum that could inhibit enzyme activity.

Following subculture, the cells are no longer a primary culture and now become a cell line. Most epithelial cell lines become dysfunctional and/or die after a few subculturings or passages. However, in some cases, cell lines spontaneously undergo a transformation and become continuous cell lines characterized by increased cell growth, an apparently infinite life span, highly viable cells and tumorigenicity (although this characteristic is not a feature of all cell lines).

11.2.5 Dilution cloning

Homogeneous cell lines can be derived from continuous cell lines by dilution plating. In this technique, cell suspensions are diluted with a sufficient volume of medium to allow addition of single cells to culture dishes. The cells then divide to form isolated colonies derived from a single parental cell.

11.3 ADVANTAGES AND DISADVANTAGES OF PRIMARY CELL CULTURE AND CONTINUOUS CELL LINES

It is clear from the above description that there are distinct differences between primary cell cultures and continuous cell lines that may be advantageous or disadvantageous to the investigator, depending on the particular application. Both primary and continuous cell cultures offer the previously stated advantages of precise control of growth conditions and provide a preparation with a simple structure, i.e. an epithelium without underlying connective tissue or smooth muscle layers.

Primary cultures have the additional advantages of being composed of fully differentiated cells and, if carefully dissected, of having a known origin. Most can survive in serum-free media without fibroblast overgrowth. However, if fibroblast growth inhibitors (such as cis-proline, phentobarbitone; Freshney, 1987) are used, the effects of these compounds on epithelial function should be independently tested. A disadvantage, as mentioned above, is that cells in primary cultures are usually heterogeneous and growth usually ceases after a few divisions. Primary cultures can also demonstrate phenotypic instability, i.e. the physical, biochemical or physiological property of interest can disappear after the cells are placed in culture.

Use of continuous cell lines can eliminate most of these problems. For example, continuous cell lines, particularly subclonal cell lines, are homogeneous. Since one can, in principle, grow as many cells as needed, the supply of cells is essentially unlimited. In addition, one can freeze viable samples of cells (referred to as cell stocks) and do future experiments on the same cells. Such cell lines also give the investigator the ability to compare cellular functions in different stages of growth (i.e. actively dividing or growth-arrested stages) and to identify mutants that are defective in particular functions.

Nonetheless, there are important limitations in using continuous cell lines. First, the cells are grown from minced organs and usually their precise origin is unknown. Transformation of the cells into a continuous cell line is a relatively poorly understood process, although recent work indicates that oncogenes are responsible (Freshney, 1992). Moreover, some cell lines have abnormal numbers of chromosomes and cells clearly can differ with respect to their stage of differentiation. Often, the precise nutrient requirements of the cells are unknown and serum or other supplements are required, adding unknown factors to the growth media. Another potential problem that is associated with the need for supplements is possible differences in gene expression. Genes expressed in the continuous cell line *in vitro* may differ from those observed *in vivo*, resulting in the emergence of new phenotypes *in vitro*.

In summary, primary and continuous cell cultures can provide highly useful models for understanding fundamental principles in epithelial physiology. However, one must be aware that these cells may not retain all the normal physiological or biochemical properties of differentiated epithelial cells. Therefore, they may not provide faithful models of epithelia from particular organs, such as, for example, the kidney or trachea.

11.3.1 Overview of commonly used cell lines

Table 11.1 summarizes some of the more commonly used epithelial cell lines for which transepithelial electrical data are available. Although this is not a comprehensive list, transported substrates and agents which stimulate or inhibit transport are as indicated. The present progress in the development of immortalized cell lines using SV40 virus (Linder and Marshall, 1990), nickel and other factors (Freshney, 1987, 1992) has led to an explosion in the variety of cell lines available for study. Therefore, interested readers are encouraged to check the current literature for specific applications.

Table 11.1 Transepithelial electrical and transport properties of commonly used epithelial cell lines

Cell line	*Origin*	*Electrical properties* V_t *(mV)*	I_{sc} *(µA/cm²)*	R_t *(Ωcm²)*	*Major ion Transported*	*Stimulators*	*Inhibitors*	*Ionophores*	*Reference/source*
CaCo2	human colon carcinoma	0.3	2	150		no effect AL	no effect Am	AL	Grasset *et al.* (1984)
MDCK	canine kidney strains								Valentich (1986)
	Type 1	0	0	30–100					Paulmichi *et al.* (1986)
	Type 2	0	0	800–1000					
LLC-PK_1	male Hampshire pig, proximal tubular cells	2.8	13	200	Na^+ dependent sugar and amino acid transport, Na/H exchange, Na/Ca counter-transport	cAMP Vas Cal			cf. Meier and Insel (1989)
A6	*Xenopus laevis* kidney, distal segment	–9	1	6900	active Na^+ absorption	AL, cAMP, Ins.	Am, Ou	A, N	Perkins and Handler (1981)
					Cl^- secretion	Vas, PGE	Bu		
		–57	23	2800					cf. Wills and Millinoff (1990)
T84	human colon carcinoma	30	30	1500 (600–2000)	Cl^- secretion	VIP, Carbachol For	Iso	N	cf. McRoberts and Barrett (1989)
HCA-7	human adeno-carcinoma	–0.3	9	55	Cl^- secretion	For, Kinins			Cuthbert *et al.* (1985)

Iso = isoproterenol
cAMP = cyclic AMP
Cal = Calcium
AL = Aldosterone

VIP = Vasoactive inhibitor peptide
For = Forskolin
Vas = Vasporessin

PGE = Prosaglandin E_2
Ins = insulin
AM = Amiloride
Fu = Furosemide

Bu = Bumetanide
O = Ouabain
A = Amphotericin B
N = Nystatin

11.4 OTHER ASPECTS OF EPITHELIAL CELL CULTURES

Now that we have reviewed the basic steps of growing epithelial cells in culture, it is useful to address some specific topics such as how to select a permeable filter support, how to troubleshoot cultures that die and how to monitor epithelial properties.

11.4.1 Permeable membrane filter supports for growing epithelia

Steele *et al.* (1986) demonstrated that epithelial cells grown on impermeable supports such as plastic or glass do not fully express differentiated transport properties. Using the A6 cell line derived from the distal tubule of *Xenopus laevis* kidney, they found that epithelia grown on plastic flasks demonstrated little amiloride-sensitive Na^+ uptake across the apical membrane (a test for expression of Na^+ channels). However, when cells were grown on permeable supports, amiloride-sensitive Na^+ uptake was present. They concluded that the epithelium must receive nutrients from both the apical and basal solutions for differentiation to occur. Figure 11.2 is a representation of their design for a permeable support placed in a petri dish. The base of the support consists of a permeable filter, glued to a polycarbonate ring. It rests on three silicone rubber 'feet' which elevate the support from the bottom of the petri dish.

There are now several commercially available membrane filter supports for use with cultured epithelia. These systems can be quite expensive and the investigator will need to consider whether they are cost-effective. Some investigators prefer to make their own filter supports as described above, or by gluing thin sheets of cross-linked collagen to polycarbonate rings using the technique of Steele *et al.* (1986). The time involved in preparing and sterilizing the supports can be substantial. For the ecologically and economically-minded, we note that Bell and Quinton (1990) have reported a method for recycling filter supports.

An overview of commercially available filter supports is presented in

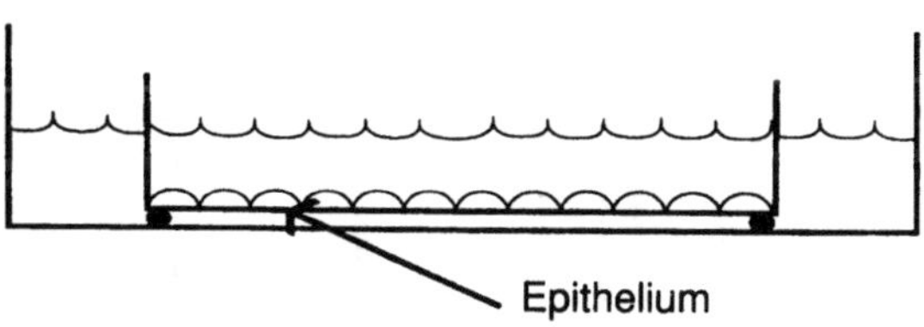

Figure 11.2 Permeable filter support for growth of cultured epithelial cells. The filter support is housed in a petri dish (as shown) or a multiwell holder and consists of a permeable filter glued to a polycarbonate ring that is elevated from the bottom of the petri dish.

Table 11.2 Summary of culture plate inserts

Specifications	*Anocell*[1,2]	*Transwell-CO1*[3]	*Transwell-*[3,e]	*Transwell*[3] *Clear*	*Millicell-HA*[4]	*Millicell-CM*[4]	*Millicell-PC*[4]	*Millicell-PCF*[4]	*Falcon*[6]	*Biocoat*	*Collagen*[5]
Membrane thickness (μm)	45	25–50	10	6–10	100–140	45–55	10	10	15±1	15±1	—
Membrane pore size	0.2–0.02[a]	0.4; 3.0	0.1; 0 4; 3; 5; 8; 12	0 4; 3.0	0 45	0 4	0.4; 3	0.4, 3	0 4; 1 0, 3 0, 8 0	0 4, 1.0, 3.0, 8.0	—
Growth area (cm^2)	0 5; 4.15	0.33; 4.71	0.33; 4 71; 7 85, 44	0.33; 4.71	0.6; 4.2	0 6, 4 2	0.6	0 6	0.3; 0.9; 4 2[f]	0 3, 0.9, 4.2	0.7, 4 5
Membrane type	inorganic[b]	Co-star membrane[d]	Polycarbonate	Polyester	Type HATF-filter[g]	Biopore membrane[h]	Polycarbonate	Polycarbonate	Polyethylene Tetraphthalate	Polyethylene Tetraphthalate	Collagen
Membrane optical properties	Transparent when wet	Transparent when wet	Translucent	Transparent	Opaque	Transparent when wet	Translucent	Translucent	Transparent; translucent	Transparent; translucent	Transparent
Substrate coating	Untreated	Type I and Type II collagen	Untreated[c]	Untreated[c]	Untreated	Untreated	Untreated[c]	Untreated[c]	Uncoated tissue culture treated	Extracellular matrix coatings (selection available	Type 1 collagen
Membrane height above base of well	1.0 mm	1.0 mm	1.0 mm	1.0 mm	1 0 mm	1.0 mm	1.0 mm	1.0 mm	0.9; 0 8	0 9, 0.8	1 0 mm

Comments

a. At membrane growth surface and at lower surface, respectively
b. Gamma aluminium oxide membrane
c. Requires coating with extracellular attachment factors
d. Composition unknown
e. Tissue culture treated
f. Special insert system available for large-scale cell culture (100 cm^2 area)
g. Surfactant-free mixed esters of cellulose nitrate and acetate
h. Hydrophilized PTFE biopore membrane

NOTE: Due to rapid development of these products, the accuracy of the above information cannot be guaranteed. Readers are advised to contact suppliers before placing orders

Suppliers

1. Whatman, Inc., 9 Bridgewell Place, Clifton, NJ 07014 (Tele: 800-922-0361) (FAX: 201-472-6949)
2. Nunc Intermed., Als Nunc, Postbox 280, DK-4000, Roskilde, Denmark (Tele: 45 42 359065) (FAX: 45 42 350105). USA 800-288-6862)
3. CoStar Corp., 205 Broadway, Cambridge, MA 02139 (Tele: 800-492-1110)
4. Millipore Corp., 80 Ashby Road, Bedford, MA 01730 (Tele: 800-645-5476)
5. ICN Biochemical, Inc., Biomedicals Div., P.O. 28050, Cleveland, OH 44128 (Tele: 800-845-0530)
6. Becton Dickinson, Labware, Two Oak Park, Bedford, MA 01730 (Tele: 800-343-2035)

Table 11.2. The filters differ with respect to their optical transparency, their composition and the size and distribution of pores. Some supports are pre-coated with attachment factors such as collagen (e.g. Transwell-Col) whereas others require treatment before cell seeding (e.g. Millicell-CM). Untreated filters (specifically Millicell-HA and Anocell) can be used for various epithelial cell lines (e.g. A6 and T84), although it is usually necessary to soak the filter supports in serum-containing culture medium for approximately one hour before seeding. A summary of attachment factors is given in Table 11.3.

11.4.2 Troubleshooting epithelial cell cultures: failure to thrive

Many factors can cause cell cultures to fail. Some may be related to dissociation procedures. Cells can be irreversibly damaged during tissue dissection, or cell disruption can occur as a result of trituration or pipetting, centrifugation or enzyme treatment and these cells can die before plating. The dissociation medium may have an inappropriate pH, temperature or osmolarity. Toxic compounds may be present in the medium or the isolation procedure may be too slow and tissue drying or anoxia may have occurred. Special care should be taken to use appropriately filtered water. Volatile organic substances are not removed by distillation, thus a system based on a battery of filters and exchange resins, such as the MilliporeQ system, is preferable.

If cells appear to be viable at the time of plating but die within 24 hours, there are again several possible explanations. It may be that the cells did not attach to the substratum or that, as above, there are toxic compounds in the culture medium or serum. Dead cells or heme may be present and toxic to living cells; there may be too many cells for the volume of medium; or the culture may be contaminated.

Cells which live for several days after plating but then die or fail to grow may indicate a lack of appropriate growth factors, or that the medium or serum is old or of an inappropriate composition. The cells may have been exposed to inappropriate conditions in the incubator such as extremes of pH, temperature or humidity. Contamination is also a possible problem and usually takes about a week before it becomes noticeable as cloudiness or foreign growths in the medium. We have experienced a mycoplasma infection of T84 cells which dramatically slowed cell growth. Symptoms of mycoplasma infection include a slow cell growth rate and an altered cell metabolism and cellular RNA profile. Because of other unknown effects of mycoplasma infections on cell properties, it is sensible to have cell stocks tested regularly for this pathogen; suitable kits are commercially available (Mycotrim TC, Hana Biologics, Inc., Berkeley, CA) and most epithelial cell lines can be tested by sending samples to American Type Culture Collection (Grand Island, NY).

Table 11.3 Summary of cell attachment factors

Factor	*Source*	*Cell type*	*Coating solution*
Collagen, Type 1[1]	Bovine tendon	Epithelial cells	1.2 mg/ml in culture medium
Collagen, Type 1[2]	Bovine skin	Primary cultures of epithelial cells and other cell types	0.6 mg/35 mm dish
Collagen, Type 1[2]	Rat tail	Most normal and transformed mammalian cells	0.4 mg/35 mm dish
Collagen, Type IV[2]	Mouse tumor	Epithelial, endothelial, muscle and nerve cells	1 mg/0.75–1.5 ml in culture medium
Poly-D-Lysine[2]	Synthetic	Many cell types, e.g. fibroblasts, chick embryo	2.5–5 μg/cm^2
Cell-Tak[2]	Marine mussel	Normal and transformed cells	3.5 μg/cm^2
Human extracellular matrix[2]	Human placenta	Epithelioid cells (especially of human origin)	1.25–10 μg/cm^2
Laminin[2]	Mouse tumor	Neoplastic cells of ectodermal or endodermal origin	1–2 μg/cm^2 1–5 μg/ml in culture medium
Human fibronectin[2]	Human fibronectin	Many cell types (especially fibroblasts and sarcomas)	1–5 μg/cm^2 8–40 μg/35 mm dish 5 μg/ml in culture medium

Source:
1. Organogenesis, 83 Rogers Street, Cambridge, MA 02142
2. Collaborative Res. Inc., Two Oak Park, Bedford, MA 01730

11.4.3 Methods for monitoring the condition of epithelial monolayers

Most investigations of cultured epithelia begin after the cells have reached confluence and tight junction formation has been completed. Usually, these properties are determined from electrical measurements of the transepithelial voltage (V_T) and conductance (G_T). Measurements of G_T are more useful than measuring only V_T, since non-transporting or electroneutral transporting epithelia can have a low conductance without an appreciable V_T.

Some caveats concerning electrical measurement systems

A few Ussing-type chambers have been designed for measuring transepithelial electrical properties under sterile conditions. However, it is crucial to consider the configuration of voltage and current electrodes when using such chambers. Calculation of G_T from measurements of voltage deflections across the epithelium in response to transepithelial current pulses generally assume that the current field is uniform across the tissue. If the current electrodes are too close to the tissue, inhomogenities in the current field can result. Voltage deflections measured near the current source will be artifactually high, resulting in underestimates of G_T.

An example of this problem is presented in Figure 11.3a which illustrates a commercially available measurement system. The system consists of an electrometer and two chopstick-like devices which contain voltage and current electrodes. The voltage electrodes consist of Ag-AgCl pellets at the end of each 'chopstick' and the current electrodes are Ag-AgCl wires wound around the tip of the sticks. The advantage of such a system is that the electrodes can be easily sterilized. However, the basic flaw is that the current-passing electrodes are too close to the epithelium. We initially observed that measured voltage deflections varied considerably in relation to the depth of immersion of the electrodes in the culture media. Even when this variable was carefully controlled, the G_T measurements obtained for A6 epithelium, for example, were approximately one third of those measured for the same preparations in a standard Ussing-type chamber.

The problem of the non-uniform current field is illustrated in the upper panel of Figure 11.3b, which shows measurements from a pair of voltage-measuring electrodes that were placed across the epithelium and moved progressively from 0 to 180° around the circumference of the petri dish (shown in Figure 11.3a). The magnitude of this voltage deflection was then compared to the magnitude of the voltage deflection at position 0°. As shown in the graph at the bottom of Figure 11.3b, the voltage deflection progressively decreased as the two pairs of voltage-sensing electrodes became more separated. The ratio of the voltage deflections decreased

(a)

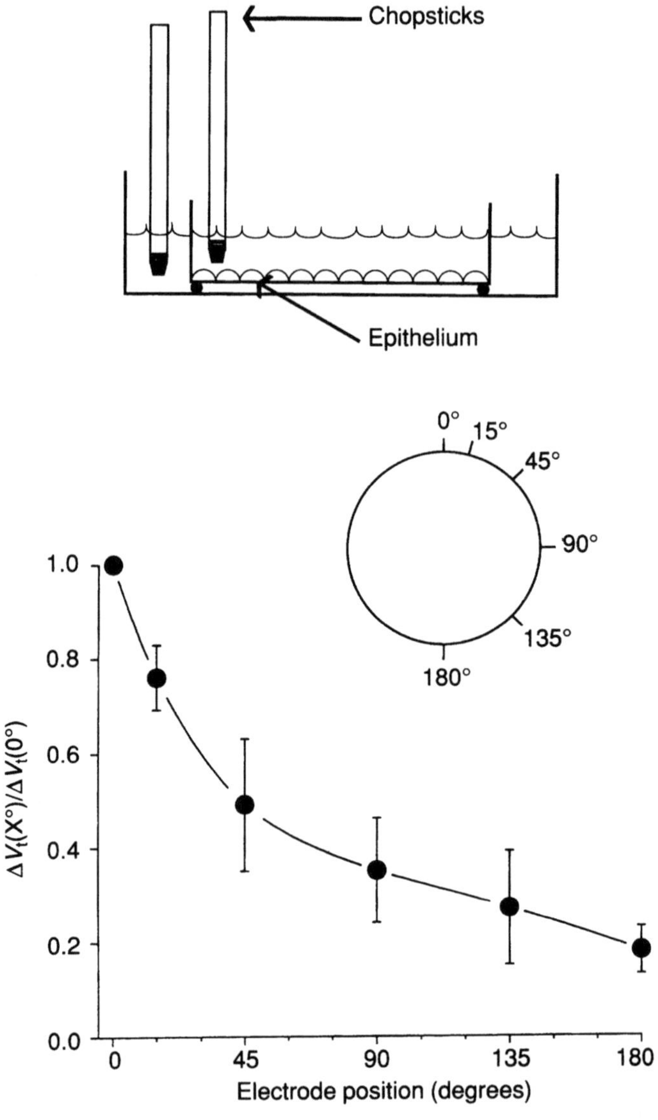

(b)

Figure 11.3 (a) Cross-section of a filterwell support showing 'chopstick' electrodes on either side of the epithelium. (b) Top: view of filter support and position of current-passing electrodes and an independent set of voltage sensing electrodes. Below: fractional decrease in voltage response to a square current pulse measured by voltage-sensing electrodes [$\Delta V_t(X°)$] positioned at various locations around the circumference of the filterwell support (see insert above). The potential is compared with the voltage deflection measured at position 0° defined as the location of the current-passing electrodes [$\Delta V_t(X°)$]. (Reprinted from Jovov *et al.*, 1991, with permission of *Am. J. Physiol.*)

from 1.0 at position 0° to 0.2 at position 180°. Consequently, caution should be used in interpreting conductances calculated from such data. A new design is available which eliminates this problem but is not as convenient because it requires placing the filter support in a special chamber.

11.4.4 A spectroscopic method for assessing the integrity of cell cultures

Many investigators are not interested in precisely monitoring transepithelial conductance. In some cases, it is useful simply to know whether the cells have an electrical resistance or have reached confluence. Jovov *et al.* (1991) reported a method for monitoring confluence that does not involve the use of electrophysiology. The method exploits the presence of phenol red, a hydrophilic, common pH indicator that is not transported by most cells and is used in most culture media. This compound has a well-defined isobestic point at 479 nm. Therefore, the concentration of phenol red can be determined by measuring its absorbance at this wavelength using a common laboratory spectrophotometer. In this method, phenol red (15 μg/ml) is initially present on the basal or serosal side of the epithelium but is absent in the mucosal solution. (Cultured medium without phenol red can be purchased from GIBCO Laboratories, Grand Island, NY). After a period of time, the mucosal fluid is collected and its phenol red concentration is determined. This information is used to calculate the phenol red flux (J_{PR}) across the epithelium, i.e. the amount of transepithelial phenol red movement over time, using the following equation:

$$J_{PR} = (A_{479} \times Vol_m)/(t \times A \times EC)$$

where J_{PR} is phenol red flux, Vol_m is the mucosal solution volume, t is time, A is the surface area, A_{479} is the measured absorbance (at 479 nm) and EC is the extinction coefficient for phenol red at the 479 nm wavelength (8450 $l mol^{-1} cm^{-1}$). We note that in the case of low PR fluxes the absorbance signal can be amplified by performing the assay in alkaline solutions (e.g. at pH 9).

This technique has proven useful for long-term measurements of tight junction formation and epithelial integrity. Figure 11.4 illustrates an experiment in which J_{PR} was measured over a period of days following cell seeding. Note that J_{PR} decreased and was relatively constant by day 9, indicating the completion of tight junction formation during this period. These findings are consistent with electrical studies which demonstrate that the transepithelial resistance reaches a maximum value within a similar time period (7–9 days; Wills and Millinoff, 1990).

Phenol red flux measurements are also convenient for evaluating short-term effects on tight junctions or epithelial integrity. The tight junctional conductance (G_j) was measured using an independent electrical method

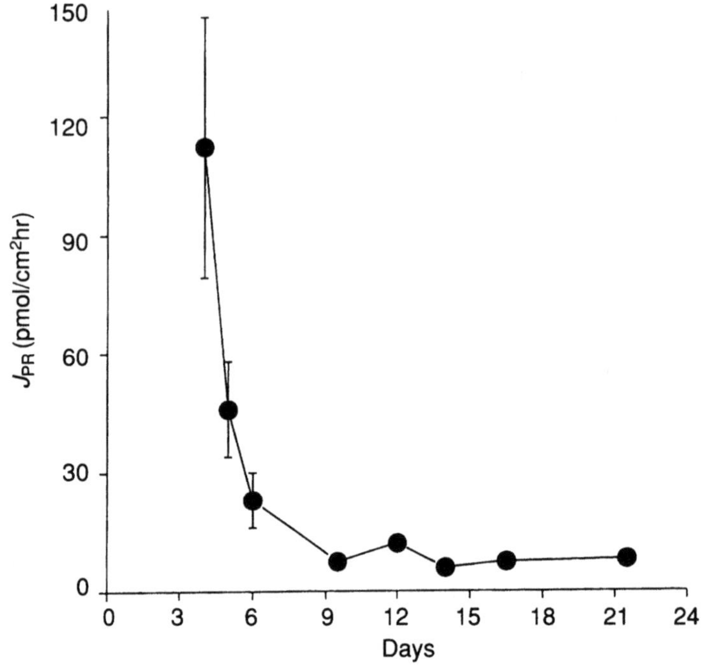

Figure 11.4 Phenol red flux measured as a function of epithelial age after cell plating on day 0. (Reprinted from Jovov *et al.*, 1991, with permission of *Am. J. Physiol.*)

described by Jovov *et al.* (1991). Figure 11.5 illustrates G_j and serosa-to-mucosa phenol red and tritiated mannitol fluxes as a function of time following calcium chelation in the serosal bathing solution. G_j, J_{PR} and mannitol fluxes all increased in parallel, consistent with the known disrupting effects of calcium removal on tight junctional integrity (Cereijido *et al.*, 1989). These findings are consistent with a paracellular route for phenol red movement across the epithelium. We have found that J_{PR} is a simple indicator of the condition of the epithelium in that contaminated epithelia showed increased phenol red fluxes before visible signs of infection were detectable. By monitoring J_{PR} flux the investigator can have an early warning of infection or other problems.

11.5 A SIMPLE METHOD FOR MONITORING WATER MOVEMENT ACROSS CULTURED EPITHELIA

In the absence of evaporative loss or condensation, water fluxes across cultured epithelia can be measured directly by comparing the volume of the mucosal and serosal bathing solutions. In practice, however, such

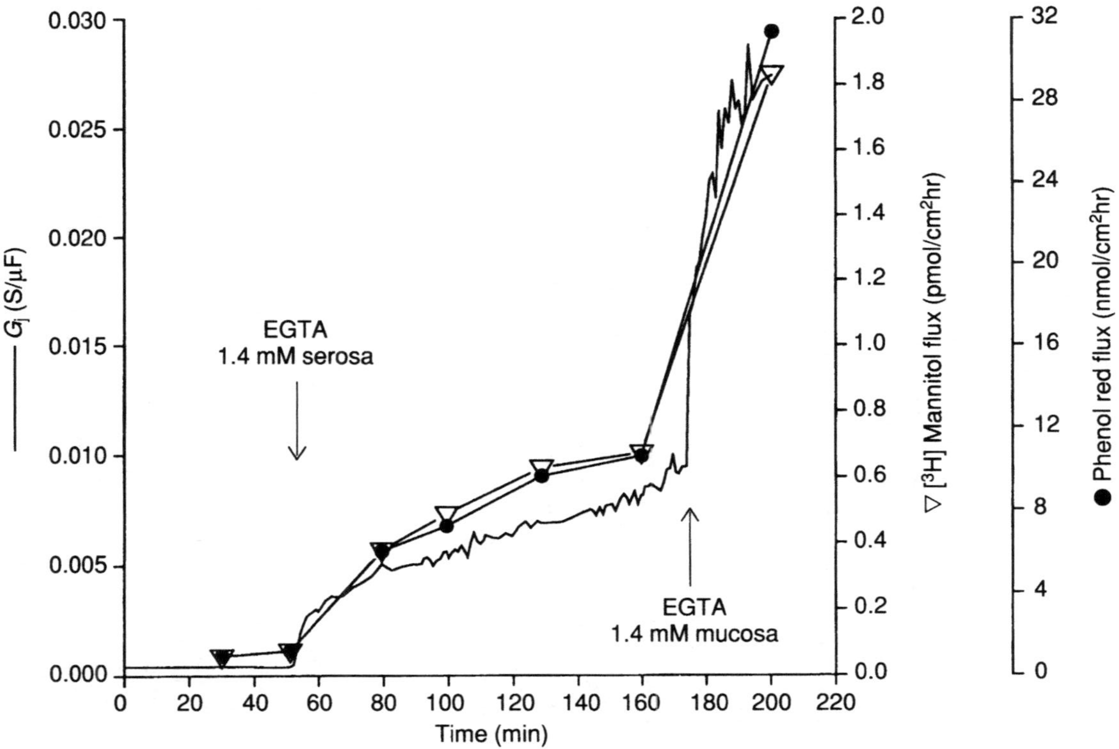

Figure 11.5 Paracellular resistance (G_J), phenol red flux and tritiated mannitol flux as a function of time following calcium chelation with EGTA as indicated (see arrows). (Reprinted from Jovov *et al.*, 1991, with permission of *Am. J. Physiol.*)

volume measurements can be hard to achieve since it is often difficult or undesirable to remove all of the bathing medium. In this regard phenol red can also serve as a useful indicator of net water movement between the serosal and mucosal compartments.

Again, in this method phenol red (PR) is initially present in one bathing compartment (usually the serosal side) and not the other. The method assumes that:

- the phenol red distribution is not in equilibrium;
- the sum of the quantity of phenol red (PR) in the mucosal and serosal solutions is equal to the original amount of PR in the initial compartment;
- a change in the volume of one compartment should result in an equal and opposite change in the volume of the other compartment.

The relationship is described as follows:

$$[PR]_o \cdot iVol_s = [PR_m] \cdot (iVol_m - x) + [PR]_s \cdot (iVol_s + x)$$

where the subscripts o, m and s refer to the initial, mucosal and serosal solutions, respectively, and [PR] is phenol red concentration calculated for each solution from PR absorbance at a wavelength of 479 nm. The initial volume of each solution is designated by *iVol*, and *x* is the volume of solution gained or lost from either compartment. Thus the amount of net transepithelial water movement can be found by solving for *x* (Jovov *et al.*, 1991).

REFERENCES

Bashor, M.M. (1979) General cell culture techniques. *Methods of Enzymol.* **58**:119–221.

Bell, C.L. and Quinton, P.M. (1990) Recycle those cell culture inserts (letter to the editor). In vitro *Cell Dev. Biol.* **26**:1123–4.

Bello-Reuss, E. and Weber, M. (1987) Electrophysiological studies of primary cultures of rabbit distal tubule cells. *Am. J. Physiol.* **252**:F899–F909.

Cereijido, M., Pounce, A. and Gonzalez-Mariscal, L. (1989) Tight junctions and apical/basolateral polarity. *J. Membr. Biol.* **110**:1–9.

Cuthbert, A.W., Kirkland, S.C. and MacVinish, L.J. (1985) Kinin effects on ion transport in monolayers of HCA-7 cells a line from a human colonic adenocarcinoma. *Br. J. Pharmacol.* **86**:3–5.

Fleming, T.P. (ed.) (1992) *Epithelial Organization and Development*, Chapman & Hall, London, England, 400 pp.

Freshney, R.I. (ed.) (1987) *Culture of Animal Cells: A Manual of Basic Technique*, 2nd edn, Alan Liss, NY, 397 pp.

Freshney, R.I. (ed.) (1992) *Culture of Epithelial Cells*, Wiley-Liss, NY, 232 pp.

Gong, J., McCarthy, K., Telford, J. *et al.* (1992) Interepithelial airway dendritic subset. *J. Exp. Med.* **175**:797–807.

Grasset, E., Pinto, M., Dussaulx, E. *et al.* (1984) Epithelial properties of human

carcinoma cell line Caco-2: electrical parameters. *Am.J.Physiol.* **247**(16):C260–C267.
Hanrahan, J.W. and Tabcharani, J.A. (1990) Inhibition of an outwardly rectifying anion channel by HEPES and related buffers. *J. Membr. Biol.* **116**:651–77.
Jakoby, W.B. and Pastan, I.H. (eds) (1979) *Cell Culture, Methods in Enzymology*, Vol. 58, Academic Press, San Diego, CA, 642 pp.
Jovov, B., Wills, N.K. and Lewis, S.A. (1991) A spectroscopic method for assessing confluence of epithelial cell cultures. *Am. J. Physiol.* **261**(30):C1196–C1203.
Linder, S. and Marshall, H. (1990) Immortalization of primary cells by DNA tumor viruses. *Exp. Cell Res.* **191**:1–7.
Matlin, K.S. and Valentich, J.D. (eds) (1989) *Functional Epithelial Cells in Culture*, Wiley, New York.
McRoberts, J.A. and Barrett, K.E. (1989) Hormone-regulated ion transport in T84 colonic cells, in *Functional Epithelial Cells in Culture*, (eds K.S. Matlin and J.D. Valentich), Plenum Press, New York, pp. 235–65.
Meier, K.E. and Insel, P.A. (1989) Hormone receptors and response in cultured renal epithelial cell lines, in *Tissue Culture of Epithelial Cells*, (ed. M. Traub), Plenum Press, New York, pp. 145–53.
Paulmichl, M., Defregger, M. and Lang, F. (1986) Effects of epinephrine on electrical properties of Madin–Darby Canine Kidney cells. *Pflügers Arch.* **406**:367–71.
Perkins, F.M. and Handler, J.S. (1981) Transport properties of toad kidney epithelia in culture. *Am. J. Physiol.* **241**:C154–C159.
Pollard J.M. and Walker J.M. (eds) (1990) *Animal Cell Culture*, Humana Press, Clifton, NJ, 713 pp.
Schaeffer, W.I. (1978) *Proposed usage of animal culture terms.* Tissue Culture Association, Inc., Manual **4**(1):779–82.
Steele, R.E., Preston, A.S., Johnson, J.P. and Handler, J.S. (1986) Porous-bottom dishes for culture of polarized cells. *Am. J. Physiol.* **251**:C136–C139.
Traub, M. (1985) *Tissue Culture of Epithelial Cells*, Plenum Press, NY, 288 pp.
Turksen, K. and Aubin, J. (1994) Positive and negative immunoselection for enrichment of two classes of osteoprogenitor cell. *J. Cell Biol.* **114**:373–84.
Valentich, J.D. (1986) Innovative approaches for the study of cultured renal epithelial. *Miner. Electrolyte Metab.* **12**:6–13.
Warrington, A., Barbarese, E. and Pfeiffer, S. (1992) Stage specific isolated oligiodendrocyte progenetors. *Dev. Neurosci.* **14**:93–7.
Wills, N.K. and Millinoff, L.P. (1990) Amiloride-sensitive Na^+ transport across cultured renal (A6) epithelium: evidence for large currents and high Na:K selectivity. *Pflügers Arch.* **416**:481–92.

12

Signaling pathways regulating ion transport in polarized cells

Norman J. Karin, Min I. N. Zhang, E. Radford Decker and Roger O'Neil

A critical function of polarized tissues is to generate and maintain differences in the composition of the milieu between the two compartments separated by the tissue. This is readily apparent in compartments separated by epithelial cells, endothelial cells and bone cells (highly specialized polarized cells). To generate and maintain this difference in composition between the compartments requires a regulated vectorial transport of solutes and water across the cells of these tissues. Hence, the regulation of vectorial transport is central to the function of all polarized cells.

The regulation of transport across epithelial cells and other polarized cells is controlled by numerous hormones and cytokines, as well as by physical forces. In epithelial cells, peptide hormones and physical stresses, such as osmotic stress, have been shown to regulate transport

Epithelial Transport: A guide to methods and experimental analysis.
Edited by Nancy K. Wills, Luis Reuss and Simon A. Lewis.
Published in 1996 by Chapman & Hall, London. ISBN 978-94-010-7168-0

processes at the plasma membrane within seconds to minutes (Bertorello and Katz, 1993; Kinne *et al.*, 1993; Reuss *et al.*, 1991; Widdicombe *et al.*, 1991). The mechanisms controlling this rapid regulation of transport have not been fully elucidated. However, in the past two decades investigators have made tremendous progress in unraveling a number of important biochemical pathways and their interactions that play a role in regulating plasma membrane transporters. The conceptual framework that has developed is that the binding of first messengers, such as peptide hormones or cytokines, to plasma membrane receptors, or the application of physical stresses, such as osmotic stress or membrane stretch, either directly modify the plasma membrane transport process or activate one or more signal transduction pathways (signaling pathways) that give rise to second and third messengers within the cell (Hepler and Gilman, 1992; Lambert, 1993). These messengers, in turn, modify the regulatory biochemical pathways controlling specific transport processes at the plasma membrane.

This chapter will focus on the methods for studying the dominant signaling pathways underlying the rapid regulation of ion transport in epithelial cells controlled by peptide hormones (receptor-coupled regulation) or physical stresses (osmotic stress or mechanical stretch). Three signaling pathways will be emphasized: the adenylate cyclase/cyclic AMP (cAMP)/protein kinase A pathway, the phospholipase C/diacylglycerol (DAG)/1,4,5-inositol trisphosphate (IP_3)/protein kinase C pathway, and the calcium signaling pathway (Figure 12.1). While other pathways assuredly are involved in regulating transport (Mills and Mandell, 1994; Frazier and Yorio, 1992; Whatley *et al.*, 1990), the three pathways noted are the most fully understood at the present time in epithelial cells. Also, because of space limitations, the current discussion will be brief, but with appropriate references to more detailed discussions of the methods.

12.1 RECEPTOR-COUPLED REGULATION OF ION TRANSPORT

Peptide hormones exert their actions on cells by binding to plasma membrane receptors which in turn are coupled to signal transduction pathways within the cell. The coupling typically involves one or more G proteins acting as intermediates (Figure 12.1). Peptide hormone receptors coupled through G proteins are part of a family of receptors which share many structural and functional characteristics (Birnbaumer, 1993; Hepler and Gilman, 1992; Lambert, 1993). This chapter will not discuss the receptor or G protein functions, but will focus on the signaling pathways coupled to these receptors.

The actions of peptide hormones on ion transport can readily be studied in epithelial cells using standard ion transport monitoring procedures as

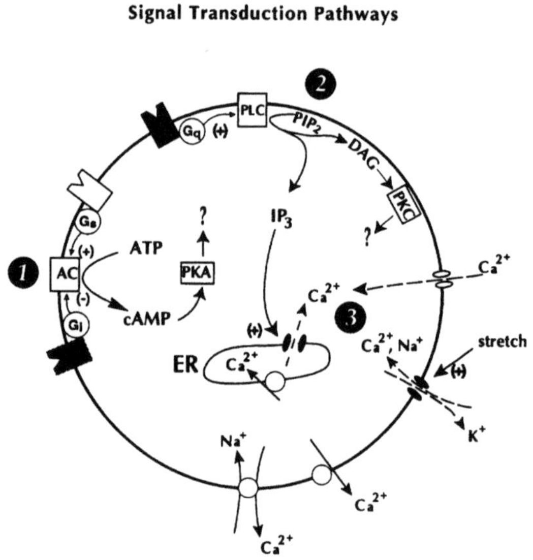

Figure 12.1 Cellular model of the signaling pathways controling ion transport in polarized cells. Components of the pathways are described in the text. Pathway 1 is the adenylate cyclase (AC), cyclic AMP (cAMP), protein kinase A (PKA) pathway which is typically coupled to a peptide-hormone receptor through a G protein, G_i or G_s. Pathway 2 is the phospholipase C (PLC), diacylglycerol (DAG), 1,4,5-inositol trisphosphate (IP_3) pathway which is typically coupled to a peptide-hormone receptor through G_q or a related G protein. Pathway 3 is the calcium signaling pathway depicting the various calcium transport mechanisms controling calcium signaling including three calcium-permeable channels: an IP_3-sensitive calcium release channel of the endoplasmic reticulum (ER), a plasma membrane calcium-selective channel, and a plasma membrane stretch-activated non-selective cation channel. Also depicted are the ER and plasma membrane calcium pumps and a plasma membrane Na:Ca exchanger, all of which can modify the calcium signal.

described in Chapters 5, 6 and 10. The role of signaling pathways in regulating the transport processes can be assessed by studying the components of the pathways.

12.1.1. Cyclic AMP/cyclic AMP-dependent protein kinase

Many peptide hormone receptors in polarized cells are coupled through G_s and G_i proteins to stimulate or inhibit adenylate cyclase, respectively, and the generation of the second messenger, cyclic AMP (cAMP). Hence, studies of this pathway have often focused on evaluating alterations in the cytoplasmic levels of cAMP in response to hormones and other effectors. Cyclic AMP-dependent protein kinase, or protein kinase A (PKA), is postulated to be the sole acceptor for cAMP in eukaryotic cells and consists of a cAMP-binding regulatory subunit and a catalytic subunit. While the holoenzyme is inactive, the binding of cAMP to the regulatory subunit dissociates the protein–protein interactions and releases the active catalytic subunit. This system recently gained a great deal of attention in polarized epithelia since cAMP-dependent phosphorylation was shown to regulate the activity of the cystic fibrosis gene product (CFTR) in tracheal and intestinal epithelia (Anderson *et al.*, 1992; Welsh *et al.*, 1992).

The earliest assays for cAMP involved determinations of binding of [^{3}H]cAMP to purified PKA in the presence and absence of unlabeled cAMP as a specific competitor (Gilman, 1970). This method has largely been replaced by radioimmunoassays for cAMP (Brooker *et al.*, 1979).

PKA traditionally was assayed by the transfer of ^{32}P from [γ-^{32}P]ATP into histones (Rubin *et al.*, 1974). More recently, synthetic substrates for phosphorylation have been employed since, among other reasons, histones can be phosphorylated by a variety of protein kinases and actually constitute a relatively poor substrate for PKA. Kemptide is a seven-amino acid peptide (Leu-Arg-Arg-Ala-Ser*-Leu-Gly) which can be phosphorylated at a serine residue by PKA. While it was first described as a PKA substrate in 1977 (Kemp *et al.*, 1977), it has only recently gained widespread use. Non-radioactive assays for protein kinase activity are also available commercially (Technical Bulletin 132, Promega Corporation, Madison, WI), and typically include a dye-linked phosphate acceptor (such as Kemptide) whose isoelectric point is monitored electrophoretically: as phosphate is transferred to the colored substrate, its P_I becomes more negative, as evidenced by its increased mobility toward the anode. A dye-linked peptide substrate is also available for the assay of protein kinase C (see below).

Inhibitors of PKA have often been employed to assess cellular functions of PKA. Caution must be exercised when using kinase inhibitors as many inhibitors initially thought to be specific have later been shown to inhibit other kinases, as discussed below. A heat-stable inhibitor peptide exists which is an effective and specific inhibitor of PKA catalysis with a K_i of 0.5–2 nM (Scott *et al.*, 1986). This peptide will not penetrate intact cells and must be used with broken cell preparations or microinjected into intact cells, although Lefkowitz and co-workers have had success in introducing inhibitor peptides into cells that have been permeabilized with digitonin (Lohse *et al.*, 1990). Wiptide, another PKA inhibitor peptide, has also been introduced into cells via diffusion through patch electrodes (Fan *et al.*, 1993). The drug H-89 is a PKA blocker which can be used in either *in vitro* kinase assays or with intact cells (Hidaka *et al.*, 1990). While H-89 displays a K_i for PKA of 0.04 μM, it is also quite effective in blocking cyclic GMP-dependent protein kinase for which it exhibits a K_i of 0.34 μM. Another compound which is gaining widespread use is R_p-cAMPS, an analog of cAMP with an exocyclic sulfur substitution, which is an effective and specific inhibitor of PKA (Ogreid *et al.*, 1994). Similarly, R_p-(cyclic)thiophosphate cAMPS (R_p-CPT-cAMPS) is a PKA antagonist but is more permeant to intact cells than R_p-cAMPS (Ogreid *et al.*, 1994). S_p-cAMPS and S_p-CPT-cAMPS are isomers of the R_p forms which act as PKA agonists (Ogreid *et al.*, 1994).

12.1.2 1,4,5-Inositol trisphosphate/diacylglycerol

Numerous hormones, growth factors and cytokines lead to the activation of phospholipase C and the generation of 1,4,5-inositol trisphosphate (IP_3) and diacylglycerol (DAG). Extensive literature exists on the roles of IP_3 and DAG in cell signaling, and the reader is referred to two recent reviews (Nishizuka, 1992; Berridge, 1993). These two compounds arise from the agonist-stimulated hydrolysis of inositol phospholipids – primarily phosphatidylinositol 4,5-diphosphate (PIP_2) – in the plasma membrane of eukaryotic cells by phospholipase C. IP_3 binds to specific receptors to cause release of free Ca^{2+} from intracellular stores while DAG is an activator of protein kinase C which will be discussed below.

Following the hydrolysis of PIP_2, IP_3 is sequentially dephosphorylated by cytoplasmic phosphatases to form IP_2 and IP. Alternatively, some IP_3 can be phosphorylated to form IP_4. The appearance of these compounds can be monitored by incubating cells or tissue samples in medium containing myo-[2-^{3}H]inositol for several hours to label the pool of PIP_2, after which radiolabeled inositol phosphates are separated by anion-exchange chromatography (Anderson *et al.*, 1992).

The generation of DAG in response to tumor promoters has been measured by thin-layer chromatography (TLC) in extracts of tracheal epithelial cells that had been grown in [^{3}H]glycerol for 24 hours (Sesko *et al.*, 1990). Cybulsky and Cyr (1993) performed a similar DAG assay in glomerular epithelial cells that had been incubated for only 40 minutes in medium containing 1-O-[alkyl-^{3}H]2-lyso-phosphatidylcholine. DAG also can be quantified after ^{32}P-labeling by *Escherischia coli* DAG kinase, as described by Preiss *et al.* (1986). Briefly, lipids are extracted from homogenates and incubated with DAG kinase and [γ-^{32}P]ATP. The incorporation of ^{32}P into lipid is then measured after a second organic extraction. That this method does not rely upon metabolic labeling has facilitated its use with tissue samples. Pagano has pioneered the use of lipid analogues labeled with the fluorescent N-6[7-nitro-2,1,3-benzoxadiasol-4-yl]aminohexanoyl ('NBD') moieties in non-radioactive assays for lipid quantitation (Pagano and Sleight, 1985). Helvoort *et al.* (1994) used

NBD-lipids and TLC to determine that DAG is converted to phosphatidylcholine in cultured dog kidney epithelial cells.

12.1.3 Protein kinase C

Protein kinase C (PKC) is activated by the increases in cytoplasmic levels of DAG which accompany many receptor mediated events in polarized epithelial cells. At least ten members of PKC family have been described to date, many of which are expressed in epithelial cells. For example, intestinal epithelial cells and rat kidney glomerular epithelia have been reported to contain five isoforms of PKC (Saxon *et al.*, 1994; Huwiler *et al.*, 1993). The family of PKC enzymes includes both Ca^{2+}-dependent and Ca^{2+}-independent forms that also exhibit different lipid requirements (Lee and Severson, 1994). Ca^{2+} dependence is complex for those PKC forms that require this ion since the Ca^{2+} concentration for maximal PKC activity actually reflects the Ca^{2+} activity at a membrane or micelle surface (see below) and is also influenced by the ionic strength of the assay solution. In addition, exposure of PKC to high Ca^{2+} levels leads to the conversion of the enzyme into a Ca^{2+}-independent form (Bazzi and Nelsestuen, 1988). PKC isoforms phosphorylate either serine or threonine residues which are adjacent to a cluster of basic amino acids (Pearson and Kemp, 1991). The best substrates have hydrophobic regions which allow their association with the lipid bilayers which are required for maximal activity. These typically are delivered as either membranes, micelles or detergent extracts (Epand, 1994). The *in vitro* assay for PKC activity is similar to that for PKA, typically measured as the transfer of ^{32}P from [γ-^{32}P]ATP to a protein substrate. Substrate proteins include histones, protamine sulfate and myelin basic protein; however, each protein differs in its requirement for cofactors (Epand, 1994). Synthetic peptide substrates are commercially available which offer single phosphorylation sites to facilitate kinetic analyses. A newer assay method employs a fluorescent peptide substrate which permits continuous measurement of PKC catalysis (McIlroy *et al.*, 1991).

PKC activity is also stimulated by phorbol esters such as TPA, which bind to the same region of the kinase as DAG. Most PKC isoforms can be down-regulated by prolonged TPA treatment, a phenomenon which has been useful for assessing the role of PKC in cellular processes in many cell types. For example, TPA-induced PKC down-regulation in T84 epithelial cells led to a reduction in both basolateral and apical chloride transport (Matthews *et al.*, 1993). Specific inhibitors have also been useful in evaluating the physiological roles of PKC activity. While staurosporine has been used extensively to block PKC activity, it is now clear that this agent also reduces PKA activity, albeit with approximately a three-fold higher K_i, as well as inhibiting a variety of other protein phosphorylases (Tamaoki, 1991). A much more specific PKC inhibitor is calphostin C which displays a K_i for PKC of 0.05 μM (Tamaoki, 1991). PKC can also be inhibited by sphingosine, a compound which itself appears to be a physiological second messenger for $[Ca^{2+}]_i$ elevation in bone cells (Liu *et al.*, 1995); Ca^{2+} signaling in polarized cells is the topic of the next section.

12.1.4 Calcium signaling

The second messenger, calcium, plays a key role in controlling numerous enzymes, transport proteins, and cellular processes. Transient elevation in intracellular calcium levels is a signal (calcium signal) which can alter numerous processes and functions of cells including transport and the long-term expression of transport proteins.

Calcium signals typically are generated by activation of calcium release from intracellular stores, such as from the endoplasmic reticulum (ER), or by activation of plasma membrane calcium-permeable channels (Figure 12.1; see also channel descriptions below). The actual magnitude of the calcium signal, i.e. the magnitude of the rise, is a complex process determined not only by the rate of entry of calcium into the cytoplasm via calcium release from stores or calcium entry through plasma membrane channels, but also by rate of removal of calcium from the cytoplasm via the ER and plasma membrane calcium pumps and the plasma membrane Na:Ca exchanger (McCarty and O'Neil, 1992; Putney, 1993).

The most widely used approach to studying calcium signaling in polarized cells is to monitor the changes in intracellular free calcium levels, $[Ca^{2+}]_i$. Typically, cells are loaded with a calcium-sensitive fluorophore, such as fura-2, indo-1 or fluo-3, and the fluorescence emission monitored to obtain an estimate of intracellular free calcium levels (Grynkiewicz *et al.*, 1985; McCarty and O'Neil, 1991; Roe *et al.*, 1990; also see Chapter 7 for methods). Other methods utilizing the calcium-sensitive absorbance of quin-2 (Grynkiewicz *et al.*, 1985) or the calcium-sensitive luminescence of aequorin (McNeil and Taylor, 1985) have been employed with limited success. However, recent advanced methods that will greatly increase the utility of monitoring calcium levels in selected cell compartments include recombinant aequorin engineered with an ER targeting sequence (Brini *et al.*, 1994; Kendall *et al.*, 1994) and dextran-bound fura-2 with a targeting sequence for the nucleus (Allbritton *et al.*, 1994).

Fluorescent probes whose excitation or emission spectra are sensitive to calcium levels, such as fura-2 or indo-1, respectively, provide the best quantitative approach for monitoring intracellular calcium levels. The advantage of these probes is that dual excitation or emission spectral ratioing can be employed which will diminish some of the factors that have proved to be problematic for quantitation such as the intracellular fluorophore concentration and the light path length (Grynkiewicz *et al.*, 1985; Roe *et al.*, 1990). In contrast, real-time microscopic video imaging of spatial differences in intracellular calcium levels can best be obtained using single wavelength fluorescent probes with excitation and emission maxima in the visible spectral range (Fricker *et al.*, 1991). Other factors can often influence the choice of probes to be used, as discussed in the above references.

The release of calcium from intracellular stores plays a key role in initiating calcium signals. The dominant storage organelle which displays IP_3-sensitive calcium release appears to be part of the ER, although other secondary sites may be present in some cells (Fricker *et al.*, 1991; Putney, 1993). The importance of this site in calcium signaling is typically evaluated by assessing the influence of inhibitors of the ER calcium pump, such as thapsigargin, to empty passively the ER calcium storage sites prior to generating a calcium signal (Putney, 1993). Caution must be exercised in using these approaches, however, as multiple calcium ATPases of possible secondary storage sites may have varying sensitivity to the blockers (e.g. Bruce and Ullrich, 1991) and, indeed, emptying the calcium storage sites may itself regulate entry of calcium into the cells through plasma membrane calcium-permeable channels (Putney, 1993).

Hormone-mediated activation of plasma membrane calcium-permeable channels can also lead to generation or amplification of calcium signals. The role of plasma membrane channels can be specifically assessed using patch clamp techniques, as discussed below, or can quantitatively be assessed using fluorescent assessment of intracellular calcium levels. In the latter approach the magnitude of a calcium signal generated in the presence of normal extracellular calcium levels (typically 1–2 mM total calcium) can be compared with that generated upon removal of extracellular calcium (or chelation by EGTA at constant pH) to provide an index of the contribution of calcium influx to the calcium signal (Bruce and Ullrich, 1991; McCarty and O'Neil, 1991). Alternatively, Mn^{2+}, which can permeate calcium channels and quenches fura-2 fluorescence, can be used to monitor calcium influx during generation of a calcium signal (Hallam *et al.*, 1989). With Mn^{2+} in the external medium, a change in the rate of fura-2 quenching upon generation of a calcium signal would demonstrate a contribution of a calcium-permeable channel to the calcium signal. The recent synthesis of a near-membrane fura-2 probe, which is designed to monitor calcium changes near the cytoplasmic face of the plasma membrane, may provide a better index of the magnitude of a calcium influx (Etter *et al.*, 1994). Unfortunately, these methods are limited in that they only provide a qualitative measure of the activity of the plasma membrane calcium channels and do not permit a simple assessment of the magnitude of the actual calcium influx. As discussed below, direct assessment of channel activity using patch clamp methodology can overcome some of these limitations.

12.2 PHYSICAL STRESS-INDUCED REGULATION OF ION TRANSPORT

12.2.1 Osmotic stress and cell swelling

Changes in the external medium osmolality (to hypertonic or hypotonic) forms an osmotic stress condition that will immediately change cell volume due to net water flow into or out of the cells. However, cells do not respond to stress conditions passively, but manage to counteract the stress through a mechanism known as cell volume regulation in which the cells tend to recover their volume to near pre-stress conditions (for review, see McCarty and O'Neil, 1992; Hoffman and Ussing, 1992). Cell volume regulation occurs as a result of activation of plasma membrane solute transport processes, such as ion channels, to effect a net transport of solute across the plasma membrane accompanied by an obligatory coupling of water flow. Although volume regulation is present in both hypertonic and hypotonic stresses in many cells, the regulatory volume mechanisms appear to be most developed for swelling conditions and, hence, are best characterized for these conditions (McCarty and O'Neil, 1992). Osmotic swelling would appear to have, therefore, at least two actions on solute transport: regulation of net transepithelial transport (see reference above) and regulation of transport processes important in control of cell volume.

One of the consequences of osmotically-induced swelling (hypotonic conditions) is the apparent generation of membrane tension which may lead to membrane stretching. Osmotically induced swelling may therefore affect the plasma membrane transport process in a similar manner observed for mechanically induced membrane stretch (Sackin, 1987; Sachs, 1988). However, these two methods for inducing 'stretch' – osmotic versus mechanical – do differ. When ion channels are activated in membrane patches, typical pressures of 10–40 cm H_2O are used. However, in osmotic stress experiments, osmotic differences of 14–140 mOsm are typically used. These osmotic differences would exert an equivalent pressure of 360 to 3600 cm H_2O. Despite the large differences in pressure, these two types of 'stretch' are often considered to be of similar nature resulting in many common responses as discussed below. Likewise, osmotically induced cell swelling can activate many of the signaling pathways induced by peptide hormones (see above) and, hence, similarities in the cell response to osmotically induced swelling and peptide hormones may be anticipated. For example, like peptide hormone actions already noted, osmotically induced swelling in some cells has been shown to lead to rapid activation of the adenylate cyclase/cAMP pathway, the phospholipase C/IP_3/DAG pathway and calcium entry/release pathways (McCarty and O'Neil, 1992).

The effect of osmotically induced cell swelling on ion transport can be readily studied in isolated intact epithelia, epithelial cell cultures or single epithelial cells using standard ion transport monitoring procedures as noted above (also in Chapters 5, 6, 9 and 10). To assess the effect of swelling, cells are typically exposed to a hypotonic medium of defined osmolality generated by simple dilution of the extracellular bathing medium or by selective removal of solutes (normally NaCl) from the medium. Bathing medium osmolarity can be reduced by 50% or more without causing cell lysis, although moderate reductions in osmolality of 5–20% are most commonly employed. To assess the magnitude of the induced swelling, the changes in cell volume are typically monitored using any one of a variety of techniques ranging from isotopic volume markers and fluorescent dye dilution methods to computer-aided video microscopy procedures, as appropriate for the cells being studied (details of methods are given in Chapter 8).

Once the conditions for studying an intact epithelium have been established for monitoring of solute transport or cell volume, the studies can be extended to evaluate the role of signaling pathways in controlling ion transport during osmotic stress conditions. The methods described above for evaluating the various signaling pathways are directly applicable to studying the effects of osmotic stress on these pathways and the subsequent controlling of the solute transport pathways in epithelial tissues and, hence, will not be re-examined here.

Studies of stress-induced effects on single cells have focused on the regulation of ion transport across the cell membrane. Some common features have been observed on the regulation of ion transport induced by mechanical stretch versus that by osmotic swelling due to the development of membrane tension on stretch. For this reason the remainder of the chapter will focus on regulation of ion transport caused by mechanical stretch, in general, and will treat osmotic stress as a special case of 'stretching' the membrane.

12.2.2 Mechanical stretch of the cell membrane

Osmotic stress leading to cell swelling can generate membrane tension or stretch. Cells may also experience membrane stretching due to cell topological changes associated with mechanical forces altering cell shape in airway epithelium (Kim *et al.*, 1993) or urinary bladder (Lewis and DeMoura, 1982) due to mechanical force, to shear stress effects as a result of fluid flow across the surface of a cell population (Wang *et al.*, 1993; Johnson, 1994), or to hydrostatic pressure effects directly altering cell shape (Lewis and DeMoura, 1982). However, most investigations on membrane stretch have focused on single cells and the patch clamp technique to monitor single-channel activity and whole-cell currents (Morris, 1990). Details of the patch clamp technique are described in Chapter 10 and will not be discussed here.

In most studies using the patch clamp techniques, stretching was achieved by applying a low pressure in the back of the patch pipette. Methods of creating this low pressure include mouth suction (e.g. Martinac *et al.*, 1987), syringe suction (e.g. Martinac *et al.*, 1987; Sackin, 1987), in-line aspirator (e.g. Moody and Bosma, 1989), vacuum pump (e.g. Bear, 1990), or a piston (e.g. Lane *et al.*, 1991; Morris and Horn, 1991). At the most advanced end, an excellent apparatus is the pressure clamp developed by McBride and Hamill (1992) for precisely controlled stretching. The pressure clamp is capable of measuring a threshold pressure that activates channels. Its wide pressure range from positive to negative provides flexibility for studying most stretch-activated channels (McBride and Hamill 1992; Hamill and McBride 1995). The idea of the pressure clamp is depicted in Figure 12.2. However, data from stretch-activated channels using the patch clamp technique are difficult to interpret due to innate variations in the structure of each patch and possible delays in membrane movement in response to pressure (Sokabe and Sachs, 1990; Ruknudin *et al.*, 1991).

Cell membrane stretching with the patch clamp allows for direct monitoring of ion channel activity during stretching, most often in an on-cell patch where single channels can be identified. It also has the advantage of restricting the study of ion transport to a particular side of polarized cells: apical membrane side versus basolateral membrane side. One of the challenging tasks in understanding stretch-activated channels is to identify whether stretch activates a channel protein directly or indirectly through signaling pathways which interact with the channel protein (e.g. the proposed Ca^{2+} influx following internal Ca^{2+} store depletion: Putney, 1993; Roman and Harder 1993), or through disruption of the cytoskeleton system (Mills and Mandel, 1994). Patch clamp experiments should, therefore, be designed to determine if stretch has an immediate effect on channel gating or whether the effect on the channel is delayed by a few seconds or minutes (Christensen and Hoffmann, 1992; Kim *et al.*, 1993). Generally, if the channel response is delayed the role of specific signaling pathways in controlling the channel can be evaluated by blocking or activating the putative pathway in the presence of applied stretch. With the availability of the pressure clamp technique by McBride and Hamill (1992), some hint of channel activation at a mechanistic level may be obtained if multiple pressure ramp slopes are employed.

The occurrence of channel openings only at a certain pressure level in a pressure ramp (Hamill and McBride, 1995) provides strong evidence in favor of a direct stretch-activated signal for channel opening. Using the pressure ramp technique, a pressure threshold for channel opening can be defined. Caution must be exercised in defining this threshold as it can be argued that the channel actually could be triggered at an earlier stage (at less negative pressure level) and then opened after some delay due to a

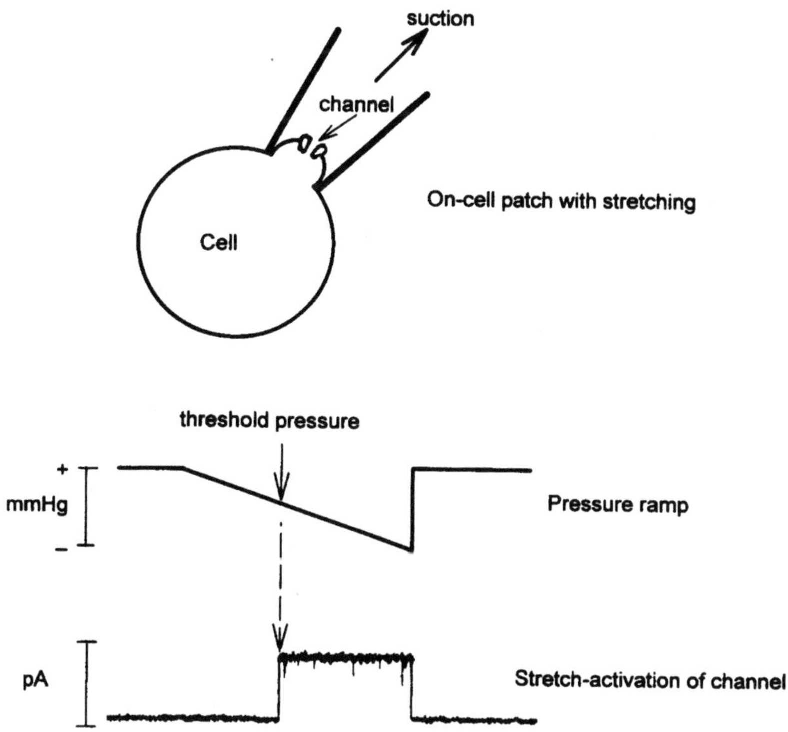

Figure 12.2 Depiction of the pressure-ramp patch clamp technique. Shown are the patch clamp arrangement with applied suction, the pressure ramp applied to the back of the patch pipette, and the current trace of a single channel activated at a threshold pressure.

secondary mechanism such as cytoskeleton breakdown or alteration of a secondary signal pathway. A detailed analysis using multiple ramp slopes will be required to assess the concept of a defined pressure threshold for each channel.

12.2.3 The stretch-activated ion channels

Osmotic swelling is known to activate a number of ion channels to bring about K^+ efflux and/or Cl^- efflux during volume regulation. The activation of these channels appears to require a Ca^{2+} signal which is generated, at least in part, by activation of calcium influx via calcium-permeable channels (Hoffmann and Ussing, 1992; McCarty and O'Neil, 1992). In some cells, however, RVD seems capable of relying on only stretch-activated K^+ efflux and/or Cl^- efflux without involving Ca^{2+} signaling (e.g. Bear, 1991;

Kim, 1992; Christensen and Hoffmann, 1992; Botchkin and Matthews, 1993; Ackerman *et al.*, 1994). It should, therefore, be noted that swelling would appear to activate multiple channels including K^+ channels, Cl^- channels and calcium-permeable channels (possibly non-selective cation channels). Since these channels appear to be similar to that activated by mechanical stretch, the procedures to study mechanical stretch-activated channels will be emphasized.

Reports on mechanical stretch-activated channels cover a wide range of channel types in both epithelial and non-epithelial cells with many common features amongst the various types of cell. These channels include K^+, Cl^- and non-selective cation channels, similar to that observed with osmotic swelling. Other stretch-activated ion channels have been reported in polarized cells, such as a non-selective anion channel in MDCK cells (Weiss and Lang, 1992) and a non-selective channel permeable to anions and cations in opossum kidney cells (Ubl *et al.*, 1988), but the available information is too limited to generalize about the physiological importance of the channels. Hence, methods assessing the general features of the more common channels will be discussed to provide a conceptual framework for studying the channels. Details of the methodology to study the channels are provided in the references.

A few studies indicate that stretch-activated K^+ channels resemble voltage gated K^+ channels in terms of common channel blockers such as externally applied Ba^{2+} (e.g. Yantorno *et al.*, 1992). Quinine and quinidine will block stretch-activated K^+ selective channels when applied extracellularly (Bear 1991; Yantorno *et al.*, 1992). Similarly, stretch-activated Cl^- channels can be inhibited by the anion channel blocker NPPB (Diener *et al.*, 1992; Yantorno *et al.*, 1992; Ackerman *et al.*, 1994). Stretch-activated Cl^- channels are also partially blocked by externally applied Gd^{3+} and La^{3+} (Ackerman *et al.*, 1994). It is evident from these results that stretch-activated channels may share characteristics with other ion channels.

The cytoskeleton is involved in the cellular response to membrane stretch (Mills and Mandel, 1994). The fungal toxins cytochalasin and phalloidin (Cooper, 1987; Schliwa, 1982), which bind to actin and alter its polymerization, have helped to elucidate the involvement of actin microfilaments in stretch-activated channels (Mills *et al.*, 1994). Addition of cytochalasin D, which depolymerizes actin filaments, activates a Cl^- channel in rabbit collecting duct cells (Schwiebert *et al*, 1994). This Cl^- channel is also activated by osmotic swelling. The activation due to cytochalasin D and osmotic stress is blocked by phalloidin, which stabilizes the actin microfilament network. Conversely, addition of cytochalasin D inactivates a K^+ channel in the rat collecting duct, while the addition of phalloidin prevented this inactivation (Wang *et al.*, 1994). In rabbit proximal tubule, cytochalasin D inactivates a Cl^- channel (Suzuki *et al.*, 1993). This channel can be reactivated in excised inside-out patches with the addition of actin polymers (F-actin), but not monomers (G-actin). In the toad kidney cell line A6, cytochalasin D activates a Na^+ channel (Cantiello *et al.*, 1991). However, like the rabbit proximal tubule Cl^- channel, the addition of short-chain actin polymers activates the Na^+ channel in excised inside-out patches. Cytochalasin D also has an effect in excised inside-out patches (Wang *et al.*, 1994; Cantiello *et al.*, 1991) which is probably due to a portion of the cell cytoskeleton being excised with the patch (Ruknudin *et al.*, 1991). Together, these results suggest a direct role of the actin cytoskeleton in the regulation of ion channels in response to membrane stretch.

Among stretch-activated channels, non-selective cation channels have received the most extensive studies. Duncan *et al.* (1992) found that parathyroid hormone promotes non-selective cation channel activity in osteosarcoma cells. Gadolinium is widely used as a blocker for non-selective cation channels (Yang and Sacks, 1989; Filipovic and Sackin, 1991; Duncan *et al.*, 1992; Kawahara and Matsuzaki, 1993) and has also been shown to block some stretch-activated Cl^- channels. Therefore Gd^{3+} does not specifically block a particular channel, which limits its usefulness. Some studies showed that Gd^{3+} also blocks the highly selective L-type and T-type Ca^{2+} channels (Biagi and Enyeart, 1990), while others found that Gd^{3+} either could not completely block stretch-activated channels as in A6 cells (Kawahara and Matsuzaki, 1993) or did not block at all as in Ehrlich ascites tumor cells (Christensen and Hoffmann, 1992).

Amiloride has also been demonstrated to be a blocker of some non-selective cation channels (Lane *et al.*, 1991). Extracellular amiloride blocks non-selective cation channels in oocytes in a voltage-dependent manner in which inward cation current is reduced but outward current is almost unaffected. At extremely negative voltages, the blocking reaches a relatively constant level. Based on their analysis, Lane and coworkers suggest that the blocking involves two amiloride binding sites associated with channel protein conformational change under voltage clamp (Lane *et al.*, 1991; Hamill *et al.*, 1992).

It seems that our understanding of the stretch-activated (swelling-activated) channels is still at an early stage. Future studies need to focus on structural/functional properties and the regulatory pathways controlling channel activity if we hope to obtain insight into control of channel activity. In particular, our understanding of the regulatory pathways controlling stretch-activated channels is in its infancy and remains to be elucidated. While stretch may directly activate some channels, still to be defined is the importance of cell signaling pathways both in short-term modulation and long-term adaptation of the channels.

ACKNOWLEDGEMENT

The authors wish to thank Leslie Sanders for preparation and editing of the manuscript.

REFERENCES

Ackerman M.J., Wickman, K.D. and Clapham, D.E. (1994) Hypotonicity activates a native chloride current in *Xenopus* oocytes. *J. Gen. Physiol.* **103**:153–79.

Allbritton, N.L., Oancea, E., Kuhn, M.A. and Meyer, T. (1994) Source of nuclear calcium signals. *Proc. Natl. Acad. Sci. USA* **91**:12458–62.

Anderson, M.P., Sheppard, D.N., Berger, H.A. and Welsh, M.J. (1992) Chloride channels in the apical membrane of normal and cystic fibrosis airway and intestinal epithelia. *Am. J. Physiol.* **263**:L1–L14.

Bazzi, M.D. and Nelsestuen, G.L. (1988) Constitutive activity of membrane-inserted protein kinase C. *Biochem. Biophys. Res. Comm.* **152**:336–43.

Bear, C.E. (1990) A non-selective cation channel in rat liver cells is activated by membrane stretch. *Am. J. Physiol.* **259**:C421–C428.

Bear, C.E. (1991) A K^+-selective channel in the colonic carcinoma cell line: CaCo-2 is activated with membrane stretch. *Biochim. Biophys. Acta* **1069**:267–72.

Berridge, M.J. (1993) Inositol trisphosphate and calcium signalling. *Nature* **361**:315–25.

Bertorello, A.M. and Katz, A.I. (1993) Short-term regulation of renal Na-K-ATPase activity: physiological relevance and cellular mechanisms. *Am. J. Physiol.* **265**:F743–F755.

Biagi, B.A. and Enyeart, J.J. (1990) Gadolinium blocks low- and high-threshold calcium currents in pituitary cells. *Am. J. Physiol.* **259**:C515–C520.

Birnbaumer, L. (1993) Heterotrimeric G proteins. Molecular diversity and function correlates. *J. Recent Res.* **13**:19–26.

Botchkin, L.M. and Matthews, G. (1993) Chloride current activated by swelling in retinal pigment epithelium cells. *Am. J. Physiol.* **265**:C1037–C1045.

Brini, M., Pasti, L., Bastianutto, C. *et al.* (1994) Targeting of aequorin for calcium monitoring in intracellular compartments. *J. Biolumin. Chemilumin.* **9**(3):177–84.

Brooker, G., Harper, J.F., Terasaki, W.L. and Moylan, R.D. (1979) Radioimmunoassay for cyclic AMP and cyclic GMP. *Adv. Cyclic Nucleotide Res.* **10**:1–33.

Bruce, B. and Ullrich, V. (1991) Calcium mobilization in human platelets by receptor agonists and calcium-ATPase inhibitors. *FEBS Lett.* **284**(1):1–4.

Cantiello, H.F., Stow, J.L., Prat, A.G. and Ausiello, D.A. (1991) Actin filaments regulate epithelial Na^+ channel activity. *Am. J. Physiol.* **261**:C882–C888.

Christensen, O. and Hoffmann, E.K. (1992) Cell swelling activates K^+ and Cl^- channels as well as non-selective, stretch-activated cation channels in Ehrlich ascites tumor cells. *J. Membrane Biol.* **129**:13–36.

Cooper, J.A. (1987) Effect of cytochalasin and phalloidin on actin. *J. Cell Biol.* **105**:1473–8.

Cybulsky, A.V. and Cyr, M.D. (1993) Phosphatidylcholine-directed phospholipase C: activation by complement C5b-9. *Am. J. Physiol.* **265**:F551–F560.

Diener, M., Nobles, M. and Rummel, W. (1992) Activation of basolateral Cl^- channels in the rat colonic epithelium during regulatory volume decrease. *Pflügers Arch.* **421**:530–8.

Duncan, R., Hruska, K.A. and Misler, S. (1992) Parathyroid hormone activation of stretch-activated cation channels in osteosarcoma cells. (URM-106.01) *FEBS* **307**:219–23.

Epand, R.M. (1994) *In vitro* assays of protein kinase C activity. *Anal. Biochem.* **218**:241–7.

Etter, E.F., Kuhn, M.A. and Fay, F.S. (1994) Detection of changes in near-membrane Ca^{2+} concentration using a novel membrane-associated Ca^{2+} indicator. *J. Biol. Chem.* **269**:10141–9.

Fan, S.F., Wang, S. and Kao, C.Y. (1993) The transduction system in the isoproterenol activation of the Ca^{2+}-activated K^{+} channel in guinea pig taenia coli myocytes. *J. Gen. Physiol.* **102**:257–75.

Filipovic, D. and Sackin, H. (1991) A calcium-permeable stretch-activated cation channel in renal proximal tubule. *Am. J. Physiol.* **260**:F119–F129.

Frazier, L.W. and Yorio, T. (1992) Eicosanoids: their function in renal epithelia ion transport. *Proc. Soc. Exp. Biol. Med.* **201**(3):229–43.

Fricker, M.D., Gilroy, S., Read, N.D. and Trewavas, A.J. (1991) Visualization and measurement of the calcium message in guard cells. *Symp. Soc. Exp. Biol.* **45**:177–90.

Gilman, A.G. (1970) A protein binding assay for adenosine 3′:5′-cyclic monophosphate. *Proc. Natl. Acad. Sci. USA* **67**:305–12.

Grynkiewicz, G., Poenie, M. and Tsien, R.Y. (1985) A new generation of Ca^{2+} indicators with greatly improved fluorescence properties. *J. Biol. Chem.* **260**:3440–50.

Hallam, T.J., Jacob, R. and Merritt, J.E. (1989) Influx of bivalent cations can be independent of receptor stimulation in human endothelial cells. *Biochem. J.* **259**:125–9.

Hamill, O.P., Lane, J.W. and McBride, D.W. Jr (1992) Amiloride: a molecular probe for mechanosensitive channels. *Trends Pharmacol. Sci.* **13**:373–6.

Hamill, O.P. and McBride, D.W. Jr (1995) Mechanoreceptive membrane channels. *Am. Sci.* **83**:30–7.

Helvoort, A.V., van't Hof, W., Ritsema, T. *et al.* (1994) Conversion of diacylglycerol to phosphatidylcholine on the basolateral surface of epithelial (Madin–Darby Canine Kidney) cells. Evidence for the reverse action of a sphingomyelin synthase. *J. Biol. Chem.* **269**:1763–9.

Hepler, J.R. and Gilman, A.G. (1992) G proteins. *TIBS* **17**:383–8.

Hidaka, H., Watanabe, M. and Tokumitsu, H. (1990) Search for the functional substrate proteins of protein kinases and their selective inhibitors. *Adv. Second Messenger Phosphoprotein Res.* **24**:485–90.

Hoffmann, E.K. and Ussing, H.H. (1992) Membrane mechanisms in volume regulation in vertebrate cells and epithelia, in *Membrane Transport in Biology*, Vol. 5, (eds G.H. Giebisch, J.A. Schafer, H.H. Ussing and P. Kristensen), Springer-Verlag, Heidelberg, pp. 317–99.

Huwiler, A., Schulze-Lohoff, E., Fabbro, D. and Pfeilschifter, J. (1993) Immunocharacterization of protein kinase C isoenzymes in rat kidney glomeruli, and cultured glomerular epithelial and mesangial cells. *Exp. Nephrol.* **1**:19–25.

Johnson, R.M. (1994) Membrane stress increases cation permeability in red cells. *Biophys. J.* **67**:1876–81.

Kawahara, K. and Matsuzaki, K. (1993) A stretch-activated cation channel in the apical membrane of A6 cells. *Japanese J. Physiol.* **43**:817–32.

Kemp, B.E., Graves, D.J., Benjamini, E. and Krebs, E.G. (1977) Role of multiple basic substrate residues in determining the substrate specificity of cyclic AMP-dependent protein kinase. *J. Biol. Chem.* **252**:4888–94.

Kendall, J.M., Badminton, M.N., Dormer, R.L. and Campbell, A.K. (1994) Changes

in free calcium in the endoplasmic reticulum of living cells detected using targeted aequorin. *Anal. Biochem.* **221**(1):173–81.
Kim, D. (1992) A mechanosensitive K^+ channel in heart cells: activation by arachidonic acid. *J. Gen. Physiol.* **100**:1–20.
Kim, Y.-K., Dirksen, E.R. and Sanderson, M.J. (1993) Stretch-activated channels in airway epithelial cells. *Am. J. Physiol.* **265**:C1306–C1318.
Kinne, R.K., Czekay, R.P., Grunewald, J.M. *et al.* (1993) Hypotonicity-evoked release of organic osmolytes from distal renal cells: Systems, signals, and sidedness. *Renal Physiol. Biochem.* **16**:66–78.
Lambert, D.G. (1993) Signal transduction: G proteins and second messengers. *Br. J. Anaesth.* **71**:86–95.
Lane, J.W., McBride, D.W. Jr and Hamill, O.P. (1991) Amiloride block of the mechanosensitive cation channels in *Xenopus* oocytes. *J. Physiol.* (Lond.) **441**:347–66.
Lee, M.W. and Severson, D.L. (1994) Signal transduction in vascular smooth muscle: diacylglycerol second messengers and PKC action. *Am. J. Physiol.* **267**:C659–C678.
Lewis, S.A. and DeMoura, J.L.C. (1982) Incorporation of cytoplasmic vesicles into the apical membrane of mammalian urinary bladder epithelium. *Nature* **297**:685–8.
Liu, R., Farach-Carson, M.C. and Karin, N.J. (1996) Effects of sphingosine derivatives on MC3T3-E1 pre-osteoblasts: psychosine elicits release of calcium from intracellular stores. *Biochem. Biophys. Res. Comm.* **214**:676–84.
Lohse, M.J., Benovic, J.L., Caron, M.G. and Lefkowitz, R.J. (1990) Multiple pathways of rapid beta 2-adrenergic receptor desensitization. Delineation with specific inhibitors. *J. Biol. Chem.* **265**:3202–11.
Martinac, B., Buechner, M., Delcour, A.H. *et al.* (1987) Pressure-sensitive ion channels in *Escherichia coli. Proc. Natl. Acad. Sci. USA* **84**:2297–301.
Matthews, J.B., Awtrey, C.S., Hecht, G. *et al.* (1993) Phorbol ester sequentially downregulates cAMP-regulated basolateral and apical Cl^- transport pathways in T84 cells. *Am. J. Physiol.* **265**:C1109–C1117.
McBride, D.W. Jr and Hamill, O.P. (1992) Pressure-clamp: a method for rapid step perturbation of mechanosensitive channels. *Pflügers Arch.* **421**:606–12.
McCarty, N.A. and O'Neil, R.G. (1991) Calcium-dependent control of volume regulation in renal proximal tubule cells: I. Swelling-activated Ca^{2+} entry and release. *J. Membrane Biol.* **123**:149–60.
McCarty, N.A. and O'Neil, R.G. (1992) Calcium signaling in cell volume regulation. *Physiol. Rev.* **72**:1037–61.
McIlroy, B.K., Walters, J.D. and Johnson, J.D. (1991) A continuous fluorescence assay for protein kinase C. *Anal. Biochem.* **195**:148–52.
McNeil, P.L. and Taylor, D.L. (1985) Aequorin entrapment in mammalian cells. *Cell Calcium* **6**:83–93.
Mills, J.W. and Mandel, L.J. (1994) Cytoskeletal regulation of membrane transport events. *FASEB J.* **8**:1161–5.
Mills, J.W., Schwiebert, E.M. and Stanton, B.A. (1994) Evidence for the role of actin filaments in regulating cell swelling. *J. Exp. Zoo.* **268**:111–20.
Moody, W.J. and Bosma, M.M. (1989) A non-selective cation channel activated by membrane deformation in oocytes of the ascidian *Boltenia villosa. J. Membrane Biol.* **107**:179–88.
Morris, C.E. (1990) Mechanosensitive ion channels. *J. Membrane Biol.* **113**:93–107.
Morris, C.E. and Horn, R. (1991) Failure to elicit neuronal macroscopic mechanosensitive currents anticipated by single channel studies. *Science* **251**:1246–9.

Nishizuka, Y. (1992) Intracellular signaling by hydrolysis of phospholipids and activation of protein kinase C. *Science* **258**:607–14.

Ogreid, D., Dostmann, W., Genieser, H.G. *et al.* (1994) (Rp)- and (Sp)-8-piperidino-adenosine 3′,5′-(cyclic)thiophosphates discriminate completely between site A and B of the regulatory subunits of cAMP-dependent protein kinase type I and II. *Eur. J. Biochem.* **221**:1089–94.

Pagano, R.E. and Sleight, R.G. (1985) Defining lipid transport pathways in animal cells. *Science* **229**:1051–7.

Pearson, R.B. and Kemp, B.E. (1991) Protein kinase phosphorylation site sequences and consensus specificity motifs: tabulations. *Meth. Enzymol.* **200**:62–81.

Preiss, J., Loomis, C.R., Bishop, W.R. *et al.* (1986) Quantitative measurement of sn-1,2-diacylglycerols present in platelets, hepatocytes, and ras- and sis-transformed normal rat kidney cells. *J. Biol. Chem.* **261**:8597–600.

Putney, J.W. Jr (1993) Excitement about calcium signaling in inexcitable cells. *Science* **262**:676–678.

Reuss, L., Segal, Y. and Altenberg, G. (1991) Regulation of ion transport across gallbladder epithelium. *Annu. Rev. Physiol.* **53**:361–73.

Roe, M.W., Lemasters, J.J. and Herman, B. (1990) Assessment of fura-2 for measurement of cytoplasmic free calcium. *Cell Calcium* **11**:63–73.

Roman, R.J. and Harder, D.R. (1993) Cellular and ionic signal transduction mechanisms for the mechanical activation of renal arterial vascular smooth muscle. *J. Am. Soc. Nephrol.* **4**:986–96.

Rubin, C.S., Erlichman, J. and Rosen, O.M. (1974) Cyclic AMP-dependent protein kinase from bovine heart muscle. *Meth. Enzymol.* **38**:308–15.

Ruknudin, A., Song, M.I. and Sachs, F. (1991) The ultrastructure of patch-clamped membranes: a study using high voltage electron microscopy. *J. Cell Biol.* **112**(1):125–34.

Sachs, F. (1988) Mechanical transduction in biological systems. *Crit. Rev. Biomed. Eng.* **16**:141–69.

Sackin, H. (1987) Stretch-activated potassium channels in renal proximal tubule. *Am. J. Physiol.* **253**:F1253–F1262.

Saxon, M.L., Zhao, X. and Black, J.D. (1994) Activation of protein kinase C isozymes is associated with post-mitotic events in intestinal epithelial cells *in situ*. *J. Cell Biol.* **126**:747–73.

Schliwa, M. (1982) Action of cytochalasin D on cytoskeletal networks. *J. Cell Biol.* **92**:79–91.

Schwiebert, E.M., Mills, J.W. and Stanton, B.A. (1994) Actin-based cytoskeleton regulates a chloride channel and cell volume in a renal cortical collecting duct cell line. *J. Biol. Chem.* **269**:7081–9.

Scott, J.D., Glaccum, M.B., Fischer, E.H. and Krebs, E.G. (1986) Primary-structure requirements for inhibition by the heat-stable inhibitor of the cAMP-dependent protein kinase. *Proc. Natl. Acad. Sci. USA* **83**:1613–16.

Sesko, A., Cabot, M. and Mossman, B. (1990) Hydrolysis of inositol phospholipids precedes cellular proliferation in asbestos-stimulated tracheobronchial epithelial cells. *Proc. Natl. Acad. Sci. USA* **87**:7385–9.

Sokabe, M. and Sachs, F. (1990) The structure and dynamics of patch-clamped membranes: a study using differential interference contrast light microscopy. *J. Cell Biol.* **111**:599–606.

Suzuki, M., Miyazaki, K., Ikeda, M. *et al.* (1993) F-actin network may regulate a Cl^- channel in renal proximal tubule cells. *J. Membrane Biol.* **134**:31–9.

Tamaoki, T. (1991) Use and specificity of staurosporine, UCN-01, and calphostin C as protein kinase inhibitors. *Meth. Enzymol.* **201**:340–7.

Ubl, J., Murer, H. and Kolb, H.-A. (1988) Ion channels activated by osmotic and mechanical stress in membranes of opossum kidney cells. *J. Membrane Biol.* **104**:223–32.

Wang, N., Butler J.P. and Ingber, D.E. (1993) Mechanotransduction across the cell surface and through the cytoskeleton. *Science* **260**:1124–7.

Wang, W.-H., Cassola, A. and Giebisch, G. (1994) Involvement of actin cytoskeleton in modulation of apical K^+ channel activity in rat collecting duct. *Am. J. Physiol.* **267**:F592–F598.

Weiss, M.J. and Lang, F. (1992) Ion channels activated by swelling of Madin–Darby Canine Kidney (MDCK) cells. *J. Membrane Biol.* **126**:109–14.

Welsh, M.J., Anderson, M.P., Rich, D.P. *et al.* (1992) Cystic fibrosis transmembrane conductance regulator: a chloride channel with novel regulation. *Neuron* **8**:821–9.

Whatley, R.E., Zimmerman, G.A., McIntyre, T.M. and Prescott, S.M. (1990) Lipid metabolism and signal transduction in endothelial cells. *Prog. Lipid Res.* **29**(1):45–63.

Widdicombe, J.H., Kondo, M. and Mochizuki, H. (1991) Regulation of airway mucosal ion transport. *Int. Arch. Allergy Appl. Immunol.* **94**:56–61.

Yang, X.C. and Sacks, F. (1989) Block of stretch-activated ion channels in *Xenopus* oocytes by gadolinium and calcium ions. *Science* **243**:1068–71.

Yantorno, R.E., Carre, D.A., Coca-Prados, M. *et al.* (1992) Whole cell patch clamping of ciliary epithelial cells during anisosmotic swelling. *Am. J. Physiol.* **262**:C501–C509.

13

The cytoskeleton and epithelial function

John W. Mills

The number of known and proposed functions of the cytoskeleton have grown dramatically since the first descriptions of these filamentous components of the cytoplasm. Clear evidence to support a role in cell shape and structure (a true expression of function based on the term skeleton) has been repeatedly produced. Other functions include motile processes (whether the cell or its parts are locomoting or specific cytoplasmic constituents are moving inside a static cell) and the anchoring of the cells to substrates and to each other or the gelling of the cytoplasm so that movement within the cell space is restricted. With regard to epithelial cells some of the most significant functions for the different cytoskeletal

Epithelial Transport: A guide to methods and experimental analysis.
Edited by Nancy K. Wills, Luis Reuss and Simon A. Lewis.
Published in 1996 by Chapman & Hall, London. ISBN 978-94-010-7168-0

elements are the establishment of epithelial polarity (Chapter 3), the maintenance of cell and substrate junctions, the trafficking of cell membrane proteins to and from membrane domains, the anchoring of these transport proteins in the cell membrane and the role that the cytoskeleton could play in signal transduction and regulation of transport events. The purpose of this chapter is to describe selected examples of the wide array of proteins grouped under the broad heading of cytoskeleton, review the use of drugs that act on the cytoskeleton and discuss selected epithelial models where a picture is emerging as to the role that the cytoskeleton plays in transport.

13.1 CYTOSKELETON

The term cytoskeleton covers three types of filamentous protein: actin filaments (microfilaments), intermediate filaments and microtubules. For each of these types there are also other proteins that associate with them and are therefore placed in the same group. For example, proteins that cap or bind to the sides of actin filaments are 'actin-associated proteins' and can be designated as part of the actin cytoskeleton. This terminology will be used throughout this chapter, i.e. 'actin cytoskeleton' will be used broadly to cover actin in its various states of polymerization and secondary organization (bundling) as well as the proteins that interact with it either directly or with other actin-binding proteins.

13.1.1 Actin

Actin is a globular, 43 kD protein (G-actin) that assembles into 8 nm thick filaments (F-actin). This polymerization requires ATP, a monovalent cation and a divalent cation. In the cell these are most likely K^+ and Mg^{2+} although other combinations of divalents and monovalents can produce filaments. The G-actin combines with ATP prior to addition to the filament and then the ATP is hydrolyzed to ADP which remains attached to the filament. Addition of G-actin units to the filament can occur at both ends. However, there is a marked difference in the polymerization rate at the two ends with the **barbed end** (so named from the arrowhead appearance of actin filaments after decoration with myosin heads) growing much faster than the **pointed end**. These ends therefore are also termed the plus (+) and minus (–) ends, respectively. This difference in rate of addition is due to a difference in the critical concentration at the two ends, i.e. the concentration of actin at which addition of actin subunits is matched by loss. Since there is a difference in the critical concentration at the two ends it is possible that the concentration of G-actin available for assembly will only support net addition of units at one end (+) and net loss at the other

(–). This results in **treadmilling** of actin subunits through the filament. Thus an energy-dependent (since ATP hydrolysis is occurring) steady state can be established.

13.1.2 Actin-associated proteins

In most cases when we deal with actin filaments we must consider and understand the role that the very broad class of actin-associated proteins plays in filament dynamics. In several cases the actin-binding proteins have become such a center of research that they are considered as separate cytoskeletal components. Thus the actin-binding protein spectrin, which assembles into filaments as well, can be considered as the spectrin skeleton (Bennett and Lambert, 1991). A list of these proteins would soon need to be modified as the number is growing rapidly. Only a few of these proteins, limited to those implicated as playing a role in epithelial transport, will be discussed.

(a) Spectrin (fodrin)

The spectrin family of plasma membrane-associated proteins are rod-shaped molecules assembled as heterodimers (α,β spectrin) in an antiparallel fashion and then head to head into tetramers. The assembled short (200 nm) rods have the ability to bind to F-actin at the paired amino and carboxy terminal ends. Thus the assembled rod can serve as a cross-linker of actin filaments. The spectrin rod can also bind to ankyrin and to several other membrane skeleton proteins such as protein 4.1 and adducin.

The spectrin skeleton is assembled in the cortical cytoplasm of epithelial cells (Figure 13.1) and is distributed around the entire periphery in some cell types (Drenckhahn *et al.*; 1985, Glenney *et al.*, 1983; Mercier *et al.*, 1989). In others, spectrin is localized specifically to the basolateral domain once the epithelial sheet is formed (Morrow *et al.*, 1989). The functional significance of this difference is not known. The distribution in the cortical cytoplasm is related to an interaction with either ankyrin or actin (Glenney *et al.*, 1983; Morrow *et al.*, 1989).

Spectrin appears to play a central role in the establishment of specific membrane domains in epithelia. The evidence is quite convincing that the placing and maintenance of membrane transport proteins (the Na^+ pump, the anion exchanger AE1) to the specific membrane domain is associated with an interaction with spectrin. Thus spectrin plays a role in the maintenance of the vectorial nature of epithelial transport. Investigation into a regulatory role in epithelial transport for the proteins in the spectrin family has been hindered by the lack of specific inhibitors of spectrin assembly and steady-state dynamics. The possibility for such a role arises from the fact that the α subunit of spectrin contains a sequence homology

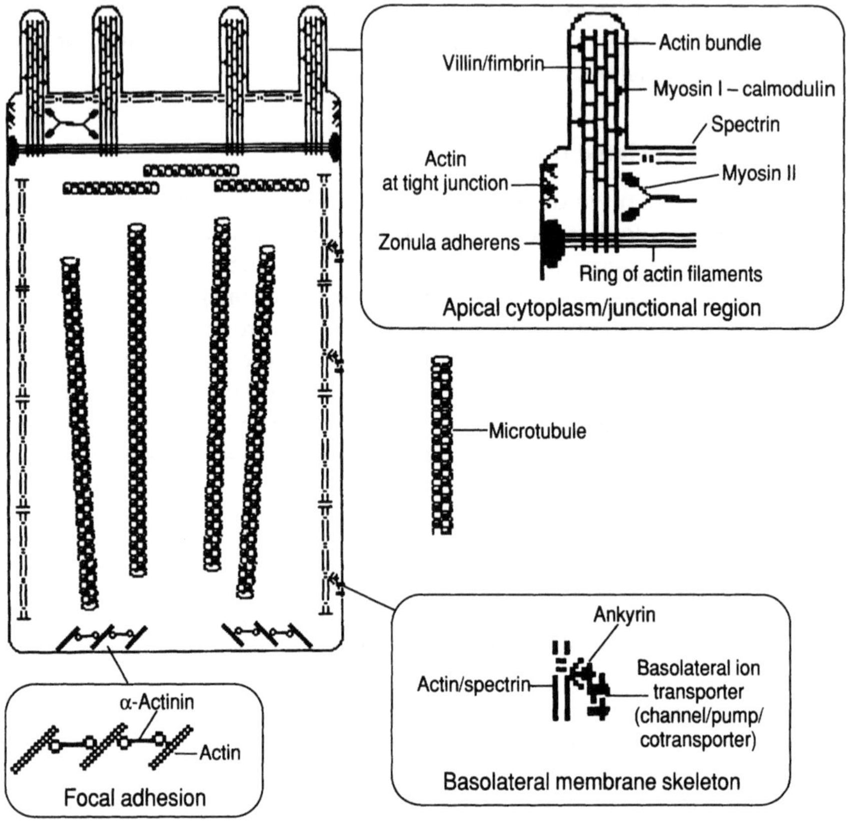

Figure 13.1 Simplified schematic of the actin and microtubule organization in a differentiated epithelial cell. The apical pole is characterized by several separate organizational domains for actin. These include the microvillar actin complex, assumed to exist wherever apical microvilli are found irrespective of whether a true 'brush border' is present. In addition, actin filaments are shown associated with the tight junction as well as the adhering junction, based on the data presented by Madara (1987). The apical spectrin is shown with the actin oligomers associated with the ends of the coiled molecule although this has not been verified. At the basolateral domain the actin/spectrin complex is shown associated with ankyrin and this links to membrane ion transporters. This complex is assumed to follow the basolateral membrane faithfully, including those epithelial cells where the membrane is arranged in extensive folds or microplicae (Mills *et al.*, 1993). Finally, at the base of the cells actin is found associated with a complex of proteins at the focal adhesion. Microtubules are shown running in an apical/basal direction (with plus ends at the base, minus ends at the apical surface) in the central core of the cytoplasm and at the apical surface just beneath the circumferential band of actin filaments. (Diagram modified from various sources including Drenckhahn and Dermietzel, 1988; Fath *et al.*, 1993; Luna and Hitt, 1992; Mays *et al.*, 1994.)

with the SH3 domain of the src-tyrosine kinases (Bennett and Gilligan, 1993). This domain is implicated in intracellular signaling.

(b) Ankyrin

Ankyrin, a globular protein first associated with the spectrin cytoskeleton of red blood cells, is characterized as having multiple binding sites for both membrane transport proteins and cytoskeletal proteins. These include the anion exchanger (Band 3), the Na^+-pump, spectrin and tubulin. Two forms of ankyrin are characterized: $ankyrin_B$ expressed in brain and $ankyrin_R$ expressed in the erythrocyte (Bennett and Gilligan, 1993). Interestingly, both forms of ankyrin can be expressed in epithelial cells. For example, $ankyrin_R$ is found only in the distal tubules and collecting ducts of rat kidney cells whereas $ankyrin_B$ is found throughout the nephron (Davis *et al.*, 1989; Drenckhahn *et al.*, 1985).

The importance of ankyrin with regard to epithelial transport has grown significantly with the repeated demonstration that it associates with many different classes of membrane transporter. Besides the pump and the exchanger, both voltage and amiloride-sensitive Na^+-channels and the H^+-K^+-ATPase associate with ankyrin. The fact that (at least for the anion exchanger, the voltage-dependent Na^+-channel and the Na^+-pump) the binding sites on ankyrin are distinct indicates that this transport protein–ankyrin interaction has evolved specifically for each protein and is not the result of a general binding to membrane proteins (Bennett, 1992).

13.1.3 Distribution of actin and associated proteins in epithelial cells

Actin filaments are arranged into specific domains within epithelial cells arranged in a confluent and therefore truly epithelioid sheet (Figure 13.1). At the base of the cell (i.e. that portion attached to the substrate: for cultured cells this can be glass, plastic or connective tissue proteins such as collagen or fibronectin), actin is assembled into fiber bundles (stress fibers) with one end attached to a dense plaque of material (the focal contact) containing a group of proteins, including actin-binding proteins and membrane-spanning proteins, that play a role in attachment to the substrate (Luna and Hitt, 1992). The other end of the stress fiber is either attached to another focal contact or ends in a zone of intermediate filaments in the vicinity of the nucleus (Alberts *et al.*, 1994).

F-actin is also distributed in the cortical cytoplasm all around the periphery of the cell in the membrane domains classified as basolateral. The filaments make up what is termed the cortical actin web and are closely associated with the cell membrane, including the long lateral interdigitations present in some epithelia (Mills *et al.*, 1993). F-actin is also concentrated in the apical domain in two areas. A band of actin filaments, associated with the adherens junction and containing myosin II, forms a complete ring around the cell (Fath *et al.*, 1993). Bundles of actin filaments associated with actin binding proteins including myosin I are arranged in the central core of the microvilli, whether densely packed on the surface as in intestinal epithelial cells or sparse as in toad bladder or MDCK cells (Mills, 1987; Simon *et al.*, 1993). Finally there appear to be some actin filaments associated with the cytoplasmic surface of the occluding junction (Madara, 1987).

13.1.4 Microtubules

Microtubules (MTs) are polymers formed from the subunits α and β tubulin which assemble in a sequential manner as a set of 13 protofilaments forming a cylindrical structure. The α and β tubulin are products of different genes and there are several different genes for each of the two subunits. *In vitro* the polymerization of tubulin requires divalent cation (usually Mg^{2+}) and GTP. As with actin polymerization there is a kinetic barrier that must be overcome (the **lag phase**) before elongation can proceed. Once nucleation sites are established, polymerization occurs rapidly. If the tubulin monomer pool is not replenished, a **plateau phase** is reached where the addition and removal of subunits are at equilibrium. Again, as with actin, the concentration of tubulin where this equilibrium is achieved is known as the **critical concentration**. Another similarity between MTs and actin filaments is that the formed polymer has a polarity with regard to the rate of addition of subunits with the fast growing end designated plus (+) and the slow end designated minus (–).

The polymerization of MTs *in vivo* is further regulated by the presence of the centrosome (microtubule-organizing center or **MTOC**). This is an area in the cytoplasm, usually characterized by the presence of centrioles (but not in epithelia – see below), that has a distinct set of proteins, including the variant form of tubulin known as γ tubulin. Experimental studies have demonstrated that the MTs grow out from the centrosome with the (+) end leading and the (–) end remaining in the centrosome matrix. The centrosome plays a role also as a cap on the (–) end so that little loss of subunits occurs. The subsequent growth or shortening of the MTs thus takes place at the (+) end via a process known as **dynamic instability** (Wordeman and Mitchison, 1994). Thus one can observe, *in vivo*, MTs that go through sequential elongation and shortening phases at one end. The phenomenon of dynamic instability is not observed in all MTs and one explanation for this is that the (+) end gets capped so that net addition or loss does not occur.

The microtubule array in epithelial cells is also arranged into specific domains (Figure 13.1). There is an apical web of microtubules, just deep to the actin filament domain. Deep to this and arranged in a pattern around the nucleus is a group of microtubules arranged in an apical–basal direction. The (+) ends of the microtubules are located at the basal pole and the (–) ends are at the apical end (Bacallao *et al.*, 1989; Gilbert *et al.*, 1991). This arrangement appears to play a critical role in the movement of vesicles from the Golgi to the apical domain and possibly with the movement of vesicular contents during transcytosis (Gilbert *et al.*, 1991; Hunziker *et al.*, 1990).

13.1.5 Microtubule-associated proteins (MAPs)

There is a wide class of proteins designated as MAPs. Several of these have been described biochemically and structurally but their function, especially *in vivo* , is not well understood (Matus, 1994; Muller *et al.*, 1994; Olmstead, 1991). Within the MAP1 and MAP2 family the proteins appear to play a role in assembly, stabilization and cross-linking of microtubules. There does not appear to be any information regarding a possible role for these proteins in epithelial ion tranport function.

The motor proteins kinesin and dynein comprise another class of MAPs. Important for consideration of epithelia is the fact that these two motor proteins direct movement of 'cargo' (membrane vesicles) in opposite directions on the microtubule. Kinesin directs vesicle movement towards the (+) end while cytoplasmic dynein directs movement to the (–) end. As mentioned above, in epithelia the (+) and (–) ends of the perinuclear microtubule domain are directed toward the basal and apical membranes respectively. Thus, apically directed targeting of vesicles containing transport proteins, as is hypothesized to account for the increased transporter activity observed in several epithelia (Bacskai and Friedman, 1990; Fuller *et al.*, 1994; Hansch *et al.*, 1993) could involve dynein. Although there is some evidence to indicate that microtubule motor proteins are needed for membrane fusion in MDCK cells, the relevance of this to vesicles containing transport proteins has not been investigated. It is obviously of importance to determine what role the motor proteins play in this aspect of epithelial transport regulation, especially with regard to the movement of the transporters into the domain of the apical actin web (Fath *et al.*, 1993).

13.1.6 Intermediate filaments

Intermediate filaments (IFs) are quite different from MTs and actin filaments. The filaments are assembled from fibrous proteins that are distinguished by a variable NH_2-terminus head region, a COOH-terminus tail and central α-helical rod segment. The α-helical rod contains a heptad repeat which is characterized by the close approximation of hydrophobic amino acids. This hydrophobic region promotes the assembly of monomers into coiled-coiled dimers (with the hydrophobic regions associating). Further assembly into a tetramer occurs as the dimers associate (Shoeman and Traub, 1993). The intermediate filament forms from the association of tetramers into higher order oligomers. Importantly this assembly occurs in an antiparallel manner so that the subsequent tetramer is non-polarized. This association then can be repeated to form IFs of various lengths.

The intermediate filaments are divided into five major types (Stewart, 1993), one of which encompasses the diverse group of keratins (or cytokeratins) that are an almost ubiquitous component of epithelial cells. Other types, sometimes called cytoplasmic IFs, are components of mesenchymal cells (vimentin) and neuronal cells (neurofilaments and glial fibrillary acidic protein). A special type of IF composes the nuclear lamins, thought to be the evolutionary precursor of the other types (Weber *et al.*, 1989), and the only type expressed in all eukaryotic cell types.

Intermediate filaments are major components of epithelial cells. They course throughout the cytoplasm and are organized into dense bundles in the areas of cell–cell attachment (desmosomes) and cell–substrate attachment (hemidesmosomes). The interaction with adhesion zones appears to be a lateral association of the filaments with adhesion proteins (Fuchs and Weber, 1994). The filaments thus form a continuous network connecting adhesion zones in a three-dimensional array. There is also evidence indicating that IFs can interact directly with the cell membrane and interact with integral membrane proteins via the actin-associated membrane skeleton (Asch *et al.*, 1990; Horkovics-Kovats and Traub, 1990). In addition the intermediate filaments have been shown to interact with the nuclear envelope, specifically in the area of the nuclear pore. This association with the nucleus and the potential to interact directly with the nuclear lamins (proteins of the IF family that appear to play a critical role in chromosome structure) led to the hypothesis that the IFs may serve a role in gene activation and/or signal transduction from cell membrane to nucleus (Georgatos and Blobel, 1987).

A clear picture of the function of intermediate filaments has not yet arisen although genetic studies are begining to provide important insight (Fuchs, 1994). An essential role in cell physiology is lacking since there are cultured cell lines that have no IFs (Venetianer *et al.*, 1983). What is emerging is the concept that IFs serve a tissue specific role or a developmental (differentiation) specific role (Fuchs and Weber, 1994). With regard to epithelia and transport function, several results indicate that a potential role for IFs still needs to be considered. For example, spectrin and ankyrin have a binding domain for IFs, cAMP-dependent protein kinase associates with IFs *in vitro*, and disruption of microtubules leads to a collapse of the IF network (Dosemeci and Pant, 1992; Frappier *et al.*,

1992; Ishikawa *et al.*, 1968). Thus an indirect effect on transport may occur via interaction with other components of the cytoskeleton or the signal mechanism.

13.1.7 Intermediate filament associated proteins (IFAPs)

There is a growing body of information on the proteins that associate with IFs. The definition of an IFAP is much less stringent than for the actin and microtubule-associated proteins, and thus some proteins may not be directly involved with the filaments themselves (Foisner and Wiche, 1991). Not enough is known about these proteins to develop a functional description, especially with any relevance to epithelial transport. However, since one of these, plectin, has been shown to serve a cross-linking function for IFs (Foisner and Wiche, 1991) and interact with microtubule-associated proteins and spectrin (Herrmann and Wiche, 1987), a role in regulating epithelial transport cannot be ruled out.

13.2 FUNCTIONS OF THE CYTOSKELETON IN EPITHELIA

The cytoskeleton is involved in many functions within epithelial cells. First and foremost is the role that the various filaments play in maintaining the shape or structure of the epithelium. Some of these functions have been covered in the sections above. For example, actin filaments form the core of the microvilli on the apical surface of epithelia (Figure 13.1). Disruption of the filaments leads to a change in shape of the microvillar surface with the breakdown of microvilli into pleomorphic structures or the appearance of large membrane protrusions usually called blebs. IFs also play a structural role, serving to anchor cell adhesion areas and provide the mechanical support for the cell when undergoing stress. These functions as determinants or maintainers of shape are critical to epithelial function, since it is a hallmark of epithelia that the polarization of the cell into distinct apical and basolateral membrane domains and subdomains occurs. It is not surprising that massive alterations of the cytoskeletal components, especially those that lead to a breakdown in adhesive properties and barrier function, result in a loss of any or all functions unique to the epithelium. Clearly, more detailed and well-controlled experiments are needed to correlate less drastic changes in the state of organization of the cytoskeleton, both biochemical and structural, with changes in epithelial or individual transport protein function. With the development of more sophisticated techniques for analyzing transport and structure (e.g. patch clamp methods, confocal microscopy), better definition of the role of the cytoskeleton in specific transport events is now possible (Cantiello *et al.*, 1993; Holmgren *et al.*, 1992; Matthews *et al.*, 1994).

A list of proposed functions of the cytoskeleton in epithelial transport is

Table 13.1 Possible functions of the cytoskeleton in epithelial transport

Actin filament/actin-associated skeleton	*Microtubules*
Maintain epithelial barrier	Delivery retrieval of transport proteins to/from membrane domains
Modulate transcellular permeability	
Maintain polarity of transport proteins in membrane specific domains	
Delivery and removal of transport proteins to the membrane domains	
Regulate activity via direct interaction with transporter	
Participate in signal transduction/second messenger control of transporter activity or number	

given in Table 13.1. Depending on the particular function, the cytoskeleton may be directly involved with the transporter (e.g. anchoring of the Na^+-pump in the basolateral domain by the actin cytoskeleton) or it could be indirect (a gel–sol transition in the apical cytoplasmic domain that mediates movement and eventual docking of cytoplasmic vesicles with the apical membrane). In almost all of the cases listed, evidence supporting the role of the cytoskeleton comes from experiments where drugs/chemicals have been used either to disrupt or to alter the steady-state dynamics of the cytoskeleton. Use of any foreign agent always needs to be approached with caution. However, with appropriate controls and careful regard for possible non-specific effects, much useful information can be gained.

13.3 DRUGS INTERACTING WITH ACTIN

There are many drugs and xenobiotics that appear to interact with the actin filament system and alter filament dynamics and distribution. Two groups have been used much more than any others as tools to elucidate the role that actin filaments plays in cell function: cytochalasins, and the heptapeptide phallotoxins from mushrooms.

13.3.1 Cytochalasins

The fungal metabolites known as cytochalasins (Figure 13.2) are the most commonly used drugs for studying the actin filament system. *In vitro*, cytochalasins bind to the barbed end of the actin filament (the fast-growing end) and this prevents the association of G-actin as well as the dissociation from the filament tip. There is also some evidence for actin

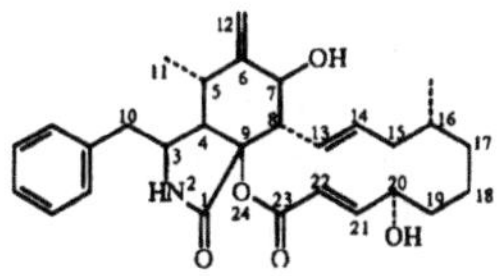

Cytochalasin B

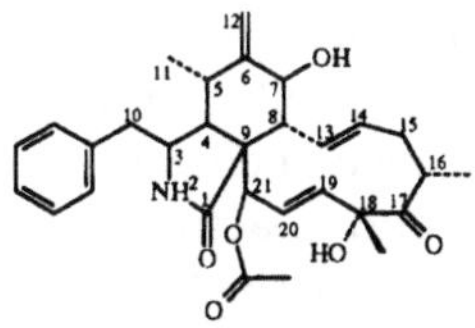

Cytochalasin D

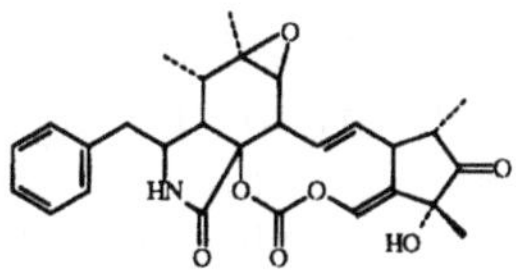

Cytochalasin E

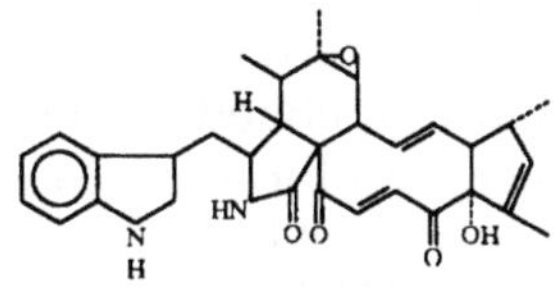

Chaetoglobosin

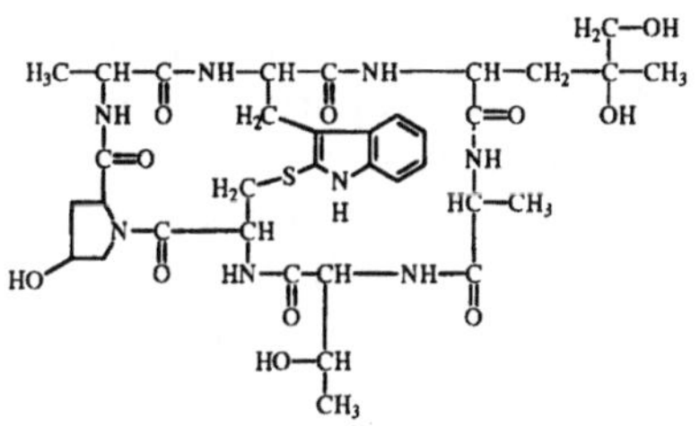

Phalloidin

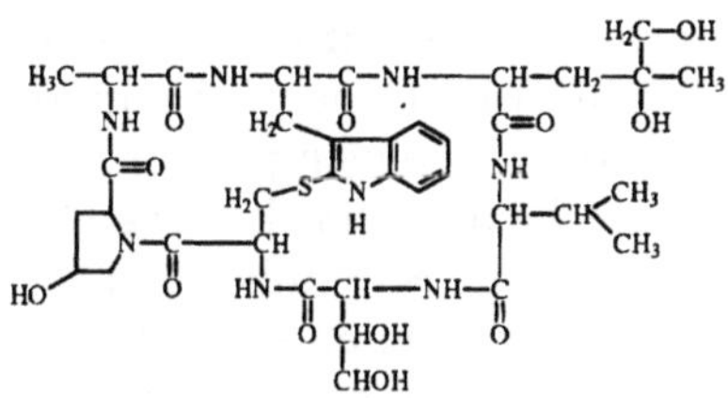

Phallacidin

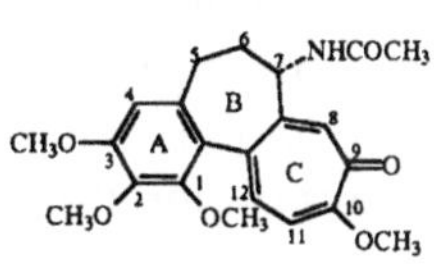

Colchicine

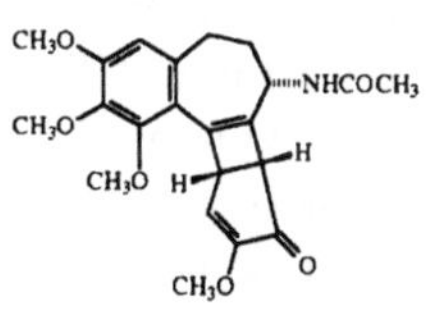

β-Lumicolchicine

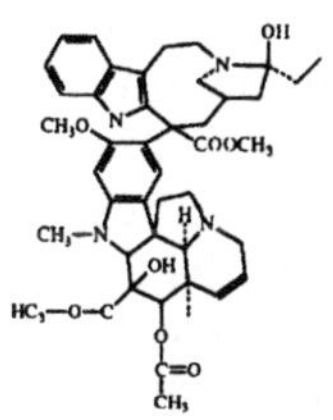

Vinblastine

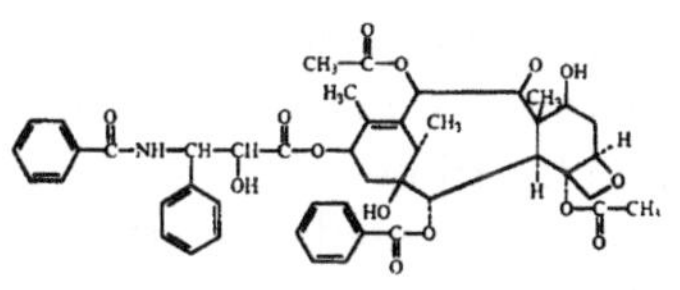

Taxol

Figure 13.2 Structures of the most commonly used drugs that are used to perturb actin filaments and microtubules. The conventional numbering and lettering systems are shown.

filament severing activity (Cooper, 1987). The net result of prolonged cytochalasin treatment in intact cells is the breakdown of actin filament organization and the appearance of clumps of actin which is apparently a mix of short actin filaments and G-actin. Removal of the cytochalasin by washing leads to rapid recovery of the filamentous array.

The most commonly used are cytochalasin B (CB), cytochalasin D (CD), cytochalasin E (CE) and, more recently, 21,22-dihydrocytochalasin B (DHCB). A serious consideration when working with CB is that hexose sugar transport is also inhibited by this drug. The affinity for the sugar transporter is higher than for actin (Mookerjee *et al.*, 1981) and thus one cannot control for this effect by a titration. However, available data indicates that CD, CE and DHCB do not inhibit sugar transport (Rampal *et al.*, 1980) and have no affinity for the binding site associated with the hexose transporter (Mookerjee *et al.*, 1981). Thus, use of CB should be approached only as an alternative to use of CD or DHCB, or the results should be verified by use of one of the more specific types. An important control to consider when using CD or DHCB is the application of chaetoglobosin C, a cytochalasin that has membrane perturbing effects but does not cap actin filaments (Feuilloley *et al.*, 1993; Rampal *et al.*, 1980). Thus when an effect of CD on an epithelial transport parameter has been determined, control experiments with chaetoglobosin C will allow analysis of effects on membrane structure versus actin binding.

The mechanism of action of cytochalasin on the actin filament system in intact cells is not clearly understood. *In vitro* analysis of CD on actin has produced data that indicates cytochalasins can have multiple effects. These depend on the concentration of CD, the divalent cation present and the availability of ATP versus ADP as the nucleotide (Cooper, 1987). For example, at low concentrations (< 1 μM) CD should bind preferentially to the barbed end of filaments. This could lead to a shortening of filament length and disruption of filament networks with a net loss of F-actin. At concentrations greater than 1 μM evidence from *in vitro* studies indicates that the CD would bind to actin monomers and enhance dimer formation. A dimer then can serve as a nucleation site for filament formation. The filament formation may not be rapid or even occur. This is due to the fact that the CD would have capped the dimer at the barbed end and thus the filament growth would occur at the pointed end – the end with the higher critical concentration. Another possibility is that the filament growth, induced by the CD–actin dimerization, occurs at the expense of other filaments rather than a recruitment of G-actin from the cellular pool.

Experiments to determine the effect of CD on the state of polymerization of cellular actin have produced some conflicting results. In platelets, exposure to CD concentrations < 1 μM inhibited the increase in F-actin that occurs upon stimulation of the platelet (Casella *et al.*; 1981, Fox and Philipps, 1981). However, concentrations of CD up to 25 μM had no effect on the resting F-actin content. Likewise in HEp-2 cells, 2 μM CD had no effect on F-actin content (Morris and Tannenbaum, 1980). At very high concentrations of CD (40 μM) an increase in F-actin was detected in mouse lymphocytes (Wilder and Ashman, 1991).

The structural organization of actin filaments shows a concentration-dependent response to CD as well. In the MDCK epithelial cell line treatment with 4 μM CD disrupted the actin filament arrays associated with junctional complexes in the apical region and those associated with stress fibers at the basal membrane domain. At a concentration of 40 μM CD, the filament bundles at the base of the cell were disrupted whereas the apical, junctional F-actin appeared relatively intact (Stevenson and Begg, 1994). The actin filaments associated with the apical microvilli (Mills, 1987) also appeared normal at this higher concentration. In this case no measurement of the F-actin or G-actin content was made so it is difficult to know if the concentration-dependent effects were related to polymerization effects as well.

A detailed analysis of the effects of CD on actin content, correlated with a physiological response, has been done in the toad urinary bladder (Franki *et al.*, 1992). The basis for the study was the observation that exposure of the toad bladder epithelial cells to vasopressin resulted in a decrease in the F-actin content correlated with the vasopressin-sensitive increase in osmotic water flux (Ding *et al.*, 1991). However, earlier studies had reported that exposure to CD inhibits vasopressin-induced water flow (Kachadorian *et al.*, 1979). Thus it was important to determine if the CD effect on actin was a net depolymerization or polymerization. Using the phalloidin binding assay to determine relative F-actin content, Franki *et al.* (1992) determined that there was a biphasic effect of CD on actin (Figure 13.3). At concentrations of 0.25 μM there was a significant reduction in F-actin content. From 0.5 μM to 2.0 μM there was no change in F-actin. Interestingly the depolymerizing action of vasopressin could be completely blocked by 1.0 μM CD. At 20 μM CD there was a significant increase in F-actin. The polymerizing effect of CD was correlated with a change in microvillar length and area, indicating an effect on the apical pool of actin. These results point out the importance of consideration of dose when using the cytochalasins. Low (< 1.0 μM) and high concentrations (> 2.0 μM) should be examined in order to cover the range of possible depolymerizing and polymerizing effects. In addition, since the apical actin pool showed differential sensitivity to CD and the basal and apical pools in the MDCK cell have been shown to respond differently to changing doses of CD (see above) one must also consider morphological analysis in order to determine the site of action of CD.

There are several other important considerations in the use of cytochalasins. Exposure to 0.05 μg/ml CE (or 5 mg/ml CB) for 2–4 hours results in DNA fragmentation in some, but not all, cell lines (Kolber *et al.*, 1990).

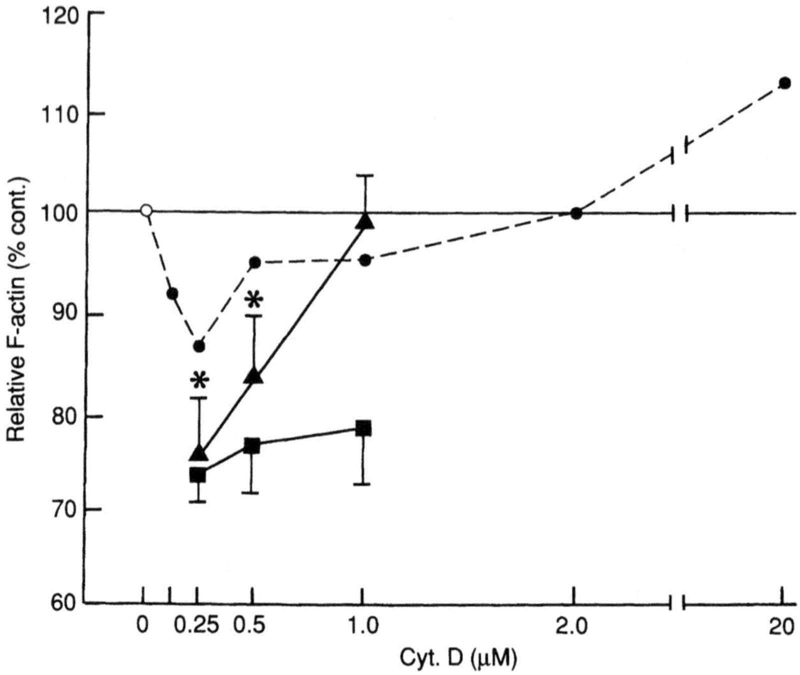

Figure 13.3 Effect of cytochalasin D (CD) alone (●), vasopressin alone (■) and vasopressin plus CD (▲) on relative F-actin content of bladder epithelial cells at different concentrations of CD. Value for vasopressin alone is that obtained in all three CD plus vasopressin experiments. * indicates significant decrease from unstimulated CD-treated bladder. (Reprinted from Franki *et al.*, 1992, with permission of *Am. J. Physiol.*)

Thus long-term exposure to cytochalasins needs to be avoided unless the effect on DNA is ruled out by analysis. This effect on DNA is probably not a non-specific one, however: first, CE was much more effective than CB in fragmenting DNA, indicating that actin was the primary target for the cytochalasin effect; and second, F-actin has been associated with the polytene chromosomes of *Drosophila* (Sauman and Berry, 1994) and exposure of salivary gland cells to 10 μM CD resulted in a change in the morphology of the chromosomes.

Cytochalasins alter protein synthesis in cells. Again this could be a non-specific effect and raises concern as to the specific interpretation of a cytochalasin effect as being directly related to interaction with actin (Cooper, 1987). However, this effect may still be due to the alteration of actin filaments in the presence of the drug: mRNA is known to associate with the actin filament system and when the filament network is perturbed by cytochalasin the mRNA is released. The result is a decrease in protein synthesis (Ornelles *et al.*, 1986). A role for the structural integrity of the actin filament system in biosynthetic processes has been emphasized recently in studies of steroid secretion and polyphosphoinositide metabolism in adrenocortical cells (Feuilloley *et al.*, 1993). In these cells, treatment with CB disrupted actin filaments, inhibited steroid secretion and the formation of polyphosphoinositides, and blocked the angiotensin-II stimulation of phosphoinositide breakdown and possibly phospholipase C activation. None of these effects were seen with chaetoglobosin C, indicating a CB effect mediated via the changes in actin filaments.

Finally, there is the possibility that cytochalasins, even though highly specific for actin, may have an indirect effect on other components of the cytoskeleton. For example, in quail oviduct and the flagella of *Chlamydomonas*, CD (at very high doses) alters microtubule dynamics (Boisvieux-Ulrich *et al.*, 1990; Dentler and Adams, 1992). Although the mechanism of the CD effect is not known, the evidence indicates that it was not due to any previously unknown direct effect of CD on tubulin nor was there an indication that there were metabolic/toxic effects. Thus, a simple precaution in the use of CD to dissect a role of actin in a physiological process is the application of colchicine as a control.

In summary, the use of cytochalasins as a tool for analyzing a potential role for actin filaments in epithelial function needs to be approached with caution. There is abundant evidence to indicate that cytochalasin B should never be used. This is due to its ability to bind to the glucose transporter and to the fact that other cytochalasins have significantly higher affinities for actin than CB does. A second important caveat is that, whenever cytochalasins are used, a dose/response analysis should be combined with the physiological measurements and, where possible (de rigueur in cultured cell systems), a biochemical and structural determination of the state of the actin filament system (F-actin and G-actin levels). Relative F-actin levels can be determined by use of the phalloidin binding assay (Ding *et al.*, 1991); G-actin and total cellular actin (and therefore F-actin) can be measured by DNase inhibition assay (Fox *et al.*, 1981). A bimodal effect of CD on actin filament polymerization is well enough established that a single dose analysis, without a determination of the effect on the filament content and organization, will not allow any insight into the possible mechanism other than to conclude that perturbation of the actin steady state is involved.

13.3.2 Phalloidin

The phallotoxins – cyclic heptapeptides produced by the Amanita mushroom – bind with high specificity to filamentous actin (Faulstich, 1982). The result of filament binding is a stabilization of the actin filament including the depolymerizing effect of CB (Low *et al.*, 1975). *In vitro*, once phalloidin binds, there is a drop in the dissociation constant for the actin subunits. The net effect is a drop in the critical concentration for polymerization and subsequently an increase in the amount of actin assembled into filaments.

There has been no report of a non-specific or alternative specific (whether of high or low affinity) binding site for phalloidin. Some concern has been raised about the possibility of alternative sites since phalloidin treatment can prove toxic and lead to cell death (Cooper, 1987). Although the toxicity still could be mediated by an effect on actin, it remains problematic since the phallotoxins do not readily cross most cell membranes. Therefore one important area of research with regard to these compounds is to determine the mechanism of the toxic effect. This will allow a more accurate interpretation of results.

13.4 DRUGS INTERACTING WITH MICROTUBULES

There is a wide variety of drugs that act on microtubules. Of this group a few have been used most extensively to elucidate the functional role of microtubules in cells and specifically the role that microtubules may play in epithelial transport function. By far the most widely used is colchicine and its derivatives (colcemid, podophyllotoxin) and, most recently, taxol.

13.4.1 Colchicine

Colchicine is an alkaloid (Figure 13.2) found in the plants of the lily family. It is one of the oldest drugs in use, recommended to treat gout since the sixth century. Colchicine binds to the tubulin heterodimer, although the exact binding site and whether it involves both α and β tubulin subunits is not known. This binding is relatively slow and has varying affinity, depending on the cellular source. For example, yeast microtubules are resistant to the effects of colchicine, as are cells from the Syrian hamster, and even within individual cells some microtubules appear to be resistant to the effect of colchicine (Eilers *et al.*, 1989; Thatte *et al.*, 1994).

Exposure of microtubules to colchicine *in vivo* has long been thought to cause a depolymerization of the microtubules. This is due to the binding to the tubulin dimers. Colchicine–tubulin becomes incorporated at the end of the microtubule. If the concentration of this colchicine–tubulin complex is high enough, the further addition of unbound tubulin to the microtubule end is blocked. However, at low concentrations (nanomoles) this does not prevent further addition of tubulin *in vitro* and in fact the ends can grow, with incorporation of the colchicine–tubulin complex within the microtubule. At the concentrations used in most epithelial studies (> 10^{-6} M) this may not be of concern, especially since cells can accumulate microtubule-active drugs from the medium (Wilson and Jordan, 1994).

Although it is widely held that most of the effects of colchicine are due to binding to tubulin (Hastie, 1991) and although colchicine binding assays have been used to determine the presence of tubulin (Fulton and Simpson, 1979), there is some evidence that there are binding sites on cell membranes unrelated to tubulin (Stadler and Franke, 1974) and that nucleoside transport is also inhibited by colchicine. One way to overcome the problem of a microtubule specific effect is to use an analog of colchicine, β-lumicolchicine (Figure 13.2) as a control. This photochemical product of colchicine has a very low affinity for tubulin (McClure and Paulson, 1977), although many investigators treat it as if it had none. However, β-lumicolchicine does have affinity for the colchicine binding sites on membranes that are unrelated to tubulin, including inhibition of nucleoside transport and the $GABA_A$ receptor-mediated Cl^- currents (Mihic *et al.*, 1994).

13.4.2 Vinca alkaloids

Another class of compounds that is used widely to analyze the role of microtubules in cell function is the vinca alkaloids, vinblastine (Figure 13.2) and vincristine. These drugs also inhibit microtubule polymerization, and exposure in intact cells leads to a loss of microtubules. The mechanism of action appears to be slightly different from colchicine in that the binding occurs directly to the microtubule end rather than the free tubulin. However, the result is the same: at sufficiently high concentrations tubulin incorporation is inhibited and microtubules depolymerize.

13.4.3 Taxol

Taxol (Figure 13.2) is a diterpene that is isolated from the bark of the yew *Taxus brevifolia*. It has only recently been used in studies of epithelial function and in most cases has been employed as a counter treatment to the use of colchicine or vinca alkaloids. This is due to the stabilizing action of this drug on microtubules. Taxol binds to microtubules and does not appear to have any affinity for free tubulin (Ringel and Horwitz, 1991). Taxol-treated microtubules are very resistant to depolymerization, including the actions of colchicine and Ca^{2+} ions (Wilson and Jordan, 1994). Besides stabilizing microtubules, taxol stimulates microtubule formation (Schiff *et al.*, 1979) and can lead to microtubule bundling (Schiff and Horwitz, 1980). The fundamental effect appears related to the reduction of the dissociation rate constant at each end of the microtubule. With no alteration in the association rate, net growth can occur. The mechanism by which the effect of colchicine is inhibited by taxol is not known.

All of the effects of taxol appear to be related to its microtubule binding properties. Thus it could become a drug of wide use in epithelial studies where a role of microtubules is under investigation but some precautions need to exercised in its use. Exposure of microtubules to taxol in the steady state may stabilize the array but it can lead to net change in microtubule mass and to changes in the array unrelated to a change in the depolymerization/polymerization state (e.g. bundling). If microtubule depolymerization is suspected as playing a role, taxol can be used to inhibit this action; however, the taxol treatment might mediate an effect via an increase in microtubule mass rather than by preventing depolymerization. In addition, more careful attention must be paid to dose as there may be multiple effects *in vivo* that only now are being elucidated.

As pointed out by Wilson and Jordan (1994), almost all of the microtubule-active drugs are used in cells at high concentrations. This leads to a depolymerization or stabilization, as expected. *In vitro* studies show that each of the drug classes discussed above have different effects depending on the concentration used. For example, at sub-stoichiometric doses vinblastine does not change the length of microtubules nor does it alter microtubule mass. The same is true for colchicine and taxol. In most epithelial studies, exposure to these drugs has occurred at a single dose. Appropriate controls included carrier alone or inactive analogs. Finally, in many cases (but not all) the morphological detection of microtubules is included to assess the effectiveness of the drug on the microtubule array. Recent studies on the mitotic apparatus of HeLa cells indicate that inhibition of a microtubule-driven function (mitosis) can occur at doses below those that were previously used to inhibit mitosis and either depolymerize or stabilize the microtubules. Most importantly, for all of the drugs, except colchicine, the dose that was effective for mitotic arrest did not appreciably alter the microtubule mass. For vinblastine, the maximum inhibition of mitosis was achieved at a concentration that had no effect on microtubule mass (Figure 13.4). Rather, at the lowest effective doses for mitotic arrest, morphological analysis showed subtle changes in the organization of the microtubule array of the spindle, not a loss (or, for taxol, a gain) of microtubules. Thus a microtubule function was inhibited by altering the dynamics of the microtubule system without producing a detectable change in microtubule polymerization. Similar dose studies combined with careful analysis of microtubule mass and organization should be attempted in analyzing microtubule function in epithelia in order to analyze more clearly the role that the microtubules play.

13.5 ROLE OF THE CYTOSKELETON IN EPITHELIAL TRANSPORT

There is abundant evidence indicating that elements of the cytoskeleton may play a key role in several epithelial transport functions (Mills and

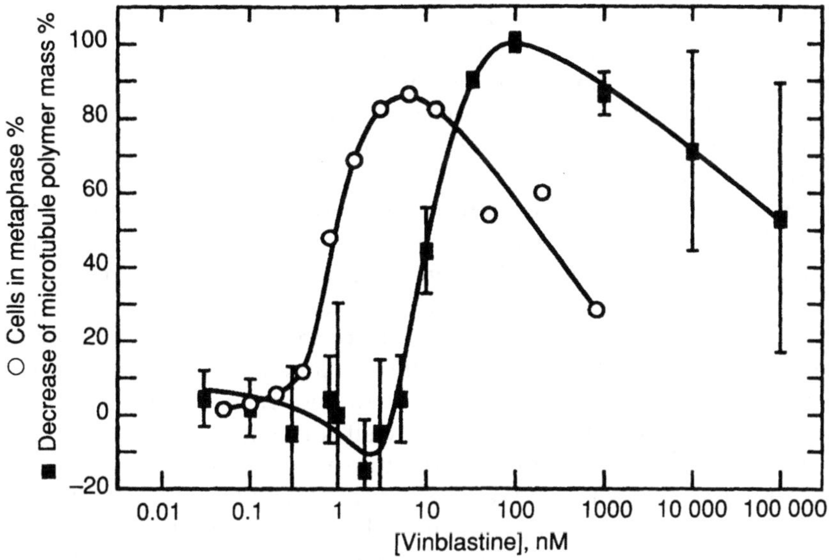

Figure 13.4 Relationship between metaphase arrest and microtubule depolymerization in HeLa cells incubated with increasing concentrations of vinblastine. Percentage of cells in metaphase (circles) and percentage decrease in mass of polymerized microtubules (squares) compared with control cells. Metaphase arrest was assayed by immunofluorescence microscopy of fixed cells and microtubule polymer mass was measured by ELISA for tubulin in stabilized cytoskeletons. At high concentrations of vinblastine (10–100 μM) the microtubule polymer mass increased as a result of vinblastine–tubulin paracrystal formation. (Reprinted from Jordan *et al.*, 1992, with permission of the publisher.)

Mandel, 1994). A selection of these studies will be reviewed here with special emphasis on the use of methods to investigate the mechanism by which the cytoskeleton may mediate the functions and how inhibitors of cytoskeletal processes were used to elucidate the mechanism.

13.5.1 Epithelial barrier function

A well-established role for the actin filament system is in the establishment of the epithelial barrier. Actin filaments and associated proteins are attached to the adherens junctions of epithelial cells – those that attach both to the substrate and to the zonula adherens – and are closely associated with the membrane comprising the tight junction. In addition, myosin has been shown to be present in the circumferential ring close to the tight junction (Figure 13.1). That this organization plays a role in the permeability function of the epithelial barrier is demonstrated by the fact that treatment with cytochalasins alters the structure of the junctions as well as reducing selective permeability (Madara *et al.*, 1986). The mechanism by which actin interacts with the junctions to control permeability is not known. At the primary permeability barrier (the tight junction) it is thought that pores, for ions or molecules that may permeate the barrier, are located in the region of the tight junction. The presence of these pores has not been verified.

A possible mechanism by which the cytoskeleton may regulate the epithelial transport function at the junction is via an increase in the tension generated by actin–myosin skeleton (Madara, 1989). This results in an increase in tight junction permeability, presumably due to a reduction in sieving/selective permeability properties of the intercellular domain of the junction brought about by a laterally developed tension acting to separate or perturb junction structure (Pitelka and Taggart, 1983). Primary evidence to support this comes from well-correlated physiological and morphological studies of junction structure and transepithelial resistance (Madara *et al.*, 1986, 1987a; Stevenson and Begg, 1994) during different physiological states or after treatment with inhibitors of cytoskeletal organization. A key element in the interpretation of the data dealing with junction function and cytoskeletal involvement is that the peripheral actin–myosin ring has contractile properties. This has been demonstrated in isolated brush border preparations. The equivalent in an intact cell could therefore be the generation of tension.

What complicates this model is the recent observation that CD treatment of MDCK cells has a different effect on junctional cytoskeleton and epithelial resistance depending on the concentration of CD used. At 4 μM CD, resistance dropped more than 50% and junctional actin was altered. At 40 μM CD, resistance dropped approximately 20% and junctional actin was not disrupted (Stevenson and Begg, 1994). This is in contradiction to what was seen in intestinal epithelia exposed to 20 μM CD (Madara *et al.*, 1986). The reason for the discrepancy is not known. Both of these high concentrations are in the range where actin polymerization may have occurred and, therefore, insight into this differential effect could be obtained by analysis of the F-actin content and domain-specific effects of CD as discussed above for cytochalasin effects. The simplest conclusion is that, whether net actin depolymerization or polymerization occurs, the steady-state organization of the tight junction cytoskeleton is disturbed by CD and the junction opens.

There is no evidence that CD treatment can lead to increases in tension within the actin–myosin ring although it may be that disruption leads to localized contraction (Stevenson and Begg, 1994). However, treatment of smooth muscle with CD relaxes the smooth muscle (Adler *et al.*, 1983). Further support for a contractile response is that treatment of intestinal epithelium with the uncoupler DNP prevents the CD effect on transepithelial permeability and junctional structure (Madara *et al.*, 1987b).

Significantly, CD treatment, at all concentrations, inhibits the drop in resistance mediated by Ca^{2+} chelation while the differential effect of CD on the junctional cytoskeleton at low Ca^{2+} was verified (Stevenson and Begg, 1994). Since removal of Ca^{2+} causes detachment of the actin ring from the junction (Volberg *et al.*, 1986) this would imply that tension on the junctional belt is disrupted. How CD prevents this is not known.

13.5.2 Establishment of membrane domains

Another function for the actin/spectrin cytoskeleton is in the establishment and maintenance of epithelial domains, especially with regard to the polarized distribution of transport proteins (Chapter 3). This interaction with membrane proteins can serve an important function in the maintenance of vectorial transport across the epithelium. There is abundant evidence demonstrating a link between components of the actin cytoskeleton and transport proteins (Mills and Mandel, 1994). This linkage has been shown to be membrane domain specific and can even be altered for the individual specific transporters based on the direction of the transport in different cell types (Gundersen *et al.*, 1991). What is not clear is whether this interaction, besides having an anchoring or stabilizing function, can serve a regulatory function with regard to transporter activity.

13.5.3 Regulation of membrane transport

Several studies from Cantiello and co-workers have provided important new insights into a possible regulatory role for the actin cytoskeleton in regulating ion transporters. Using patch clamped membranes from A6 cells, Cantiello *et al.* (1991) showed that exposure to cytochalasin D increased the activity of Na^+ channels. Exposure of patches to 5 μM actin for short periods mimicked the CD effect, whereas long exposures had no effect. This time treatment is thought to lead to the formation of short filaments and then long filaments. Treatment with DNase I, which will sequester actin monomer, prevented the actin stimulation. Further analysis demonstrated that DNase I could inhibit protein kinase A activation of the Na^+ channel (Prat *et al.*, 1993). Thus it was concluded that 'short' actin filaments play a role in regulating the channel activity. One interesting aspect to these studies was the observation that other cytochalasins, including CE, added over a 10-fold range, did not alter channel activity. It is not clear why these other cytochalasins did not work, since they should also produce a change in the actin filaments. Such disparities point out the usefulness of microscopic analysis of the cytoskeleton to verify drug and biochemical treatments to alter actin filaments.

We have used cytochalasins and phalloidin in order to investigate the role that the actin filament system plays in regulating Cl^- channel activity (Schwiebert *et al.*, 1994). When RCCT-28A cells, a renal cell culture line, are exposed to a hypotonic medium they undergo a classic swelling and then volume recovery phase (RVD). The RVD response is associated with an increased activity of K^+ and Cl^- channels. The Cl^- channels are voltage sensitive and can also be activated by stretch in the patch excised mode. When excised patches of RCCT-28A apical membrane are exposed to 1.0 μM concentrations of DHCB, CD or CB there is an increase in Cl^- channel activity (Figure 13.5), indicating that disruption of the actin filament array within the patch leads to activation of the channel.

When the patches are activated by DHCB treatment, subsequent application of suction does not lead to further channel activity. Most important was the demonstration that cells pretreated with phalloidin did not elicit an RVD and patches pretreated with phalloidin could not be activated by the application of maximum-activating suction. That the channels were present and still capable of functioning in the phalloidin treated patch was shown by subsequent activation with depolarizing voltage. Finally, in whole-cell configurations, DHCB elicited an increase in Cl^- conductance

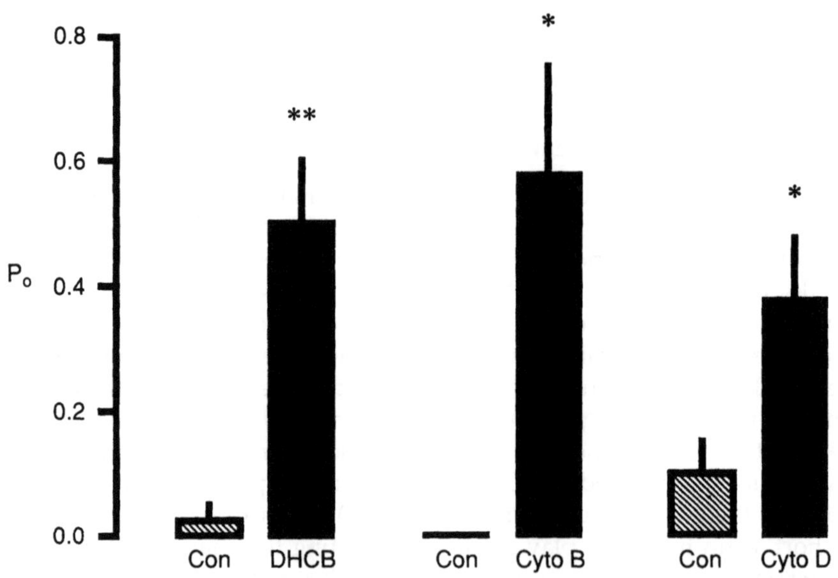

Figure 13.5 Effect of cytochalasin on the P_o of the 305 pS Cl^- channel in paired experiments: DHCB (10^{-6} M, n = 10), cytochalasin B (Cyto B: 10^{-6} M, n = 4) and cytochalasin D (cyto D: 10^{-6} M, n = 4). Each cytochalasin increased the P_o after a delay of 3–5 min. The voltage across the membrane patch was –20 mV. * = $P < 0.05$ versus control; ** = $P < 0.01$ versus control. (Reprinted from Schwiebert *et al.*, 1994, with permission of the publisher.)

in the isotonic condition whereas phalloidin significantly reduced Cl^- conductance in the isotonic state and prevented the hypotonic-sensitive increase (Table 13.2). Thus activation of the volume sensitive (stretch-activated) Cl^- channel in RCCT-28A cells was sensitive to the state of the actin filaments in the whole cell and in the membrane patch excised from the apical membrane. We do not know the effect of the DHCB or phalloidin treatment on the actin filaments in the cell or the patch. At the concentrations used we might predict a net decrease of F-actin in the the case of cytochalasin treatment and an increase with phalloidin. Thus 'short' versus long filament arrays might be a key element in this regulation. The advantage of being able to duplicate the effects on patches in the whole-cell configuration means that the actual changes in F-actin can be analyzed both biochemically and by confocal microscopy to determine the domain specific and biochemical outcome of the inhibitor treatments (Franki *et al.*, 1992; Holmgren *et al.*, 1992).

An elegant molecular biology approach was used in further dissecting the role of the cytoskeleton in regulating ion transport (Cantiello *et al.*, 1993). Melanoma cell lines deficient in actin-binding protein (ABP) or transfected to express ABP were exposed to hypotonic conditions and the K^+ channel activity monitored during regulatory volume decrease (RVD). In ABP^- there was no RVD and no increase in K^+ channel activity, as is seen in ABP replete cells. Important also was the demonstration that in ABP^- cells there was an unregulated K^+ conductance. Since ABP can link actin to integral membrane proteins, this protein could serve as a connection between the K^+ channel present in the membrane and the actin cytoskeleton. Under conditions that lead to activation of the channel this link is necessary in order for normal activation to occur. The fact that baseline K^+ channel activity was altered in ABP^- cells also implies that normal steady-state activity of the channel is regulated by an interaction with the

Table 13.2 Whole cell chloride conductance

Condition	*Isotonic*	*Hypotonic*	n	p *values*
Control	15.0±1.3	29.5±1.5	14	$p < 0.001$
DHCB (10^{-6} M)	34.8±5.7[a]	37.8±5.4[a]	7	NS[b]
Phalloidin (2×10^{-10}M)	8.1±2.2[a]	8.7±2.2[a]	7	NS
Colchicine (10^{-6} M)	12.9±1.6	19.7±2.4	4	$p < 0.05$

The p values indicate the level of significance comparing isotonic with hypotonic data by paired t test.

[a]Significantly different from isotonic or hypotonic data in the same column by analysis of variance and Student–Neuman–Keuls test ($p < 0.05$).
[b]NS, not significant
(Reprinted from Schwiebert *et al.*, 1994, with permission)

cytoskeleton. Thus, as pointed out by Cantiello *et al.* (1993), the actin cytoskeleton may serve to both 'trigger and control' membrane transporters.

In T84 cells both microfilaments and microtubules may play a role in regulating ion transport. In this cell line Cl^- secretion is stimulated by either the cAMP mediated pathway or a Ca^{2+} pathway. When T84 cells are exposed to cAMP, Cl^- secretion increases and this is associated with a change in the actin filaments at the basal pole of the cell. The change is similar to that seen in MDCK cells exposed to cAMP (Mills, 1987). In a clever experiment, Shapiro *et al.* (1991) exposed the T84 cells to phalloidin for a period of time long enough for the cells to take up the toxin and have it bind to the actin filaments. This did not produce any noticeable toxicity. When phalloidin-treated cells were exposed to cAMP the actin filaments did not alter and the Cl^- secretory response was blocked. In addition, phalloidin treatment inhibited ^{86}Rb uptake through the basally located $Na^+/K^+/2Cl^-$ transporter (Figure 13.6). However, bumetanide binding was unaffected indicating that the transporter was present in the cell membrane but inhibited. Treatment with 2 μM CD, while markedly reducing transepithelial resistance, did not block the cAMP-mediated rise in secretion (Matthews *et al.*, 1994). The results indicated the possibility of a regulatory step in the cAMP-mediated pathway that involved a linkage between the actin filaments and the activity of the cotranporter (Matthews *et al.*, 1992). What is not known is whether the inhibitor treatments lead to net changes in actin filaments. This would be significant since phalloidin treatment may have led to an increase in F-actin. Thus the inhibitory effect could have been related to increases in actin filaments and linkages or a change in G-actin content.

A possible regulatory linkage between actin and the $Na^+/K^+/2Cl^-$ cotransporter has also been demonstrated in Ehrlich ascites tumor cells (Jessen and Hoffmann, 1992). In these cells, exposure to hypotonicity results in a stimulation of cotransporter activity. This activation can be inhibited in Ca^{2+}-free medium. When cells in Ca^{2+}-free hypotonic medium are pretreated with CB (42 μM) the activation is recovered. Thus a regulation of the transporter, linking a rise in Ca^{2+} to changes in the organization of the actin filaments, was proposed. Interestingly, the cotransporter can also be activated during recovery from cell shrinkage and in this case CB treatment actually leads to inhibition of the response. Whether the CB treatment resulted in a depolymerization or polymerization response in the cells under the different experimental conditions is not known. However, the results do indicate that several regulatory pathways, responding to different stimuli, may involve the actin filament system.

Further studies into the role of the cytoskeleton in regulating the cAMP-stimulated Cl^- secretory response in T84 cells has implicated microtubules as well (Fuller *et al.*, 1994). When MTs are depolymerized in the presence

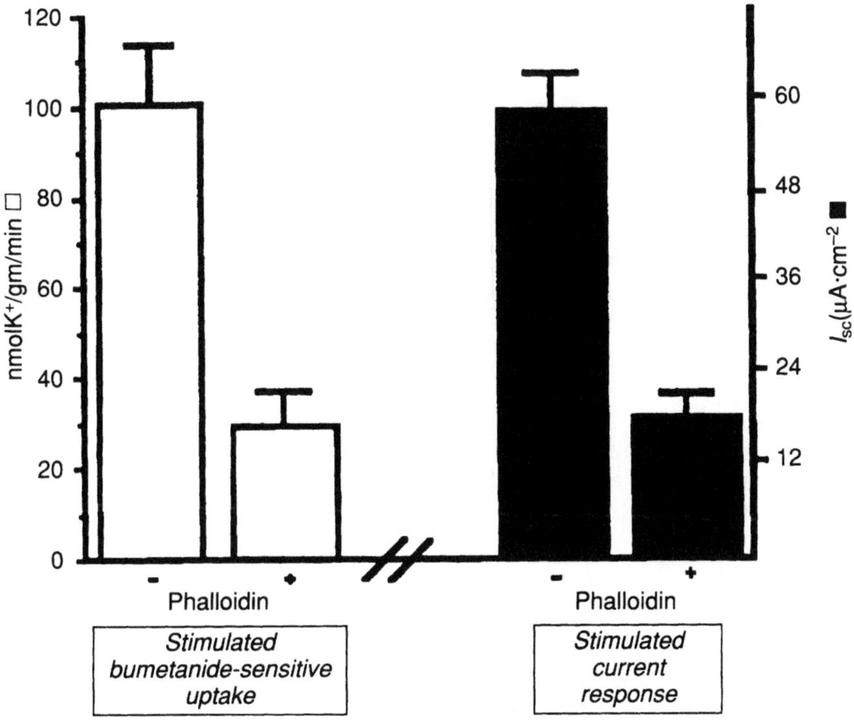

Figure 13.6 Phalloidin loading inhibits cAMP stimulation of $Na^+/K^+/2Cl^-$ cotransporter activity in T84 cells. The left-hand side of the figure (white bars) indicates that $Na^+/K^+/2Cl^-$ cotransporter activity, as measured by bumetanide-sensitive ^{86}RB uptake under forskolin-stimulated conditions, is inhibited under phalloidin-loaded conditions compared with controls (n = 12 for each, $P < 0.001$); the percentage reduction in bumetanide-sensitive ^{86}Rb uptake is comparable to the inhibition of the forskolin-stimulated peak I_{sc} (black bars) under these conditions. (Reprinted from Matthews *et al.*, 1992, with permission from *Am. J. Physiol.*)

of high concentrations of colchicine or nocodazole, forskolin-sensitive (cAMP-mediated) anion efflux was inhibited 40 to 50% (Figure 13.7). This effect was not seen in the presence of β-lumicolchicine. Again, as with the phalloidin effect, the microtubule active drugs did not inhibit the Ca^{2+} sensitive anion pathway. In addition, they did not appear to inhibit the transporter directly, since short-term incubation was without effect. An interesting aspect of this study was the lack of an effect of nocodozole and colchicine at doses below 33 μM and 100 μM, respectively. This is well above concentrations known not only to alter most microtubule arrays but also to lead to alteration of microtubule function without any change in microtubules.

Based on these results, a tentative model can be proposed in which two

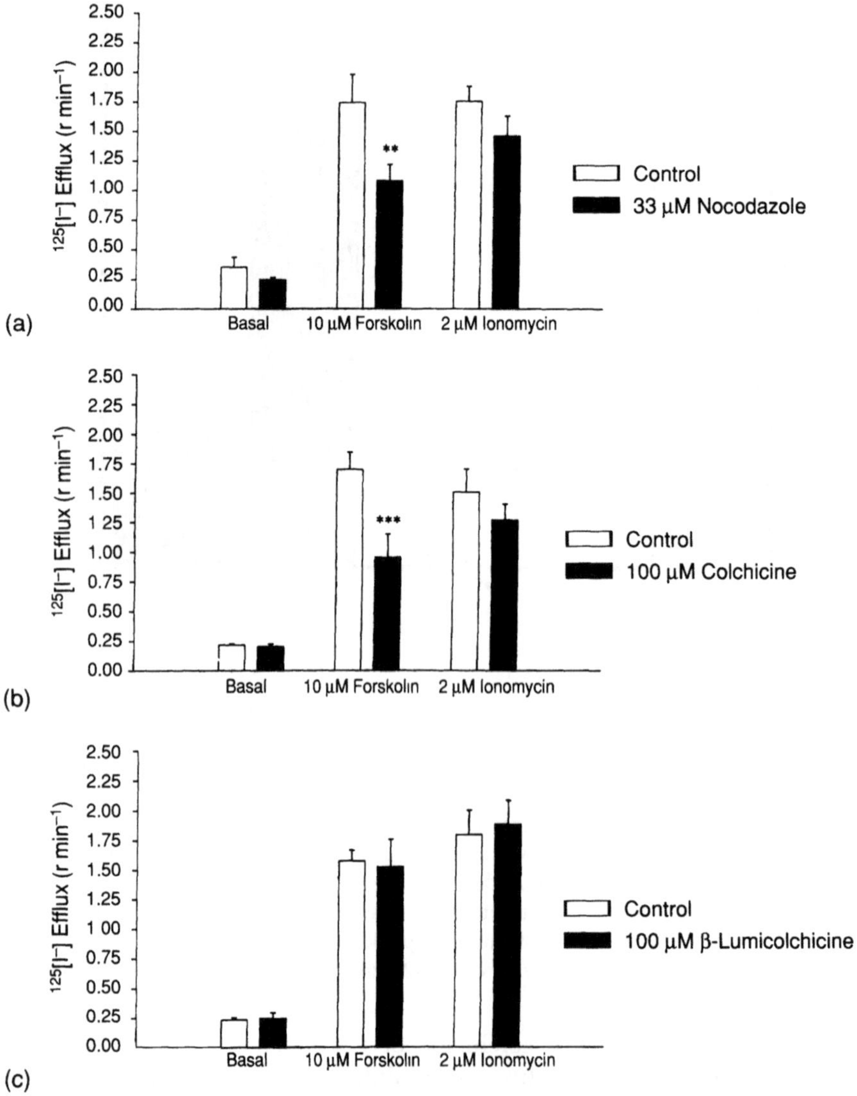

Figure 13.7 Summarized data for series of $^{125}I^-$ efflux experiments in presence of 1% DMSO (control), microtubule inhibitors nocodozole (a) or colchicine (b), or inactive analog of colchicine, β-lumicolchicine (c). Both nocodazole and colchicine significantly reduced only forskolin-sensitive I^- efflux ((a) and (b)), whereas structural analog β-lumicolchicine ((c)) was without effect on forskolin-evoked efflux of I^-. None of the compounds used had any significant effect on magnitude of ionomycin-sensitive efflux. Results represent rate of efflux of I - as measured at 20 s after addition of agonist. Results are means + SE of at least four separate experiments. Results were compared using unpaired *t* test. ** 0.02 > P > 0.01; *** 0.01 > P > 0.001. (Reprinted from Fuller *et al.*, 1994, with permissionfrom *Am. J. Physiol.*)

components of the cytoskeleton (basally situated actin filaments and apically directed microtubules) work in concert to regulate the cAMP-stimulated Cl^- efflux in T84 cells. Basally located cotransporter activity, which supplies the anion for net transport, is activated if the steady-state actin filament network is altered. Net flux across the apical membrane can be achieved through Cl^- channels already present in the membrane. However, full activation of this pathway can only occur if the microtubule array is present and mediates the movement of vesicles containing Cl^- channels to the vicinity of the apical membrane. How the vesicles move from the microtubules to the apical membrane through the apical actin network is not known but could be mediated by myosin-I interaction between vesicles and actin filaments (Fath *et al.*, 1993).

REFERENCES

Adler, K.A., Krill, J., Alberghini, T.V. and Evans, J.N. (1983) Effect of cytochalasin D on smooth muscle contraction. *Cell Motil.* **3**:545–51.

Alberts, B., Bray, D., Lewis, J. *et al.* (1994) *Molecular Biology of the Cell*, Garland Publ. Inc., New York.

Asch, H.L., Mayhew, E., Lazo, R.O. and Asch, B.B. (1990) Lipids noncovalently associated with keratins and other cytoskeletal proteins of mouse mammary epithelial cells in primary culture. *Biochim. Biophys. Acta* **1034**:303–8.

Bacallao, R., Antony, C., Dotti, C. *et al.* (1989) The subcellular organization of Madin–Darby canine kidney cells during the formation of a polarized epithelium. *J. Cell Biol.* **109**:2817–32.

Bacskai, B.J. and Friedman, P.A. (1990) Activation of latent Ca^{2+} channels in renal epithelial cells by parathyroid hormone. *Nature* **347**:388–91.

Bennett, V. (1992) Ankyrins: adaptors between diverse plasma membrane proteins and the cytoplasm. *J. Biol. Chem.* **267**:8703–6.

Bennett, V. and Gilligan, D.M. (1993) The spectrin-based membrane cytoskeleton and micron-scale organization of the plasma membrane. *Ann. Rev. Cell Biol.* **9**:27–66.

Bennett, V. and Lambert, S. (1991) The spectrin skeleton: From red cells to brain. *J. Clin. Invest.* **87**:1483–9.

Boisvieux-Ulrich, E., Laine, M.-C. and Sandoz, D. (1990) Cytochalasin D inhibits basal body migration and ciliary elongation in quail oviduct epithelium. *Cell Tissue Res.* **259**:443–54.

Cantiello, H.F., Stow, J.L., Prat, A.G. and Ausiello, D.A. (1991) Actin filaments regulate epithelial Na^+ channel activity. *Am. J. Physiol.* **261**:C882–C889.

Cantiello, H.F., Prat, A.G., Bonventre, J.V. *et al.* (1993) Actin-binding protein contributes to cell volume regulatory ion channel activation in melanoma cells. *J. Biol. Chem.* **268**:4596–9.

Casella, J.F., Flanagan, M.D. and Lin, S. (1981) Cytochalasin D inhibits actin polymerization and induces depolymerizarion of actin filaments formed during platelet shape change. *Nature* **293**:302–5.

Cooper, J. (1987) Effects of cytochalasin and phalloidin on actin. *J. Cell Biol.* **105**:1473–8.

Davis, J., Davis, L. and Bennett, V. (1989) Diversity in membrane binding sites of ankyrins. *J. Biol. Chem.* **264**:6417–26.

Dentler, W.L. and Adams, C. (1992) Flagellar microtubule dynamics in

Chlamydomonas: cytochalasin D induces periods of microtubule shortening and elongation; and colchicine induces disassembly of the distal, but not proximal, half of the flagellum. *J. Cell Biol.* **117**:1289–98.

Ding, G., Franki, N., Condeelis, J. and Hays, R.M. (1991) Vasopressin depolymerizes F-actin in toad bladder epithelial cells. *Am. J. Physiol.* **260**:C9–C16.

Dosemeci, A. and Pant, H.C. (1992) Association of cyclic-AMP-dependent protein kinase with neurofilaments. *Biochem. J.* **282**:477–81.

Drenckhahn, D. and Dermietzel, R. (1988) Organization of the actin filament cytoskeleton in the intestinal brush border: a quantitative and qualitative immunoelectron microscope study. *J. Cell Biol.* **107**:1037–48.

Drenckhahn, D., Schluter, K., Allen, D.P. and Bennett, V. (1985) Colocalization of band 3 with ankyrin and spectrin at the basal membrane of intercalated cells in the rat kidney. *Science* **230**:1287–9.

Eilers, U., Klumperman, J. and Hauri, H.-P. (1989) Nocodazole, a microtubule-active drug, interferes with apical protein delivery in cultured intestinal epithelial cells (Caco-2). *J. Cell Biol.* **108**:13–22.

Fath, K.R., Mamajiwalla, S.N. and Burgess, D.R. (1993) The cytoskeleton in development of epithelial cell polarity. *J. Cell Sci.* **17**:65–73.

Faulstich, H. (1982) Structure–activity relationship of actin-binding peptides, in *Chemistry of Peptides and Proteins*, (eds W. Voelter, E. Wunsch, Y. Ovchinnikov and V. Ivanov), W. de Gruyter, New York, pp. 279–88.

Feuilloley, M., Desrues, L. and Vaudry, H. (1993) Effect of cytochalasin-B on the metabolism of polyphosphoinositides in adrenocortical cells. *Endocrinol.* **133**:2319–26.

Foisner, R. and Wiche, G. (1991) Intermediate filament-associated proteins. *Curr. Opin. Cell Biol.* **3**:75–81.

Fox, J.E.B. and Philipps, D.R. (1981) Inhibition of actin polymerization in blood platelets by cytochalasins. *Nature* **292**:650–2.

Fox, J.E.B., Dockter, M.E. and Phillips, D.R. (1981) An improved method for determining the actin filament content of nonmuscle cells by the DNase I inhibition assay. *Anal. Biochem.* **117**:170–7.

Franki, N., Ding, G., Gao, Y. and Hays, R.M. (1992) The effect of cytochalasin D on the actin cytoskeleton of the toad bladder epithelial cell. *Am. J. Physiol.* **263**:C995–C1000.

Frappier, T., Derancourt, J. and Pradel, L.-A. (1992) Actin and neurofilament binding domain of brain spectrin B subunit. *Eur. J. Biochem.* **205**:85–91.

Fuchs, E. (1994) Intermediate filaments and disease: mutations that cripple cell strength. *J. Cell Biol.* **125**:511–16.

Fuchs, E. and Weber, K. (1994) Intermediate filaments: structure, dynamics, function and disease. *Ann. Rev. Biochem.* **63**:345–82.

Fuller, C.M., Bridges, R.J. and Benos, D.J. (1994) Forskolin – but not ionomycin – evoked Cl^- secretion in colonic epithelia depends on intact microtubules. *Am. J. Physiol.* **266**:C661–C668.

Fulton, C. and Simpson, P.A. (1979) Tubulin pools, synthesis and utilization, in *Microtubules*, (eds K. Roberts and J.S. Hyams), Academic Press, New York, pp. 118–74.

Georgatos, S.D. and Blobel, G. (1987) Two distinct attachment sites for vimentin along the plasma membrane and the nuclear envelope in avian erythrocytes: a basis for vectorial assembly of intermediate filaments. *J. Cell Biol.* **105**:117–25.

Gilbert, T., Lebivic, A., Quaroni, A. and Rodriguez-Boulan, E. (1991) Microtubule organization and its involvement in the biogenetic pathways of plasma memebrane proteins in CaCo-2 intestinal epithelial cells. *J. Cell Biol.* **113**:275–84.

Glenney, J.P., Glenney, P. and Weber, K. (1983) The spectrin related molecule TW 260/240 cross-links actin bundles of the microvillus rootlets in the brush borders of intestinal epithelial cells. *J. Cell Biol.* **96**:1491–6.

Gundersen, D., Orlowski, J. and Rodriguez-Boulan, E. (1991) Apical polarity of Na,K-ATPase in retinal pigment epithelium is linked to a reversal of the ankyrin–fodrin submembrane cytoskeleton. *J. Cell Biol.* **112**:863–72.

Hansch, E., Forgo, J., Murer, H. and Biber, J. (1993) Role of microtubules in the adaptive response to low phosphate of Na/P_i cotransport in oppossum kidney cells. *Pflügers Archiv.* **422**:516–22.

Hastie, S.B. (1991) Interactions of colchicine with tubulin. *Pharmac. Ther.* **51**:377–401.

Hays, R.M., Franki, N., Simon, H. and Gao, Y. (1994) Antidiuretic hormone and exocytosis: lessons from neurosecretion. *Am. J. Physiol.* **267**:C1507–1524.

Herrmann, H. and Wiche, G. (1987) Plectin and IFAP-300K are homologous proteins binding to microtubule-associated proteins 1 and 2 and to the 240 kilodalton subunit of spectrin. *J. Biol. Chem.* **262**:1320–5.

Holmgren, K., Magnusson, K.E., Franki, N. and Hays, R. M. (1992) ADH-induced depolymerization of F-actin in the toad bladder granular cell: a confocal microscope study. *Am. J. Physiol.* **262**:C672–C677.

Horkovics-Kovats, S. and Traub, S. (1990) Specific interaction of the intermediate filament protein vimentin and its isolated N-terminus with negatively charged phospholipids as determined by vesicle aggregation, fusion, and leakage measurements. *Biochem.* **29**:8652–7.

Hunziker, W., Vale, P. and Mellmam, I. (1990) Differential microtubule requirements for transcytosis in MDCK cells. *EMBO J.* **9**:3515–25.

Ishikawa, H., Bischoff, R. and Holtzer, H. (1968) Mitosis and intermediate-sized filaments in developing skeletal muscle. *J. Cell Biol.* **38**:538–55.

Jessen, F. and Hoffmann, E.K. (1992) Activation of the $Na^+/K^+/Cl^-$ cotransport system by reorganization of the actin filaments in Ehrlich ascites tumor cells. *Biochim. Biophys Acta* **1110**:199–201.

Jordan, M.A., Thrower, D. and Wilson, L. (1992) Effects of vinblastine, podophyllotoxin, and nocodozole on mitotic spindles: implications for the role of microtubule dynamics in mitosis. *J. Cell Sci.* **102**:401–16.

Kachadorian, W.A., Ellis, S.J. and Muller, J. (1979) Possible roles for microtubules and microfilaments in ADH action on toad urinary bladder. *Am. J. Physiol.* **236**:F14–F20.

Kolber, M.A., Broschat, K.O. and Land-Gonzalez, B. (1990) Cytochalasin B induces cellular DNA fragmentation. *FASEB. J.* **4**:3021–7.

Low, I., Dancker, P. and Wieland, T. (1975) Stabilization of F-actin by phalloidin reversal of the destabilizing effect of cytochalasin B. *FEBS Letters* **54**:263–5.

Luna, E.J. and Hitt, A.L. (1992) Cytoskeleton–plasma membrane interactions. *Science* **258**:955–64.

Madara, J.L. (1987) Intestinal absorptive cell tight junctions are linked to cytoskeleton. *Am. J. Physiol.* **253**:C171–C175.

Madara, J.L. (1989) Loosening tight junctions. *J. Clin. Invest.* **83**:1089–94.

Madara, J.L., Barenberg, D. and Carlson, S. (1986) Effects of cytochalasin D on occluding junctions of intestinal absorptive cells: further evidence that the cytoskeleton may influence paracellular permeability and junctional charge selectivity. *J. Cell Biol.* **102**:2125–36.

Madara, J.L., Moore, R. and Carlson, S. (1987a) Alteration of intestinal tight junction structure and permeability by cytoskeletal contraction. *Am. J. Physiol.* **253**:C854–C861.

Madara, J.L., Moore, R. and Carlson, S. (1987b) Alteration of intestinal tight junction structure and permeability by cytoskeletal contraction. *Am. J. Physiol.* **253**:C854–C861.

Matthews, J.B., Awtrey, C.S. and Madara, J.L. (1992) Microfilament-dependent activation of $Na^+/K^+/2Cl^-$ cotransport by cAMP in intestinal epithelial monolayers. *J. Clin. Invest.* **90**:1608–13.

Matthews, J.B., Tally, K.J., Smith, J.A. and Awtrey, C.S. (1994) F-actin differentially alters epithelial transport and barrier function. *J. Surg. Res.* **56**:505–9.

Matus, A. (1994) MAP2, in *Microtubules*, (eds J.S. Hyams and C.W. Loyd), Wiley-Liss, New York, pp. 155–66.

Mays, R.W., Beck, K.A. and Nelson, W.J. (1994) Organization and function of the cytoskeleton in polarized epithelial cells: a component of the protein sorting machinery. *Curr. Opin. Cell Biol.* **6**:16–24.

McClure, W.O. and Paulson, J.C. (1977) The interaction of colchicine and some related alkaloids with rat brain tubulin. *Molec. Pharmac.* **13**:560–75.

Mercier, F., Reggio, H., Devilliers, G. *et al.* (1989) Membrane–cytoskeleton dynamics in rat parietal cells: mobilization of actin and spectrin upon stimulation of gastric acid secretion. *J. Cell Biol.* **108**:441–53.

Mihic, S.J., Whatley, V.J., McQuilkin, S.J. and Harris, R.A. (1994) B-lumicolchicine interacts with the benzodiazepine binding site to potentiate GABAA receptor-mediated currents. *J. Neurochem.* **62**:1790–4.

Mills, J.W. (1987) The cell cytoskeleton: possible role in volume control, in *Current Topics in Membranes and Transport*, (eds R. Gilles, A. Kleinzeller and L. Boles), Academic Press, New York, pp. 75–101.

Mills, J.W. and Lubin, M.L. (1986) Effect of adenosine 3′,5′-cyclic monophosphate on volume and cytoskeleton of MDCK cells. *Am. J. Physiol.* **250**:C319–C324.

Mills, J.W. and Mandel, L.J. (1994) Cytoskeletal regulation of membrane transport events. *FASEB J.* **8**:1161–5.

Mills, J.W., Schwiebert, E.M. and Stanton, B.A. (1993) Evidence for the role of actin filaments in regulating cell swelling. *J. Exp. Zool.* **268**:111–20.

Mookerjee, B.K., Cuppoletti, J., Rampal, A.L. and Jung, C.Y. (1981) The effects of cytochalasins on lymphocytes. *J. Biol. Chem.* **256**:1290–300.

Morris, A. and Tannenbaum, J. (1980) Cytochalasin D does not produce net depolymerization of actin filaments in HEp-2 cells. *Nature* **287**:637–9.

Morrow, J.S., Cianci, C.D., Ardito, T. *et al.* (1989) Ankyrin links fodrin to the alpha subunit of Na,K-ATPase in Madin–Darby Canine Kidney cells and in intact renal tubule cells. *J. Cell Biol.* **108**:455–65.

Muller, R., Kindler, S. and Garner, C.C. (1994) The MAP1 family, in *Microtubules*, (eds J.S. Hyams and C.W. Loyd), Wiley-Liss, New York, pp. 141–54.

Olmstead, J. (1991) Non-motor microtubule-associated proteins. *Curr. Opin. Cell Biol.* **3**:52–8.

Ornelles, D.A., Fey, E.G. and Penman, S. (1986) Cytochalasin releases mRNA from the cytoskeletal framework and inhibits protein synthesis. *Mol. Cell Biol.* **6**:1650–62.

Pitelka, D.R. and Taggart, B.N. (1983) Mechanical tension induces lateral movement of intramembrane components of the tight junction: studies on mouse mammary cells in culture. *J. Cell Biol.* **96**:606–12.

Prat, A.G., Bertorello, A.M., Ausiello, D.A. and Cantiello, H.F. (1993) Activation of epithelial Na^+ channels by protein kinase A requires actin filaments. *Am. J. Physiol.* **265**:C224–C233.

Rampal, A.L., Pinokofsky, H.B. and Jung, C.Y. (1980) Structure of cytochalasins and cytochalasin B binding site in human erythrocyte membranes. *Biochem.* **259**:679–83.

Ringel, I. and Horwitz, S.B. (1991) Effect of alkaline pH on taxol-microtubule interactions. *J. Pharmacol. Exp. Ther.* **259**:855–60.

Sauman, I. and Berry, S.J. (1994) An actin infrastructure is associated with eukaryotic chromosomes: stuctural and functional significance. *Eur. J. Cell Biol.* **64**:348–56.

Schiff, P.B., Fant, J. and Horwitz, S.B. (1979) Promotion of microtubule assembly *in vitro* by taxol. *Nature* **277**:665–7.

Schiff, P.B. and Horwitz, S.B. (1980) Taxol stabilizes microtubules in mouse fibroblast cells. *Proc. Natl. Acad. Sci.* **77**:1561–5.

Schwiebert, E.M., Mills, J.W. and Stanton, B.A. (1994) Actin-based cytoskeleton regulates a chloride channel and cell volume in a renal cortical collecting duct cell line. *J. Biol Chem.* **269**:7081–9.

Shapiro, M., Matthews, J., Hecht, G. *et al.* (1991) Stabilization of F-actin prevents cAMP-elicited Cl^- secretion in T84 cells. *J. Clin. Invest.* **87**:1903–9.

Shoeman, R.L. and Traub, P. (1993) Assembly of intermediate filaments. *BioEssays* **15**:605–11.

Simon, H., Gao, Y., Franki, N. and Hays, R.M. (1993) Vasopressin depolymerizes apical F-actin in rat inner medullary collecting duct. *Am. J. Physiol.* **265**:C757–C762.

Stadler, J. and Franke, W.W. (1974) Characterization of the colchicine binding of membrane fractions from rat and mouse liver. *J. Cell Biol.* **60**:297–303.

Stevenson, B.R. and Begg, D.A. (1994) Concentration-dependent effects of cytochalasin D on tight junctions and actin filaments in MDCK epithelial cells. *J. Cell Sci.* **107**:367–75.

Stewart, M. (1993) Intermediate filament structure and assembly. *Curr. Opin. Cell Biol.* **5**:3–11.

Thatte, H.S., Bridges, K.R. and Golan, D.E. (1994) Microtubule inhibitors differentially affect translational movement, cell surface expression, and endocytosis of transferrin receptors in K562 cells. *J. Cell. Physiol.* **160**:345–57.

Venetianer, A., Schiller, D.L., Magin, T. and Franke, W.W. (1983) Cessation of cytokeratin expression in a rat hepatoma cell line lacking differentiated functions. *Nature* **305**:730–3.

Volberg, T., Geiger, B., Kartenbeck, J. and Franke, W.W. (1986) Changes in membrane–microfilament interaction in cellular adherens junctions upon removal of extracellular Ca^{2+} ions. *J. Cell Biol.* **102**:1832–42.

Weber, K., Pleismann, U. and Ulrich, W. (1989) Cytoplasmic intermediate filament proteins of invertebrates are closer to nuclear lamins than are vertebrate intermediate filament proteins; sequence characterization of two muscle proteins of a nematode. *EMBO J.* **8**:3221–7.

Wilder, J.A. and Ashman, R.F. (1991) Actin polymerization in murine B lymphocytes is stimulated by cytochalasin D but not by anti-immunoglobulin. *Cellul. Immunol.* **137**:514–28.

Wilson, L. and Jordan, M.A. (1994) Pharmacological probes of microtubule function, in *Microtubules*, (eds J.S. Hyams and C.W. Loyd), Wiley-Liss, New York, pp. 59–83.

Wordeman, L. and Mitchison, T.J. (1994) Dynamics of microtubule assembly *in vivo*, in *Microtubules*, (eds J.S. Hyams and C.W. Loyd), Wiley-Liss, New York, pp. 287–301.

14

Future perspectives: molecular biology and pathophysiology

Alicia McDonough

14.1 ROLE OF EPITHELIA IN PLASMA HOMEOSTASIS

The role of epithelia in maintaining plasma homeostasis of electrolytes and non-electrolytes was reviewed in Chapter 1. This homeostasis is maintained by a balance between transport into the plasma across absorptive epithelia, and transport out of the plasma by a number of pathways including transport into cells, filtration into the kidney, or secretion across epithelia into the urine, stools or sweat. Challenges to plasma homeostasis occur constantly and the epithelia compensate by readjusting transport rates. Sodium and potassium homeostasis provide good examples because transport of these ions into the plasma varies significantly with dietary intake; hence output from the plasma must be adjusted to match input in order to maintain homeostasis.

Sodium, as the major extracellular cation, is the primary determinant of extracellular volume. When sodium intake is low, the hormonal levels of the renin–angiotensin–aldosterone axis increase to promote sodium and

Epithelial Transport: A guide to methods and experimental analysis.
Edited by Nancy K. Wills, Luis Reuss and Simon A. Lewis.
Published in 1996 by Chapman & Hall, London. ISBN 978-94-010-7168-0

water reabsorption from that filtered across the renal glomerulus into the renal tubular fluid, as well as across other aldosterone-sensitive epithelia such as the colon. When sodium intake is high, sodium and volume reabsorption from the glomerular ultrafiltrate are decreased due to elevation of hormones such as atrial natriuretic factor and dopamine, which increase sodium excretion to match intake.

While the plasma concentration of potassium is low, it is very closely regulated because of its profound effect on membrane potential, a critical determinant of cardiac contractility. Elevated potassium intake stimulates insulin secretion which acutely transports potassium into skeletal muscle, via increased transport activity of sodium pumps. In addition, secretion of potassium from the plasma into the urine across the renal tubular epithelium increases to match intake. When potassium intake is low, potassium is actively reabsorbed from the tubule fluid into the blood via an apical K^+-ATPase, decreasing urinary loss. In addition, potassium shifts from intracellular stores in muscle to the plasma mediated, at least in part, by decreased expression of Na,K-ATPase, which buffers the fall in plasma potassium levels (McDonough *et al.*, 1992) (Figure 14.1). Interestingly, there

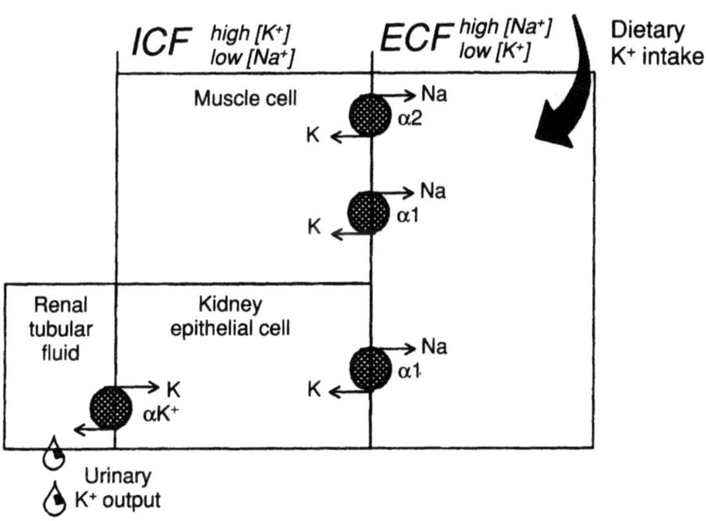

Figure 14.1 Model of factors that influence potassium homeostasis. ICF = intracellular fluid compartments; ECF = extracellular fluid. Homeostasis of potassium in ECF is maintained by a balance between dietary K^+ intake and absorption, renal K^+ excretion, and transport of K^+ between muscle ICF and ECF. α1 and α2 represent the two major isoforms of Na,K-ATPase catalytic subunit expressed in muscle; α1 is the major isoform in the kidney, expressed in basolateral membranes. αK represents the renal K^+-ATPase that drives active K^+ reabsorption during hypokalemia, expressed in apical membranes of the renal collecting duct cells. (Adapted from McDonough *et al.*, 1992.)

is an isoform specificity to the decreased sodium pump expression in hypokalemic muscle – only the α2 and β2 subunits levels decrease. Since these are preferentially expressed in muscle and fat, these tissues lose potassium preferentially, while organs that express primarily other isoforms are spared.

Changes in ion or substrate transport across an epithelium (through channels, pumps, or other transporters) can be accomplished by acute or chronic regulation of transporters (summarized in Figure 14.2). Acute (short-term) regulation of a fixed pool of transporters can be accomplished either by covalent modification of the transporters in the apical or basolateral membranes, or by redistribution of transporters between intracellular endosomal pools and the surface apical and basolateral membranes by membrane trafficking. Chronic (long-term) regulation of transport activity refers to a change in the abundance, or total cellular pool of the transporters. This can be accomplished by changing either the

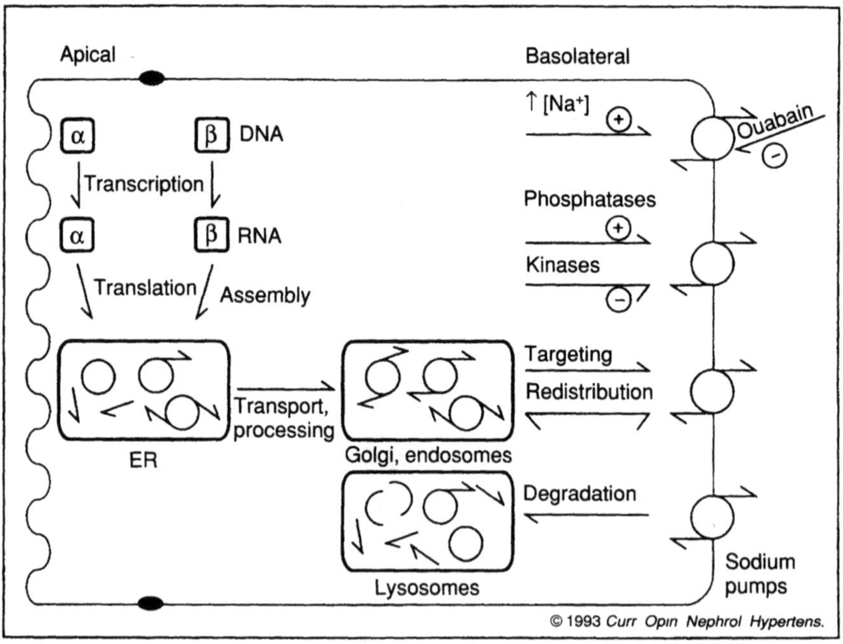

Figure 14.2 Potential sites of regulation of cellular Na,K-ATPase activity in a cell. Short-term regulation of a fixed pool of pumps can be achieved by changes in intracellular Na^+ levels, covalent modification of the pumps, redistribution of pumps to or from intracellular stores, or degradation. Long-term regulation of the pool size of the sodium pumps can be achieved by regulating transcription, translation, and assembly rates, which determine the overall synthesis rate of αβ heterodimers, or regulating degradation of Na,K-ATPase subunit mRNA or protein levels. (Reproduced from McDonough and Farley, 1993.)

synthesis or degradation rates. The goal of this chapter is to review the strategies and methods used to identify transporters and to determine how they are regulated in health and disease.

14.2 MOLECULAR METHODS USED TO ISOLATE TRANSPORTERS

14.2.1 Traditional methods

In order to study regulation of synthesis and degradation of a transporter, and to study structure–function relationships, it is important to identify the transporter's subunit protein(s), mRNA(s) and gene(s). If one of these is isolated, it can be used as a probe to isolate the other two. Traditional methods to isolate proteins, cDNAs and genes have been described in many excellent molecular biology laboratory manuals (Maniatis *et al.*, 1982; Ausubel *et al.*, 1987). These traditional strategies include three alternatives:

1. Purification of the protein of interest to homogeneity and partial sequencing of regions of the protein, followed by synthesis of a set of cDNA probes corresponding to the partial sequence, and use of these probes to screen cDNA and or genomic libraries.
2. Partial purification of a protein which is injected into mice for production of monoclonal, monospecific antibodies by myeloma hybridization; the antibodies are used to purify the protein to homogeneity by affinity chromatography, and with this the investigator can proceed as in (1).
3. Antibodies, as produced in (2), can be used to identify cDNAs by screening cDNA libraries cloned into expression vectors programmed to synthesize the proteins coded for by the cDNAs.

Blot transfer assays are often exploited to screen for clones expressing the protein or cDNA of interest. This is accomplished by placing a filter matrix on top of a field of hundreds of bacterial clones, and probing the filter with labeled cDNAs or antibodies. Blot methods are also used for further characterization of positive clones that are obtained by initial screening. In this case the starting material is fractionated by electrophoresis (Figure 14.3) to visualize specific DNA, RNA or protein producing clones (Granner, 1987).

The traditional methods described above may not be optimal for isolating transporter proteins or cDNAs which are expressed in only a very few specialized cells, or expressed in very low abundance. As an alternative strategy, epithelial transporters' mRNAs and proteins can be isolated by novel **transport expression** cloning methods which exploit the transport characteristics of the molecule. Several applications of transport expression to the cloning of transporters' cDNAs and proteins are provided below.

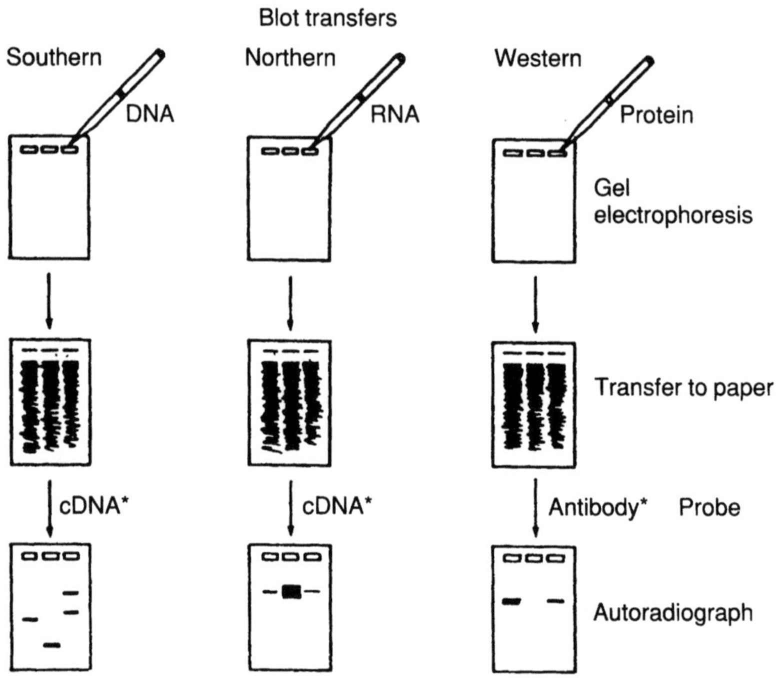

Figure 14.3 The technique of blot transfer. This is a means of visualizing a specific DNA, RNA or protein molecule among the thousands of others in a solution. DNA is first digested into smaller pieces with restriction enzymes while care is taken to retain native structure of RNA. Proteins can be resolved in their native state but are more commonly incubated with denaturants to separate multimers into individual subunits. After electrophoresis the gel is placed on a special paper, usually nitrocellulose, and the material in the gel is transferred to the paper either by simple wicking or electrophoretically. After transfer, the blot is incubated with a specific probe for the nucleic acid sequence or the protein of interest. The probes are labelled either radioactively or chemically so that they can be detected with X-ray film by autoradiography. In the case of antibodies, it is more common to add an unlabeled primary antibody and follow with a labeled secondary antibody. The autoradiographic signals can be quantitated with scanning densitometry after verification that the signal is linearly related to the amount of protein loaded. (Reproduced from Granner, 1987.)

14.2.2 Expression cloning in *Xenopus laevis* oocytes

When frog oocytes are injected with mRNAs, they are translated into proteins which can be processed, assembled as multimeric proteins and, in the case of transporter proteins, inserted into the endoplasmic reticulum and routed to the plasma membrane where transporter activity can be measured. The earliest transporters assayed in this fashion include the rat brain Na^+ channel α subunit, the Na^+/glucose cotransporter, the cardiac Na^+–Ca^{2+} exchanger, and *Torpedo* chloride channels and acetylcholine receptors (reviewed in Longoni *et al.*, 1988). This technology has also been exploited to confirm identity of transporters, such as the cardiac Na^+/Ca^{2+} exchanger, cloned by traditional methods (Nicoll *et al.*, 1990) as well as to clone transporters by screening cDNA libraries. In the latter case, the oocytes are injected with subsets of the cDNA library and screened for expression of the transport activity of interest. A cDNA for the Na^+/glucose cotransporter was isolated by screening an intestinal mucosal cDNA library for uptake of a radioactively labeled glucose analog (methyl α-D-glucopyranoside, MeGlc) that is transported by the apical Na^+/glucose cotransporter but not transported by the basolateral glucose facilitated exchanger (Hediger *et al.*, 1987). Phlorizin, a specific competitive inhibitor of the apical Na^+/glucose cotransporter, was also used to determine specificity of the transporter activity detected. Figure 14.4 summarizes the MeGlc fluxes obtained in oocytes as a function of poly A^+ RNA injection.

Both the Na^+/glucose cotransporter and the Na^+–Ca^{2+} exchanger were found to be single subunits from single genes, which simplified the cloning and functional expression of transporter activity. However, the same technique has been successfully applied to cloning the multiple subunit amiloride-sensitive epithelial Na^+ channel: Canessa *et al.* (1994) injected with RNA transcribed from cDNAs prepared from rat distal colon poly A^+ RNA of rats maintained on a low-salt diet to increase the transport activity of this channel. The oocytes were screened for expression by assay of amiloride-sensitive sodium current, measured in whole cells by the voltage clamp technique. A single clone, α, was initially isolated, but when α cRNA was injected into the oocytes, the amiloride-sensitive current was extremely low compared with that seen with the starting poly A^+ RNA. Canessa *et al.* (1994) concluded that the sodium channel was composed of multiple subunits. To find them, they rescreened the cDNA library by *functional complementation* with oocytes co-injected with α cRNA. They isolated two additional subunits, β and γ, that increased amiloride-sensitive current over that measured with α alone, but the current was still far lower than that seen with poly A^+ RNA. Finally, they determined that when the cRNAs of all three subunits were co-injected, amiloride-sensitive current was increased 10–20-fold higher than that seen with just αβ or αγ, and five times higher than that observed with poly A^+ RNA. One of the surprising discoveries was that the three subunits were actually homologous to one another structurally, sharing more than 30% protein identity.

Transporters with multiple subunits are inherently difficult to isolate. Canessa *et al.* were fortunate that a finite amiloride-sensitive current was detected when α subunit was expressed alone, and that the current was

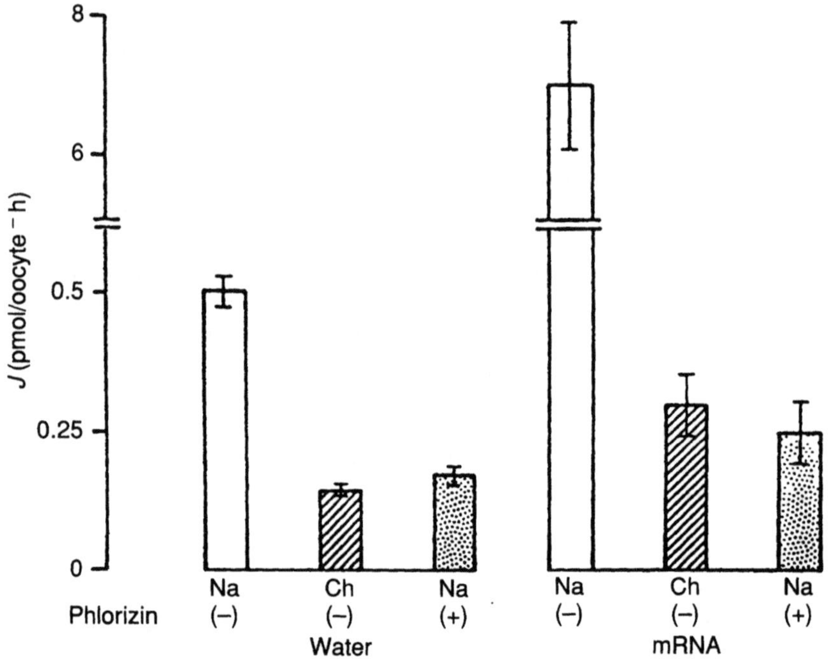

Figure 14.4 Expression of rabbit intestinal MeGlc transport in *Xenopus* oocytes. Oocytes were injected with 50 nl of water or water containing 1 μg of rabbit intestinal poly(A)$^+$ RNA per μl. After 3 days of incubation, the uptake of 50 μM MeGlc was measured in the presence of 100 mM NaCl (open bars), 100 mM choline chloride (hatched bars) and 100 mM NaCl/0.5 mM phlorizin (stippled bars). Uptakes are presented as the means obtained with 5–7 oocytes, and the bars indicate standard error of measurement. (Reproduced from Hediger *et al.*, 1987.)

increased with the co-expression of only one of the other two subunits. It is possible to imagine a transporter that is functional only when all the subunits are co-expressed. Even when a transport activity is detected after injection of a specific single clone, can an investigator conclude that the transporter consists of a single subunit? Not always. For example, when only the β subunit of Na,K-ATPase is injected into oocytes, Na,K-ATPase activity in the plasma membrane increases significantly. An investigator might conclude that β alone is the sodium pump. However, the oocytes synthesize a significant excess of Na,K-ATPase α subunit that is usually degraded unless it is stabilized by association with β. After αβ assembly, the heterodimer is transported to the plasma membrane where it operates as a sodium pump (reviewed in McDonough *et al.*, 1990). In summary, the alternative possibilities must be considered, and combinations of multiple methods should be employed to verify the structure–function connection. For example, β subunit expression alone will not produce functional sodium pumps in yeasts, which do not normally express sodium pumps; both α and β subunits must be co-expressed (McDonough *et al.*, 1990).

14.2.3 Genetic approach to isolating membrane transporters

Pouyssegur and colleagues (1987) developed an innovative strategy to clone a sodium hydrogen exchanger which also exploited its transport properties. First, they isolated fibroblasts that lacked sodium–hydrogen exchangers by selecting for cells that could survive a H^+-induced suicide protocol. Cells were first loaded with 90 mM Li^+ and then exposed to medium pH of 5.5. Under these conditions cells that have functional Na^+–H^+ exchangers take up H^+ very rapidly and cell viability drops within minutes. Following two rounds of this screening, cells lacking functional exchangers were isolated. These cells were then transfected with human DNA and screened for rescue of a functional Na^+/H^+ antiporter (Franchi and Pouyssegur, 1986). The human DNA that complemented the antiporter negative cells was subjected to three rounds of re-transfection. From these cells, they selected a clone that over-expressed sodium–hydrogen exchanger activity. They used this DNA clone to characterize tissue mRNAs by RNA blot hybridization, and proceeded to isolate and sequence the cDNA for the exchanger (Sardet *et al.*, 1989).

14.2.4 Finding isoforms by low-stringency screening

Once a transporter has been cloned, an investigator may want to determine if there are related transporters that may be expressed in a tissue-specific or development-specific pattern, or subject to differential regulation. The most successful strategy for finding such isoforms has been to screen cDNA libraries at low stringency with the cDNA to the transporter of interest. Low stringency indicates screening under conditions in which the labeled cDNA and highly related sequences will remain hybridized despite mismatches in sequences. The stringency is determined by both the hybridization temperature and the salt concentration of the hybridization wash buffer. The concept of low and high stringency is reviewed in the classic molecular biology texts cited at the beginning of the chapter. This approach has been useful for isolating isoforms of sodium pumps, sodium–hydrogen exchangers and the steroid–thyroid hormone gene family, among many other classes of proteins.

14.3 USING MOLECULAR PROBES TO STUDY REGULATION OF TRANSPORTERS

Once a transporter has been cloned, an investigator can study its biosynthesis and regulation with specific cDNA and antibody probes. There are many control points at which transport can be regulated. In assessing adaptive changes in transporter activity in health and pathologic changes in disease states, the investigator often aims to determine the mechanism(s) responsible for the change. In the final section of this chapter we will examine three diseases of altered ion transport. When approaching such studies, the investigator hypothesizes that either the transporter itself or a cellular process that regulates transport activity is altered from that found in the normal control. Transport activity can be regulated at various levels including pool size of the transport protein, cellular distribution of the transport protein, and activity per transporter. To determine which levels may be affected, the kinetic model of eukaryotic gene expression developed by Hargrove (1993) is useful (Figure 14.5). This model consists of a series of pools of intermediates from the initial transcript to the final protein, or transport protein. The pool size can be regulated at multiple levels including gene transcription (mRNA synthesis), mRNA degradation, mRNA translation into protein, and protein degradation. If the transporter is composed of multiple subunits, such as the sodium pump or epithelial sodium channel, the model can be modified to include also the kinetics of subunit assembly (reviewed in McDonough *et al.*, 1992).

The transcription rate and decay rates of nuclear and cytoplasmic mRNAs can be measured with a cDNA probe by radiolabeling nascent mRNA with ^{32}P-UTP, and isolating the labeled mRNA specific for the protein of interest by hybridizing the labeled sample with the specific cDNA. Synthesis rates of the RNAs are measured after a defined pulse labeling period (of nuclei or cells) and decay rates are measured at defined times after the label has been removed or chased. This is known as a pulse-chase protocol. A labeled cDNA can also be used to measure total mRNA pool size from cell lysates by Northern blot as described in Figure 14.3.

Protein synthesis and degradation rates can be measured with an antibody probe for the transporter protein by radiolabeling nascent proteins in intact cells with ^{35}S-methionine (using the same pulse and pulse-chase strategies described above) and then isolating the protein of interest in cell lysates by immunoprecipitation with the specific antibody. The antibody can also be used to measure relative protein pool sizes in cell lysates by Western blot as described in Figure 14.3.

With the assays described in this section, the pool sizes and rates connecting the pools, described in Figure 14.5, can be evaluated. For example, during hypokalemia there is a decrease in skeletal muscle

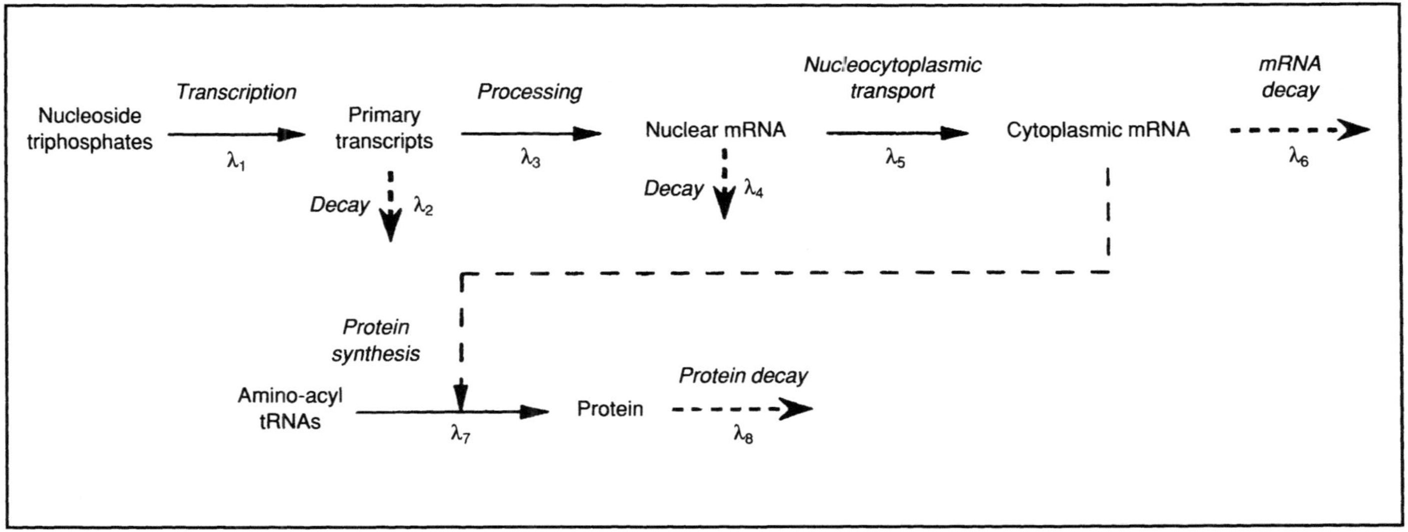

Figure 14.5 A kinetic model of eukaryotic gene expression. A model is depicted in which the rate of each subprocess involved in gene expression is equal to the concentration of each intermediate multiplied by the individual rate constant for conversion or decay, here designated λ with a subscript. The units of each first order rate constant are time $^{-1}$, and decay is assumed to be exponential. During initial conditions, the rate of transcription is assumed to be fixed (zero order) so that a constant amount of mRNA is produced per unit of time until the gene is activated to a new state, at which time a higher or lower rate of formation begins. In this figure, solid arrows indicate flow of material into a pool, and dashed arrows indicate elimination. The arrow that connects the cytoplasmic mRNA pool to the process of protein synthesis indicates that the rate of protein synthesis is proportional to the mRNA concentration. (Reproduced from Hargrove, 1993.)

sodium pump α2 and β2 subunit proteins. The levels of α2 mRNA decrease 35% relative to control pool size, while α2 protein levels decrease 80% (reviewed in McDonough *et al.*, 1992). These results demonstrate that the change in mRNA pool size cannot account for the change in protein pool size, suggesting other, perhaps more significant, control points such as mRNA translation, subunit assembly or protein degradation. With the additional measurements of transporter activity per cell or per cell protein, as described in previous chapters, an investigator can determine if transport is altered because of a change in the pool size of that transporter, because the activity per transporter is altered, or a combination of the two. If the transport activity per cell is altered but the pool size is not, the investigator will also want to determine whether the cellular location of the transporter is altered (Chapter 9).

14.4 GENETIC DISEASES OF ALTERED ION TRANSPORT

Some of the most common genetic disorders are diseases of altered ion transport, including cystic fibrosis, polycystic kidney disease and hypertension. In addition, there are infectious agents such as cholera that alter epithelial function so dramatically that the infection is fatal if not treated rapidly. Finally, there are many deadly toxins and venoms that target ion channels or pumps as their site of action. This section will examine only the category of genetic disorders of ion transport, and will outline strategies to determine the nature of the mutation in the DNA sequence, and how the mutation leads to the disease phenotype.

14.4.1 Cystic fibrosis (CF)

CF is a common and often lethal genetic abnormality in the Caucasian population. In the United States alone, one out of every 1600–2000 Caucasian babies born are affected. The major clinical symptoms of CF are elevated Na^+ and Cl^- in sweat (the primary indicator of the disease), chronic obstructive pulmonary disease and pancreatic insufficiency, and *Pseudomonas aeruginosa* infections of the airways.

Quinton (1983) aimed to determine why NaCl was elevated in sweat of CF patients. It was found that, while the secretion of NaCl by the acinar cells was the same in normal and CF patients, the reabsorption of NaCl along the reabsorptive duct was depressed because Cl^- permeability, measured in isolated sweat ductal epithelium, was almost an order of magnitude lower in CF patients compared with permeability of ducts from normal individuals. The pulmonary and pancreatic diseases are characterized by very thickened secretions in CF individuals. The reduced Cl^- permeability of these secretory epithelia will likely reduce osmotic water flow, which would contribute to thickened secretions. In studies performed on cultured cells, it was demonstrated that Cl^- channels from normal cells were activated by cAMP-dependent signal transduction pathways, whereas channels of cells from CF individuals were refractory to stimulation. In particular, it was noted that the activation of Cl^- channels by exogenous protein kinase A was defective in excised CF membrane patches (reviewed in Fuller and Benos, 1992).

Riordan *et al.* (1989) identified the mutated gene that defines this genetic abnormality by procedures, called positional cloning, that did not require knowledge of the gene product or its function. In short, the locus of the genetic defect was identified by assessing the linkage of DNA markers to the presence of the disease within the affected individuals. The gene, on chromosome 7, encodes a large membrane protein. It was named the cystic fibrosis transmembrane regulator (CFTR), since while investigators could show that it affected Cl^- transport, they could not distinguish between a transporter and a transport regulator. Introduction of the normal gene into cultured cells from a patient with CF corrected the anion transport defect of those cells.

The primary structure of CFTR predicts: an integral membrane protein that spans the membrane 12 times; two nucleotide binding domains that participate in ATP binding and hydrolysis; and a regulatory domain which contains a number of consensus sites for phosphorylation by protein kinase A and C (Figure 14.6). The most common CF mutation is in one of the nucleotide binding domains. In addition to abnormal cAMP dependent conductance properties, mutant CFTR proteins are accumulated in the endoplasmic reticulum and are not transported to the cell surface very efficiently.

As well as its role in Cl^- permeability in epithelial cells, CFTR appears to have other functions within intracellular membranes. CFTR is targeted to both apical membrane and sub-apical vesicles with marker characteristics of early endosomes (Webster *et al.*, 1994). Cell surface CFTR is rapidly and selectively internalized to this early endosomal population, and elevating cAMP levels (which activates CFTR Cl^- channel activity) inhibits CFTR internalization. In these endosomes, CFTR may play a role in acidification of intracellular organelles by acting as the counter-ion conductance for endosomal acidification by H^+ to maintain electroneutrality. These studies indicate the potential for additional roles for CFTR besides Cl^- conductance pathway on the surface. In internal membranes it may regulate vesicular trafficking and acidification (reviewed in Bradbury and Bridges, 1994). Alterations in these processes in CFTR have the potential to impair mucus production and secretion, which are characteristic of the disease. In summary, although the CF gene has been cloned and its transport characteristics have been defined, questions remain regarding the connection between the abnormal Cl^- conductance regulation observed in CF and the obstructive pulmonary disease and pancreatic insufficiency.

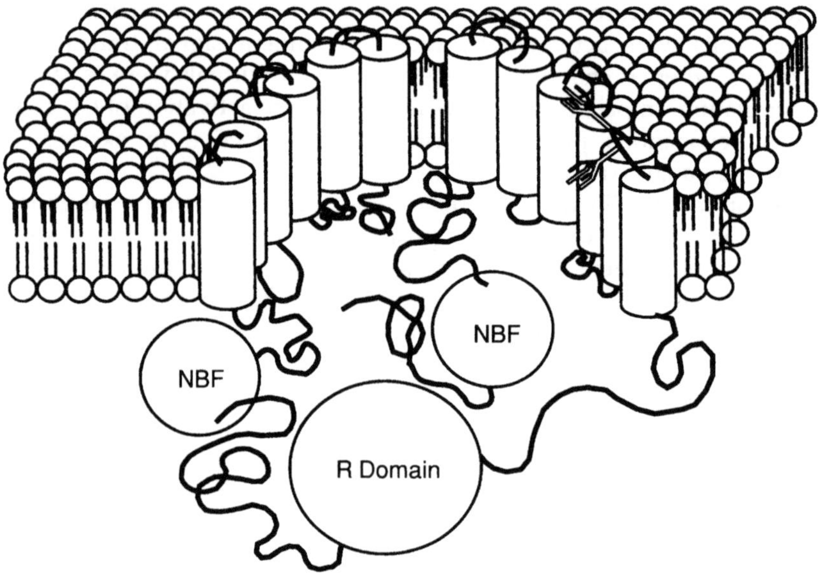

Figure 14.6 Putative model of CFTR protein as determined by hydropathy analysis of cDNA sequence. Protein consists of a doubly repeating motif comprising six membrane spanning regions and a nucleotide binding fold (NBF). The two halves of the molecule are linked by the R domain, a unique highly charged region of proposed regulatory function. The most common mutation found in cystic fibrosis is located in the first NBF. (Reproduced from Fuller and Benos, 1992, adapted from Collins, Riordan and Tsui, 1990.)

14.4.2 Autosomal dominant polycystic kidney disease (ADPKD)

ADPKD is one of the most common genetic diseases in humans, affecting approximately one in 1000 individuals. Its major feature is cystic kidneys that can lead to renal failure in adult life. Cysts in other epithelia such as liver and pancreas are also observed. Differences between normal and cystic kidneys that may be potential causes of cyst formation include changes in cell polarity, secretion, extracellular matrix and abnormal epithelial cell growth (reviewed in Wilson and Sherwood, 1991). An international collaborative team, the European Polycystic Kidney Disease Consortium (1994), has identified a gene, termed PKD1, that is responsible for producing most cases of ADPKD. The approach taken was similar to that used to isolate CFTR. The nucleotide sequence of the PKD1 sequence bears no homology to known proteins, so it remains to be determined what the gene encodes and how mutations in the gene lead to the phenotype. This situation points to the importance of further analysis of both the phenotypic characteristics of ADPKD, using methods described in this book (e.g. Chapters 4 and 5), as well as further characterization of the PKD1 gene in order to determine its role in normal cell function.

14.4.3 Hypertension

Hypertension (elevated blood pressure) is a common disorder that increases the risk of stroke, myocardial infarction and renal failure. Although the study of hypertension has yielded much information about the factors that are associated with, impact or result from hypertension, the pathophysiology of the disease is only barely understood. Many studies have concluded that a significant fraction of the variation in blood pressure is genetically determined. However, identification of genes contributing to elevated blood pressure is very complicated for three reasons:

1. Unlike the simple Mendelian genetics which facilitated the identification of CFTR and ADPKD genes, multiple environmental as well as genetic factors contribute to blood pressure and make cause–effect relationships difficult to ascertain in humans, especially once hypertension is established.
2. Blood pressure is continuously distributed in the population, making the definition of hypertension vs. normotensive artificial.
3. There is a late-onset to the elevation in blood pressure.

Thus, investigators have approached the study of hypertension by a number of routes:

- They have tested for mutations in candidate genes for potential association with hypertensive phenotypes. There are several dozen such candidate genes that may play a role in blood pressure regulation, and there are examples of single Mendelian mutations that lead to hypertension in small populations of hypertensives.
- At the other end of the spectrum is analysis for linkage of the phenotype with anonymous DNA markers. They have sought to simplify the analysis by identifying populations with intermediate phenotypes, such as salt sensitivity or insulin resistance/hyperinsulinemia, that are highly correlated with hypertension.
- They have developed animal models of hypertension that identify candidate loci or genes that can then be studied in human hypertension. The identification of genetic loci contributing to complex traits is greatly simplified in genetic crosses between inbred animals because the issues of genetic heterogeneity and environmental factors are eliminated.

Examples of human and animal studies are provided below. These strategies have been reviewed by Lifton and Jeunemaitre (1993).

One of the main determinants of blood pressure variation is the activity of the renin–angiotensin–aldosterone hormonal axis, and genes controlling the production or regulation of these hormonal systems are identified as candidate genes because of their ability to cause hypertension by promoting renal sodium reabsorption and vascular contractility. In particular, a mutation causing the Mendelian disorder known as glucocorticoid remedial hypertension (GRA) has recently been identified as a gene duplication that resulted in ectopic expression of aldosterone synthase (Lifton and Jeunemaitre, 1993). This mutation showed complete linkage to GRA: every GRA patient studied had this mutation, while none of the patients who did not have GRA had this mutation. However, as expected, some of the patients who did not have GRA were hypertensive due to other causes.

There are also examples of hypertension associated with very low levels of renin–angiotensin–aldosterone. One such subset is Liddle's syndrome, another more simple Mendelian form of hypertension. This autosomal dominant form of human hypertension affects only a small number of individuals who express phenotypes suggesting constitutive activation of the amiloride-sensitive renal epithelial sodium channel. Thus, the renal sodium channel was tested as a candidate gene that could be responsible for Liddle's syndrome. Shimkets *et al.* (1994) used the rat epithelial sodium channel cDNAs to isolate the human subunit genes. With this they first demonstrated complete linkage of the gene encoding β subunit to Liddle's syndrome, and then determined that the carboxy terminus of the β subunit was truncated. In addition to demonstrating that a mutation in the sodium channel could lead to hypertension, these results also demonstrated that the sodium channel cloned by Canessa *et al.* (1994) was indeed functional, and suggested a role of the β subunit in normal regulation of epithelial sodium channel activity. Whether other types of low renin hypertension involve mutations in this sodium channel remains to be investigated.

To simplify the study of the hypertensive phenotype, a number of strains of hypertensive rats have been developed. Typical of these strains is the Milan Hypertensive Strain (MHS) which was selected for elevated blood pressure while the Milan Normotensive Strain (MNS) was selected for normal blood pressure, both from the same inbred colony. The MHS is characterized by a generalized defect in the structure and function of plasma membranes which affects the transport of several ions in the kidney including Na^+,K^+-ATPase, Na^+, glucose co-transporter, and Na^+-K^+-Cl^- cotransporter activities which are all significantly elevated (reviewed in Bianchi *et al.*, 1994). Even the red blood cells from the MHS exhibited similarly elevated transport characteristics. Bianchi and colleagues, who developed this strain, postulated that there was a defect in a plasma membrane component that would affect all these transporters in multiple tissues, and aimed to generate antibodies to such a defective component by immunizing MNS rats with red cell membranes from

MHS. The MNS produced antibodies to the cytoskeletal protein adducin. These investigators went on to demonstrate that two point mutations within adducin are involved in the blood pressure variation in the MHS rat (Bianchi *et al.*, 1994). They also recently analyzed the α-adducin locus in hypertensive and normotensive humans, by determining the association between blood pressure variability and the distance between the α-adducin locus and four genetic markers surrounding this locus. Blood pressure variability related to the DNA marker genotype decreased exponentially with distance from the α-adducin marker, supporting the hypothesis that variability in the α-adducin locus contributes to blood pressure variability in humans (Casari *et al.*, 1994). How the mutation in adducin leads to elevated transport activity observed in the MHS remains to be determined. One possibility is that the adducin mutation increases either the stability of the ion transporters in the plasma membrane, or the percentage that is expressed in the surface vs. intracellular stores.

In summary, elevated sodium transport, and accompanying hypertension, can be the result of mutations in the renal sodium channel, in hormones controlling sodium reabsorption, in structural proteins associated with the plasma membrane, or in subtle combinations in all three. Further investigations into the genetics of hypertension will undoubtedly reveal many additional mutations in transporters, their biosynthesis, or their regulation.

REFERENCES

Ausubel, F.M., Brent, R., Kingston, R.E. *et al.* (1987) *Current Protocols in Molecular Biology*, John Wiley and Sons, New York.

Bianchi, G, Tripodi, G., Casari, G. *et al.* (1994) Two point mutations within the adducin genes are involved in blood pressure variation. *Proc. Natl. Acad. Sci. USA* **91**:3999–4003.

Bradbury, N.A. and Bridges, R.J. (1994) Role of membrane trafficking in plasma membrane solute transport. *Am. J. Physiol.* **267**:C1–C24.

Canessa, C.M., Schild, L., Buell, G. *et al.* (1994) Amiloride-sensitive epithelial Na^+ channel is made of three homologous subunits. *Nature* **367**:463–7.

Casari, G.C., Cusi, D., Stella, P. *et al.* (1994) Association of α-adducin locus to human essential hypertension. *Hypertension* **24**:387.

Collins, F.S., Riordan, J.R. and Tsui, L.-C. (1990) The cystic fibrosis gene: isolation and significance. *Hospital Practice* **25**:47–57.

European Polycystic Kidney Disease Consortium (1994) The polycystic kidney disease 1 gene encodes a 1 kb transcript and lies within a duplicated region on chromosome 16. *Cell* **77**:881–94.

Franchi, A., Perucca-Lostanlen, D. and Pouyssegur, J. (1986) Functional expression of a human Na^+/H^+ antiporter gene transfected into antiporter-deficient mouse L-cells. *Proc. Natl. Acad. Sci. USA* **83**(24):9388–92.

Fuller, C.M. and Benos, D.J. (1992) CFTR! *Am. J. Physiol.* **263**:C267–C286.

Granner, D.K. (1987) The molecular biology of insulin action on protein synthesis. *Kidney International* **32**:S82–S93.

Hargrove, J.L. (1993) Microcomputer-assisted kinetic modeling of mammalian gene expression. *FASEB Journal* **7**:1163–70.

Hediger, M.A., Ikeda, T., Coady, M. *et al.* (1987) Expression of size-selected mRNA encoding the intestinal Na^+/glucose cotransporter in *Xenopus laevis* oocytes. *Proc. Natl. Acad. Sci. USA* **84**:2634–7.

Lifton, R.P. and Jeunemaitre, X. (1993) Finding genes that cause human hypertension. *J. Hypertension* **11**:231–6.

Longoni, S., Coady, M.J., Ikeda, T. *et al.* (1988) Expression of cardiac sarcolemmal Na^+–Ca^{++} exchange activity in *Xenopus laevis* oocytes. *Am. J. Physiol.* **255**:C870–C873.

Maniatis, T., Fritsch, E.F. and Sambrook, J. (1982) *Molecular Cloning. A Laboratory Manual*, Cold Springs Harbor Laboratory, New York.

McDonough, A.A. and Farley, R.A. (1993) Regulation of Na,K-ATPase activity. *Current Op. in Nephrology* **2**:725–34.

McDonough, A.A., Geering, K. and Farley, R.A. (1990) The sodium pump needs its subunit. *FASEB Journal* **4**:1598–1605.

McDonough, A.A., Azuma, K.K., Lescale-Matys, L. *et al.* (1992) Physiologic rationale for multiple sodium pump isoforms. *Ann. NY Acad. Sci.* **671**:156–69.

Nicoll, D.A., Longoni, S. and Philipson, K.D. (1990) Molecular cloning and functional expression of the cardiac sarcolemmal Na^+-Ca^{++} exchanger. *Science* **250**:562–565.

Pouyssegur, J., Franchi, A., L'Allemain, G. *et al.* (1987) Genetic approach to structure, function and regulation of the Na^+/H^+ antiporter. *Kidney International* **32**:S144–S149.

Quinton, P. (1983) Chloride impermeability in cystic fibrosis. *Nature*, London **301**:421–2.

Riordan, J.R., Rommens, J.M., Kerem, B.-S. *et al.* (1989) Identification of the cystic fibrosis gene: cloning and characterization of complementary cDNA. *Science Wash. D.C.* **245**:1066–73.

Sardet, C., Franchi, A. and Pouyssegur, J. (1989) Molecular cloning, primary structure, and expression of the human growth factor-activatable Na^+/H^+ antiporter. *Cell* **56**:271–80.

Shimkets, R.A., Warnock, D.G., Bositis, C.M. *et al.* (1994) Liddle's syndrome: heritable human hypertension cause by mutations in the β-subunit of the epithelial sodium channel. *Cell* **79**:407–14.

Webster, P., Vanacore, L., Nairn, A.C. *et al.* (1994) Subcellular localization of CFTR to endosomes in a ductal epithelium. *Am. J. Physiol.* **267**:C340–C348.

Wilson, P.D. and Sherwood, A.C. (1991) Tubulocystic epithelium. *Kidney International* **39**:450–63.

Appendix A: Instrumentation

Instrument	*Comments*
Current/Voltage Clamp	
DVC 1000 voltage current clamp [1]	High compliance voltage, i.e. can pass large currents. Very stable. Slow response time (seconds).
EC825, EC825LV [2]	High/low compliance. Variable response time (very rapid to slow).
558C–5[3]	Rapid response Low compliance.
VCC600, VCCM6 [4]	Low compliance, multichannel (1–6 clamp modules). Rapid response time.
LNVC21 [6]	Very low noise.
IVC100 [6]	Voltage clamp for impedance analysis.
Current Clamp	
EVOM™ [1]	Used to test confluence of tissue culture cells.
Electrodes	
Ag/AgCl [1]	Silver wire or as preformed pellets.
Ag/AgCl [2]	Silver wire or as preformed pellets.
STX-2 (chop sticks) [1]	Only gives qualitative measure of tissue resistance (see Chapter 11 for details).
Chambers	
General purpose	
Ussing type (Figure 5.1a) [1]	Causes edge damage. Slow solution changes.
Ussing type (Figure 5.1b) 2	Slow solution changes. Can be used for tissue culture inserts.
Inverted microscope (Figure 5.1c) 2	Difficult to control chamber temperature. Can be used for tissue cultured cells.
Tissue culture	
Mini perfusion system [1]	Similar to Ussing chamber.
Endohm [1]	Electrode system for measuring voltage and resistance of cultured cells using EVOM.
Ussing chamber [4]	Multichannel (Precision Instrument Design)

Instrument	*Comments*
2	Same as general purpose chambers above (Figure 5.1b,c)
Data Acquisition Systems	
PP -50LAB/1 [2]	IBM based, using a Scientific Solutions™ A/D board (software developed by authors). A/D compatible with patch clamp software.
MP100WS [1]	16 bit resolution. Macintosh™ plus based system.
MacLab [5]	12 bit resolution, with software controlled gains.
Acquire & Analyse [4]	Windows 3.1, records 1–8 tissues.

Sources:

1. World Precision Instruments, 175 Sarasota Center Boulevard, Sarasota, FL 34240-9258 (Tele 813-371-1003) (FAX 813-377-5428)
2. Warner Instruments, 1125 Dixwell Avenue, Hamden, CT 06514 (Tele 203-776-0664) (FAX 203-776-1278)
3. Bioengineering Department, University of Iowa, Iowa City, Iowa 52242 (Tele 319-335-8644) (FAX 319-335-8642)
4. Physiologic Instruments, 6780 Mirmar Road, Ste. 103-226, San Diego, CA 92121. (Tele/FAX 619-451-8845).
5. AD Instruments, PO Box 845, Milford, MA 01757
6. Gert Raskin, Electronics Department, University of Leuven, Campus Gasthuisberg, B-3000 Leuven, Belgium. (Tele 32-16-345720) (FAX 32-16-345991)

Appendix B: Solving simple equivalent circuits

The purpose of this appendix is to provide ground rules for solving simple electrical circuits composed of resistors and batteries. Circuits composed only of resistors are considered first, then circuits composed of resistors and batteries in both serial and parallel combinations.

RESISTORS

Rule One

Series resistors are summed together. As an example, the total resistance (R_t) of a series arrangement of six resistors (numbered 1–6) is equal to:

$$R_t = R_1 + R_2 + R_3 + R_4 + R_5 + R_6$$

Rule Two

Parallel resistors are most easily handled as conductors ($R = 1/G$). Parallel conductors are summed together. The total conductance (G_t) of three parallel resistors is:

$$G_t = G_1 + G_2 + G_3$$

and the resistance is:

$$1/R_t = (1/R_1) + (1/R_2) + (1/R_3)$$

or:

$$R_t = \frac{R_1 R_2 R_3}{R_1 R_2 + R_1 R_3 + R_2 R_3}$$

Rule Three

In a circuit composed of a complex arrangement of resistors, reduce the complexity by summing series resistors and parallel conductors. As a rule of thumb, collapse a parallel network (without any intervening series elements) into a single resistor. Next collapse all series elements (without any intervening parallel elements) into a series resistor. Continue this process until the circuit has been reduced to a single resistive element.

Below is an example of this process. There are four (numbered) steps involved in reducing this circuit to a single resistor.

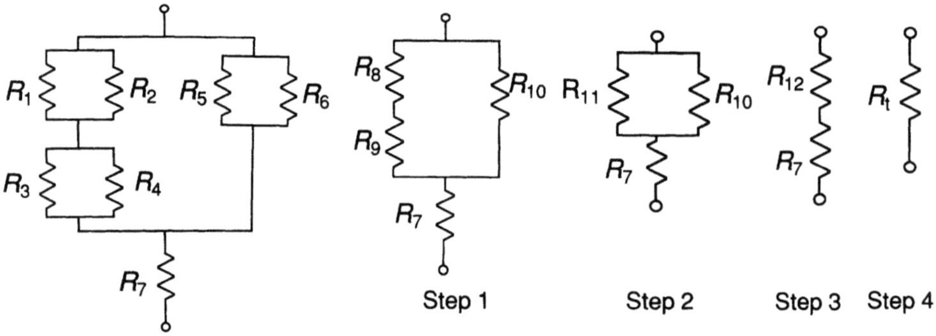

Step 1 is to collapse each of the parallel networks into three individual resistors (R_8, R_9, R_{10}):

$$R_8 = \frac{R_1 R_2}{R_1 + R_2} \qquad R_9 = \frac{R_3 R_4}{R_3 + R_4} \qquad R_{10} = \frac{R_5 R_6}{R_5 + R_6}$$

Step 2 is to collapse the left branch series circuit into a single resistor (R_{11}):

$$R_{11} = R_8 + R_9$$

Step 3 is to collapse the parallel arrangement of resistors into a single resistor (R_{12}):

$$R_{12} = \frac{R_{11} R_{10}}{R_{10} + R_{11}}$$

Step 4 is to collapse the series resistors into a single element:

$$R_t = R_{12} + R_7$$

The resistance of this circuit is equal to:

$$R_t = \frac{(R_8 + R_9)R_{10}}{R_8 + R_9 + R_{10}} + R_7$$

The final step is to substitute into the above equation the values for R_8, R_9 and R_{10}.

VOLTAGE/CURRENT SOURCES

Rule One

The ideal voltage source has no internal resistance. Thus a voltage source will produce a constant voltage independent of the load that is imposed. There is no such thing as an ideal voltage source. Thus typically a voltage source is placed in series with a resistor (this is a battery).

Rule Two

Series voltage sources (and batteries) add together.

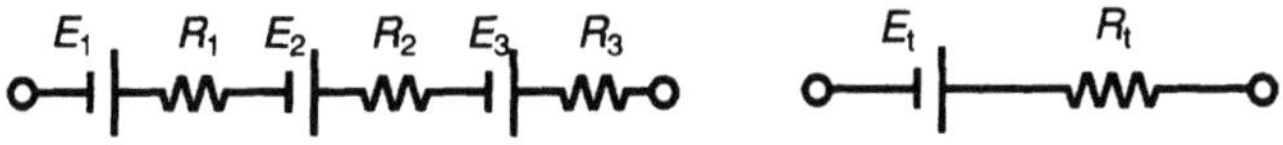

This circuit reduces to a single resistor and voltage source with values of $E_t = E_1 + E_2 + E_3$ and $R_t = R_1 + R_2 + R_3$.

Rule Three

A series combination of a resistor and a voltage source (a Thevenin equivalent circuit) can be converted to a current source in parallel with the resistor (Norton equivalent circuit).

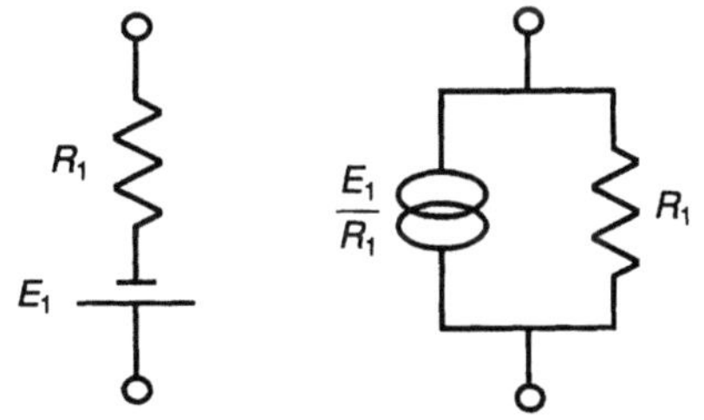

Rule Four

A current source has infinite resistance and will pass a constant current independent of the load that is imposed. Parallel current sources add.

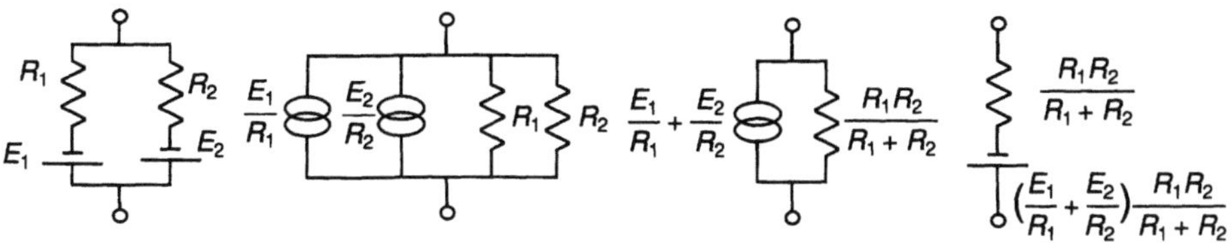

Below is an example of how to solve for an epithelial electric circuit. The process is similar to that for solving the resistor model shown above, but in this case it requires three steps.

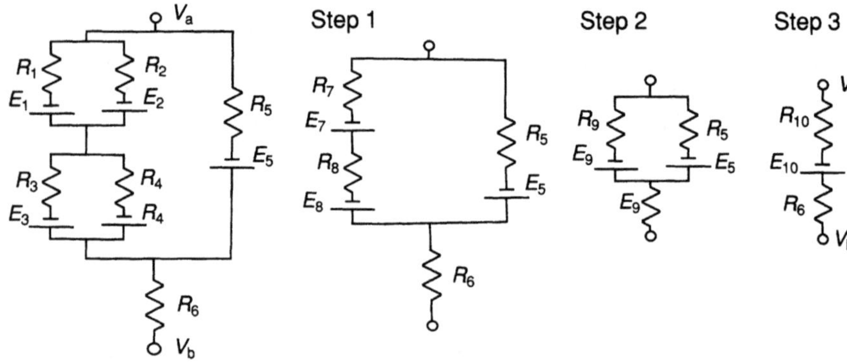

Step 1 The parallel voltage sources are first converted to Norton equivalents, the current sources summed and the resulting circuit converted to voltage sources.

$$R_7 = \frac{R_1 R_2}{R_1 + R_2} \qquad R_8 = \frac{R_3 R_4}{R_3 + R_4}$$

$$E_7 = (\frac{E_1}{R_1} + \frac{E_2}{R_2})\, R_7 \qquad E_8 = (\frac{E_3}{R_3} + \frac{E_4}{R_4})\, R_8$$

Step 2 The two series voltage sources in the left limb are summed.

$$R_9 = R_7 + R_8$$

$$E_9 = E_7 + E_8$$

Step 3 The two parallel voltage sources are converted to current sources (Norton equivalent circuits) and summed.

$$R_{10} = \frac{R_9 R_5}{R_9 + R_5}$$

$$E_{10} = (\frac{E_9}{R_9} + \frac{E_5}{R_5})\, R_{10}$$

The resulting equation is then:

$$V_a - V_b = E_{10} = \left(\frac{\left(\frac{E_1}{R_1} + \frac{E_2}{R_2}\right)\frac{R_1 R_2}{R_1 + R_2} + \left(\frac{E_3}{R_3} + \frac{E_4}{R_4}\right)\frac{R_3 R_4}{R_3 + R_4}}{\frac{R_1 R_2}{R_1 + R_2} + \frac{R_3 R_4}{R_3 + R_4}} + \frac{E_5}{R_5} \right) \left(\frac{\left(\frac{R_1 R_2}{R_1 + R_2} + \frac{R_3 R_4}{R_3 + R_4}\right) R_5}{\frac{R_1 R_2}{R_1 + R_2} + \frac{R_3 R_4}{R_3 + R_4} + R_5} \right)$$

Appendix C: A spread-sheet for marker distribution data

A portion of a spread-sheet for organizing marker distribution data is illustrated in the figure, which illustrates:

(a) **Constant parameters:** a block of cells that contain information that is used in the calculations related to all the markers that will be analyzed.
(b) **Marker 1:** a block of cells devoted to the analysis of a marker enzyme which produces a colored product with known extinction coefficient.

Similar blocks are used for other marker enzymes that produce colored products and, with obvious modifications, enzymes that produce fluorescent products, enzymes that produce radiolabeled products, receptors that bind radioligands, and immunoreactivities that yield densitometric signals after dot, slot or Western blotting.

It is convenient to have identifying information about the experiment printed on each page of hard copy. This information can be entered in response to the queries labeled **Experiment title**, **Experiment subtitle**, **Date** and **Experimental Sample**, and automatically copied to the appropriate cells for each marker. The experimental parameters that will affect all calculations include:

- **Fraction of the Tissue Analyzed**
- **Sample Volume** (total volume of each fraction)
- **Sample Dilution Factor** (factors by which samples of the fractions are diluted when they are set aside for marker determinations).

For intermediate fractions such as S_o, which gives rise to the series of pellets (**Sum** P_i) and pooled supernatants (**Sum** S_i) from the density gradient fractions, calculations of the marker recovery depend on the percentage of the total fraction volume which is carried forward to the density gradient centrifugation analysis. The spread-sheet is designed to calculate this value where required.

To calculate the total marker enzyme activity (in units of nmol/h) in a sample, it is necessary to incorporate the following values (pg. 332):

Experiment title: T1
Experiment subtitle: T2
Date: D

Experimental Sample: ES

CONSTANT PARAMETERS

Fraction of Tissue Analyzed: —

	Sample Volume (ml)	Sample Dilution Factor	Volume carried forward
S_o	—	—	—
P_o	—	—	
S_o+P_o			
Gradient Fraction			
1	—	—	
2	—	—	
3	—	—	
4	—	—	
5	—	—	
6	—	—	
7	—	—	
8	—	—	
9	—	—	
10	—	—	
11	—	—	
12	—	—	
P	—	—	
Sum P_1			
Sum S_1	—	—	

T1
T2
D

	RXN Vol (ml)	Quen vol (ml)	Ext Coef (OD/mM)	BKG
ES				
MARKER 1	—	—	—	—

	Assay Dilution Factor	Net Dilution Factor	ALIQ (ml)	RXN Time (min)	Mean O.D.	BKG	Meas. Total Act.	Adjust. Total Act. nmoles/hr	% REC wrt d.g.	% REC wrt S_o+P_o	Spec. Act.	Cumm. enrich. factor	Act./ membrane protein
S_o	—	—		—	—	—	—	—	—	—	—	—	—
P_o	—	—	—	—	—	—	—	—	—	—	—	—	—
S_o+P_o							—	—			—		
Gradient Fraction													
1	—	—	—	—	—	—	—	—	—	—	—	—	—
2	—	—	—	—	—	—	—	—	—	—	—	—	—
3	—	—	—	—	—	—	—	—	—	—	—	—	—
4	—	—	—	—	—	—	—	—	—	—	—	—	—
5	—	—	—	—	—	—	—	—	—	—	—	—	—
6	—	—	—	—	—	—	—	—	—	—	—	—	—
7	—	—	—	—	—	—	—	—	—	—	—	—	—
8	—	—	—	—	—	—	—	—	—	—	—	—	—
9	—	—	—	—	—	—	—	—	—	—	—	—	—
10	—	—	—	—	—	—	—	—	—	—	—	—	—
11	—	—	—	—	—	—	—	—	—	—	—	—	—
12	—	—	—	—	—	—	—	—	—	—	—	—	—
P	—	—	—	—	—	—	—	—	—	—	—	—	—
Sum P_i	—	—	—	—	—	—	—	—	—	—	—	—	—
Sum S_i	—	—	—	—	—	—	—	—	—	—	—	—	—
Sum P_i+S_i							—	—			—	—	
Sum P_i+S_i, Recovery from S_o:			—										

- **Net Dilution Factor** (final dilution of the sample). Samples containing marker enzyme at a very high specific activity may require secondary dilution before the assay (**Assay Dilution Factor**). The Net Dilution Factor is calculated as the product of the Sample Dilution Factor and the Assay Dilution Factor.
- **ALIQ** (volume of sample placed into each assay tube). Different fractions, depending on their marker content, may require different aliquot volumes for the reactions to yield acceptable amounts of product.
- **RXN Time** (reaction time). This is most conveniently recorded in minutes. As in selecting sample aliquots, differing amounts of enzyme in the various fractions may make it appropriate to vary reaction times in order to obtain acceptable amounts of product.
- **O.D.** (optical density; absorbance of the quenched reaction). This is routinely calculated as the mean of triplicate determinations. It is necessary to correct for the background (**BKG**), i.e. the O.D. obtained in the absence of enzyme or in the presence of enzyme at zero reaction time. The spread-sheet has been designed so that this value may be entered once and copied to the appropriate cells. If the assay requires markedly different reaction times for different samples, appropriate BKG values can be entered manually.

To calculate the total amount of product released in an assay, the measured O.D. is divided by the extinction coefficient (**Ext Coef**) and multiplied by the total volume of the quenched reaction. This volume is calculated from the sum of the volumes of the reaction medium (**RXN Vol**), the volume of the sample aliquot (ALIQ) and the volume of quenching solution (**Quen Vol**).

To calculate the total enzyme activity in a fraction, the amount of product released is divided by the aliquot (ALIQ) and reaction time (RXN Time), and this product is multiplied by the sample volume, net dilution factor, and the conversion factor for minutes to hours. The formula for this calculation is:

$$\frac{(\text{O.D.} - \text{BKG} \times (\text{RXNVol} + \text{ALIQVol} + \text{QUENvol}) \times \text{Samplevol} \times \text{Netdil} \times 60\text{min}}{\text{ALIQ} \times \text{RXNTime} \times \text{EXTCoef} \times \text{hr}}$$

When the units specified in the sample spread-sheet are employed, the activity will be calculated in units of nanomoles of product per hour (nmol/h).

Several values in addition to the measured total activity may be of interest, including:

- the total activities adjusted for the fraction of the total fraction of the tissue that was analyzed; and
- the activities in ΣP_i and ΣS_i adjusted for the percentage of S_o which was analyzed.

From the sum of the total adjusted activities one calculates the net recovery from S_o; recoveries that differ markedly from 100% will call attention to possible activation or inhibition during the course of the fractionation scheme.

It is sometimes necessary to examine only a marker's percentage distribution among the density gradient fractions. In other cases it is of interest to examine the percentage distribution with respect to the initial cell lysate (given in this example as the total amount in S_o and P_o). The percentage in each density gradient fraction is adjusted by the percentage that Sum P_i represents of the activity recovered from S_o. This is calculated as:

$$P_1(\%) = 100 \times \frac{P_i}{\Sigma P_1} \times \frac{\Sigma P_i}{\Sigma P_1 + \Sigma S_1} \times \frac{S_o}{S_o + P_o}$$

It may be convenient to calculate the marker specific activity, by dividing the total marker activity in a fraction by the total amount of protein in that fraction. It is generally more useful to calculate the cumulative enrichment factor by dividing the percentage of the total marker activity recovered in a fraction by the percentage of the total protein recovered in that fraction.

In experimental situations in which a marker is translocated from one membrane population to another, it has proven useful to normalize the marker activity in each fraction to the total membrane protein, i.e. the protein in ΣP_i. This calculation allows for a ready comparison of the distributions of the marker activity in density gradients from control and various experimental samples.

Index

Page numbers in *italics* refer to tables, those in **bold** refer to figures.

MIX
Papier aus verantwortungsvollen Quellen
Paper from responsible sources
FSC® C105338

If you have any concerns about our products,
you can contact us on
ProductSafety@springernature.com

In case Publisher is established outside the EU,
the EU authorized representative is:
Springer Nature Customer Service Center GmbH
Europaplatz 3, 69115 Heidelberg, Germany

Printed by Libri Plureos GmbH
in Hamburg, Germany